Birkhäuser

Applied and Numerical Harmonic Analysis

For further volumes:
http://www.springer.com/series/4968

Volker Michel

Lectures on Constructive Approximation

Fourier, Spline, and Wavelet Methods on the Real Line, the Sphere, and the Ball

Volker Michel
Geomathematics Group
Department of Mathematics
University of Siegen
Siegen, Germany

ISBN 978-0-8176-8402-0 ISBN 978-0-8176-8403-7 (eBook)
DOI 10.1007/978-0-8176-8403-7
Springer New York Heidelberg Dordrecht London

Library of Congress Control Number: 2012948340

Mathematics Subject Classification (2010): 33C45, 33C50, 33C55, 33F05, 41-01, 41A05, 41A10, 41A15, 41A30, 41A50, 41A52, 41A63, 42A10, 42A15, 42A82, 42A85, 42C05, 42C10, 42C40, 43A90, 65D05, 65D07, 65D20, 65D30, 65D32, 65R32, 65T40, 65T50, 65T60, 86-08, 86A22

Printed on acid-free paper

Springer is part of Springer Science+Business Media (www.birkhauser-science.com)

This book is dedicated to the Father of Geomathematics,

Prof. Dr. Willi Freeden,

a passionate teacher, a visionary scientist, and a wise mentor, in acknowledgement of his infinite support and in the hope for many further joint projects to come.

ANHA Series Preface

The *Applied and Numerical Harmonic Analysis (ANHA)* book series aims to provide the engineering, mathematical, and scientific communities with significant developments in harmonic analysis, ranging from abstract harmonic analysis to basic applications. The title of the series reflects the importance of applications and numerical implementation, but richness and relevance of applications and implementation depend fundamentally on the structure and depth of theoretical underpinnings. Thus, from our point of view, the interleaving of theory and applications and their creative symbiotic evolution is axiomatic.

Harmonic analysis is a wellspring of ideas and applicability that has flourished, developed, and deepened over time within many disciplines and by means of creative cross-fertilization with diverse areas. The intricate and fundamental relationship between harmonic analysis and fields such as signal processing, partial differential equations (PDEs), and image processing is reflected in our state-of-the-art *ANHA* series.

Our vision of modern harmonic analysis includes mathematical areas such as wavelet theory, Banach algebras, classical Fourier analysis, time-frequency analysis, and fractal geometry, as well as the diverse topics that impinge on them.

For example, wavelet theory can be considered an appropriate tool to deal with some basic problems in digital signal processing, speech and image processing, geophysics, pattern recognition, biomedical engineering, and turbulence. These areas implement the latest technology from sampling methods on surfaces to fast algorithms and computer vision methods. The underlying mathematics of wavelet theory depends not only on classical Fourier analysis but also on ideas from abstract harmonic analysis, including von Neumann algebras and the affine group. This leads to a study of the Heisenberg group and its relationship to Gabor systems and of the metaplectic group for a meaningful interaction of signal decomposition methods. The unifying influence of wavelet theory in the aforementioned topics illustrates the justification for providing a means for centralizing and disseminating information from the broader, but still focused, area of harmonic analysis. This will be a key role of *ANHA*. We intend to publish the scope and interaction that such a host of issues demands.

Along with our commitment to publish mathematically significant works at the frontiers of harmonic analysis, we have a comparably strong commitment to publish major advances in the following applicable topics in which harmonic analysis plays a substantial role:

Antenna theory	*Prediction theory*
Biomedical signal processing	*Radar applications*
Digital signal processing	*Sampling theory*
Fast algorithms	*Spectral estimation*
Gabor theory and applications	*Speech processing*
Image processing	*Time-frequency and time-scale analysis*
Numerical partial differential equations	*Wavelet theory*

The above point of view for the *ANHA* book series is inspired by the history of Fourier analysis itself, whose tentacles reach into so many fields.

In the last two centuries, Fourier analysis has had a major impact on the development of mathematics, on the understanding of many engineering and scientific phenomena, and on the solution of some of the most important problems in mathematics and the sciences. Historically, Fourier series were developed in the analysis of some of the classical PDEs of mathematical physics; these series were used to solve such equations. In order to understand Fourier series and the kinds of solutions they could represent, some of the most basic notions of analysis were defined, e.g., the concept of “function”. Since the coefficients of Fourier series are integrals, it is no surprise that Riemann integrals were conceived to deal with uniqueness properties of trigonometric series. Cantor’s set theory was also developed because of such uniqueness questions.

A basic problem in Fourier analysis is to show how complicated phenomena, such as sound waves, can be described in terms of elementary harmonics. There are two aspects of this problem: first, to find, or even define properly, the harmonics or spectrum of a given phenomenon, e.g., the spectroscopy problem in optics; second, to determine which phenomena can be constructed from given classes of harmonics, as done, e.g., by the mechanical synthesizers in tidal analysis.

Fourier analysis is also the natural setting for many other problems in engineering, mathematics, and the sciences. For example, Wiener’s Tauberian theorem in Fourier analysis not only characterizes the behavior of the prime numbers, but it also provides the proper notion of spectrum for phenomena such as white light; this latter process leads to the Fourier analysis associated with correlation functions in filtering and prediction problems, and these problems, in turn, deal naturally with Hardy spaces in the theory of complex variables.

Nowadays, some of the theory of PDEs has given way to the study of Fourier integral operators. Problems in antenna theory are studied in terms of unimodular trigonometric polynomials. Applications of Fourier analysis abound in signal processing, whether with the fast Fourier transform (FFT), or filter design, or the adaptive modeling inherent in time-frequency-scale methods such as wavelet theory.

The coherent states of mathematical physics are translated and modulated Fourier transforms, and these are used, in conjunctionwith the uncertainty principle, for dealing with signal reconstruction in communications theory. We are back to the raison d'être of the *ANHA* series!

College Park, MD

John J. Benedetto
Series Editor

Preface

This book is the result of numerous courses titled "Constructive Approximation" and taught at the Universities of Kaiserslautern and Siegen. More and more students have encouraged me to turn my lecture notes into a textbook; hence, when Birkhäuser asked me if I had plans to write a book, I decided to accept the students' advice. The more I thought about the project, the more ideas I had about what else could be added and what could be presented in a different way than in my lectures. Although I ran the risk of turning the book project into a never-ending story, I finally managed to finish it. So, here it is.

As a textbook, this book cannot cover the whole area of constructive approximation on the real line, the sphere, and the ball. The one-dimensional part is kept brief because there are many books that already present this area in detail. The main purpose is to demonstrate the features of Fourier, spline, and wavelet methods so that analogues in the multivariate case become clear. The parts on the sphere and the ball concentrate on tools that have already been successfully applied to geophysical problems or problems of medical imaging. Alternative approaches and additional theoretical and practical achievements are also mentioned, and corresponding references are given.

This book addresses several distinct readers:

- Undergraduate students in their last year as well as graduate students of mathematics will hopefully find this book a helpful companion while they are attending courses in constructive approximation, harmonic analysis, numerical analysis, or spline and wavelet methods. I have purposely not written as a research monograph. Advanced experts will consider some explanations trivial, but students who are not familiar with spherical analysis, in particular, might profit from additional help given in the derivations.
- Students in geoscientific studies with a focus on mathematical methodologies are provided with an introduction to the fundamental numerical methods for approximating functions on the sphere and the ball. This book treats classical

global approximations by orthogonal polynomials (spherical harmonics) as well as modern localized methods based on splines, wavelets, and Slepian functions with a view to sparse regularization.
- Geoscientists who realize that they need to learn more about advanced approximation methods for their problems.
- Mathematicians facing an application where they, for example, need to approximate an unknown function on a sphere or a ball can use this textbook to learn how the presented approximation methods work and what the current state of the art of these tools is.
- Geomathematicians who are already familiar with constructive approximation for geoscientific applications will, hopefully, find some new insights into well-known concepts. They will also find references to further advanced results and additional publications.

This book would not exist without the help of several people, listed in no particular order. The first courses that I taught on constructive approximation basically summarized Willi Freeden's achievements on spherical approximation methods at that point in time. So, without him, there wouldn't have been such a course and, consequently, there wouldn't be the book that you are holding in your hands. For the last ten years, I have added further topics. Besides the fact that I am starting now with a brief introduction to one-dimensional approximation, I have added some of the results of constructive approximation on the three-dimensional ball that I obtained in cooperation with my own research group. For this reason, I want to thank Nahid Akhtar, Muhammad Akram, Abel Amirbekyan, Paula Berkel, Doreen Fischer, and Dominik Michel for their courage to tackle complicated inverse problems in three dimensions and their valuable contributions to my own long-range research project. They all left their footprints in this book. Successful research in geomathematics and many other scientific areas is nowadays only possible with a highly qualified and ambitious group of scientists.

I also want to thank those who proofread this book and (in addition to catching typing errors) gave numerous suggestions for improvements. They are Nicole Dröge, Doreen Fischer, Willi Freeden, and Roger Telschow. I am also thankful to the anonymous reviewers for giving useful comments and catching some more minor errors. Furthermore, I would like to express my gratitude to Frederik J. Simons for his comments regarding the section on Slepian functions. Moreover, I am grateful to so many students who have attended my lectures for more than ten years. Their feedback was always valuable, and without their encouragement I would have never written this textbook. In this context, my gratitude also goes to a series of PhD students who organized tutorials that accompanied my lectures and assisted the students in learning the subject. Furthermore, I also want to thank Tom Grasso and Ben Cronin from Birkhäuser for their advice and their patience with an author who didn't meet the deadline. Above all, special thanks go to my wife, Bärbel Michel, for assisting me by typing the whole book, deciphering my handwriting, catching errors during the typing, and coping with stylistic requirements.

I hope that this book will be helpful to many students and scientists, and I appreciate any feedback.

Siegen, Germany Volker Michel

Contents

Chapter 1
Introduction: The Problem to be Solved

In practice, a function of interest is only given pointwise, that is, there exists a point grid $\{x_j\}_{j=1,\ldots,G}$ where $\{F(x_j)\}_{j=1,\ldots,G}$ is known. Usually, two types of tasks occur in this context:

(T1) We want to know approximately what F looks like elsewhere.
(T2) We want to analyze F, that is, we want to extract some characteristic properties of F, for example, in order to find certain phenomena or in order to compare F with some reference functions.

For a better understanding, let us consider the following examples of given data $\{F(x_j)\}_{j=1,\ldots,G}$.

(D1) The temperature measurements at a particular meteorological station: This station provides us with one value per hour, and data are available for the period of the last 10 years, that is, we have almost 88,000 data. However, the device was temporarily broken.
(D2) Somebody said something to a speech recognition system. Now we have an acoustic signal of several seconds with measurements every few milliseconds. What did this guy say?
(D3) The temperature measured at all weather stations around the globe at 8:00 UTC this morning.
(D4) The gravity field and the magnetic field of the Earth derived from measurements of a satellite mission such as CHAMP (*Cha*llenging *M*inisatellite *P*ayload [117, 118, 154]) during several months.
(D5) EEG of a human brain: Data of the electric potential collected at a grid of sensors which looks approximately like a hemisphere.
(D6) A model of the Earth's interior: At some points inside the Earth, characteristic values like the density or some elastic parameters were calculated out of data collected at the surface.
(D7) Tomographic data of the human brain: For instance, from computerized tomography (CT) or other methods, a characteristic anatomic or physiological parameter was calculated at a point grid inside the head of a patient.

V. Michel, *Lectures on Constructive Approximation*, Applied and Numerical Harmonic Analysis, DOI 10.1007/978-0-8176-8403-7_1,

The functions of interest differ here with respect to their domain. In the case of (D1) and (D2), the domain is an interval on the real line. (D3) to (D5) refer to a sphere (or a part of it), and (D6) and (D7) correspond to functions on a ball (or a fragment of a ball). All three cases are treated in this book. However, since on the one hand the interpolation of functions on a real interval is a part of a basic course on numerical analysis, and on the other hand, advanced methods for analyzing such functions by wavelets can cover several terms of maths courses and have already been treated in many textbooks, the 1D part is kept very short—only these topics which are necessary for the following chapters are treated.

The main focus of the book is on teaching Fourier, spline, and wavelet methods on the sphere and the ball. For this purpose, the sphere will be treated in detail. Then the differences and analogues with respect to the ball are discussed. If you understand how it works on the interval and the sphere, you will understand the step to the "next dimension." Furthermore, the previous discussion of the analysis of functions on real intervals allows us to combine the analysis on a sphere or a ball with the investigation of time-dependent data sets.

Let us have a closer look at our seven data sets.

(D1) Remember the number 88,000 when we discuss polynomial methods later on. This will become a huge problem. Besides this, what might we be interested in? For instance, we want to observe the global warming. Did the temperatures rise during the observation period? In order to find an answer to this question, we need not (or better should not) take care of the fact that last year at July 15 the temperature fell from 1 pm to 2 pm. Anyway, the temperature apparently fell almost every day from the evening to the next morning. So, imagine you have a huge plot of the whole data pinned up on your wall. Your perception is dominated by these effects, which you are actually not interested in, if you stand too closely to the plot. So step back. Now you observe something else. Every year, the temperature fell approximately from August to the next February. Well, that's not a surprise, either, is it? (For the people on the Southern hemisphere: replace "fell" by "rose"). Take another step back. Now you can see the trend over 10 years!

The question for us now is as follows: How can we perform this stepping back mathematically? The keyword is "multiscale analysis." We look at the observable at different scales, that is, at different resolutions. You will learn how data on a real line, a sphere, or a ball can be analyzed by decomposing them into parts which correspond to temporal or spatial domains of different size.

By the way, (D1) is connected to another problem: The data grid is not equidistant since occasionally the instrument did not work.

(D2) The audio signal is, from the mathematical point of view, a function of the time. Of course, the speech recognition system has some reference signals. Some test persons said relevant phrases like "please switch on the light" in advance. However, the audio signals of this phrase uttered ten times will all look different. Therefore, we need a way to analyze similarities and differences.

Let us compare the phrase with a symphony played by a whole orchestra. The ideal way of identifying the symphony is to extract all notes played by all instruments—certainly including the correct points of time, the precise length of each note, and the loudness. From the physical point of view, a musical tone consists of a superposition of a fundamental frequency and some overtones. The occurring frequency defines the note (or the pitch).[1]

Thus, we need a way to extract the contained frequencies and their amplitudes. However, we also need to know how long each frequency occurred. For speech recognition, this is similar. Vowels have some characteristic frequencies which do not change during the utterance of the vowels. Consonants, however, have characteristic modulations of the frequencies; see [89] and the references therein for further details.

For the extraction of different frequencies, a representation of the signal as

$$F(t) = \sum_{n=0}^{\infty} a_n \cos(nt) + \sum_{n=1}^{\infty} b_n \sin(nt), \quad t \in [0, 2\pi], \tag{1.1}$$

would be a first choice, since the trigonometric polynomials describe oscillations of different frequencies. This imagination of an approximation by (or better, decomposition into) oscillations includes that, for increasing n, the oscillations increase and the amplitudes decrease and tend to 0 [the latter is necessary for the convergence of (1.1)]. However, this will not suffice for our purposes, since the amplitudes a_n, b_n represent temporal means over the whole time period. For example, playing the whole symphony in reverse order would only change the sign of the b_n and nothing else. This means that we will find out for the symphony that the note C occurred with a certain percentage during the whole concert. This might suffice to distinguish the symphony from another one, if we are lucky, but it doesn't appear to be an appropriate analysis of the symphony.

It would be much better to subdivide the interval into many small subintervals and to do the decomposition (1.1) in each subinterval. This allows us to analyze the temporal variations of the notes. One problem remains: The fixed length of the time intervals is a handicap for the determination of the length of each note, since we subdivide the symphony, for instance, in fragments of the length of halfnotes. So, we need different levels (we will call them scales) of analyses. We should see that at some parts of the signal (where eighth notes occur) a finer subdivision is needed than elsewhere.

Now you might ask the following question: Why don't we use the same finest-level subdivision everywhere? The problem is the uncertainty principle. What is meant here is not precisely Heisenberg's uncertainty principle from quantum mechanics. However, there is a strong mathematical similarity between Heisenberg's uncertainty principle and the uncertainty principle of

[1] This is certainly a simplification. For further details, see [58, 98, 145, 171, 202].

signal analysis (see, e.g., [27, p. 56] for the 1D case and [66, p. 102] as well as [68, 142] for the sphere). The latter one says that the precise frequency and the precise point of time can never be measured together. More precisely, the product of the inaccuracy of the frequency and the inaccuracy of the time cannot be less than a certain positive constant. This means, if we choose a very fine temporal subdivision, then we will be able to detect roughly changes of pitches during a short period of time quite well, which is good for consonants and eighth notes, respectively, but we are in trouble, if it was a vowel where the frequencies have to be known exactly (and it might be difficult to distinguish a C from a C sharp, respectively). Vice versa, a coarse subdivision allows us to detect vowels much better because the frequencies can be detected more accurately, but we are in trouble in the case of a consonant and we can better distinguish a C sharp from a C, but we don't know the length of the note, respectively.

(D3) First of all, you can imagine that we will have to deal with quite a lot of data. However, there is another problem. The data are extremely nonuniformly distributed over the sphere. In North America, Europe, and Australia, we can expect a high density of stations. On the other continents, the density is significantly smaller, and over the oceans the density is very poor. Thus, we would like to develop a model which takes into account this inhomogeneity. The model should have a high resolution in regions with a high density of the data points and a low resolution in poorly covered areas. In other words, the resolution should be spatially dependent according to the distribution of the point grid.

Remember example (D2). Instead of an expansion in trigonometric polynomials, the sphere provides us with the possibility of expanding functions with respect to certain algebraic polynomials, the so-called spherical harmonics, as we will see later on. The expansion will look as follows:

$$F(\xi) \approx \sum_{n=0}^{N} \sum_{j=-n}^{n} F_{n,j} Y_{n,j}(\xi), \quad |\xi| = 1. \tag{1.2}$$

Here, the truncation of the series has already been performed as it is unavoidable in practice. The truncation parameter N is in this case our only possibility to change the resolution. We have, for example, globally either a model of degree 100, 200, or 300. Imagine what the graph of a polynomial looks like. A high-degree model over Central Europe might be a good idea, but it is definitely a bad choice in regions with only a few data. You will learn how alternative basis functions—they will be called spline basis functions—can be constructed that allow a spatially varying adaptation of the resolution.

(D4) In comparison to (D3), the data of (D4) are rather uniformly distributed. However, the quality of the data varies. For example, there is the South-Atlantic anomaly (see, e.g., [81, 90, 118, 151]): an area of weakness in the Earth's magnetic field. Since this field acts as a protecting shield against the particles emitted by the sun, the satellite is heavily disturbed over the South-Atlantic such that the data are worse there. This is another reason for a spatial adaptation of the resolution.

However, there are some further interesting tasks. For instance, there are areas in our planet where the gravity field shows a complicated structure and changes at a rather small spatial scale. This is, for example, the case at the Cordilleras and the Himalayas. In addition, there are regions such as the Amazon area where the gravity field seasonally changes due to the transport of water masses [52, 54, 55, 57]. Other areas are rather uninteresting because the gravity field only changes slightly over a large spatial scale. These are further reasons for a multiscale analysis. We need a tool that allows us to look at a spherical function at different spatial scales: What happens at continental size? Are there effects at smaller spatial extension? Is there a global trend? This can be solved by wavelets. Moreover, major earthquakes like the Sumatra event at December 26, 2004 are connected to a huge transport of masses. Hence, the gravity field around the epicenter is different after the earthquake. Can we observe this? A polynomial representation (1.2) would have the consequence that the changes contribute to all coefficients $F_{n,j}$, since these coefficients are—similar to the temporal means a_n and b_n in (1.1)—spatial means. A localized basis system such as splines and wavelets provides us with coefficients that refer to a spatial parameter. Only a few will change after such an earthquake.

Now you might say, who cares? Well, the scientist working in practice cares. No data are exact. This noise propagates to the coefficients in the model, which all have a certain inaccuracy. Note that the change of the gravity field due to an earthquake is, globally regarded, rather small, if a model with global basis functions such as spherical harmonics is used. All coefficients might be affected by the event, but each only gets a small glimpse of the effect. More precisely, if we represent the same change by two models where model 1 reflects the phenomenon in more coefficients than model 2, then the probability that the changes of the coefficients are lower than the noise level (and are, consequently, not significant) is higher for model 1.

By the way, there is another problem connected to this data set: the so-called downward continuation of the gravity model from a sphere at orbit height to the spherical Earth's surface. However, this is beyond the scope of this book; see [73, 128, 166] for further details.

(D5) In comparison to (D3) and (D4), the size of (D5) is much smaller, but we have again nonuniformly distributed grid points—in a very special sense. There is a subset of the sphere covering a bit more than a hemisphere where the points are rather uniformly distributed, and at the complement of this subset there are no data at all. Polynomial basis systems are not a good choice here, either. Spherical splines perform much better.

(D6), (D7) The problems and tasks connected to these data sets are related to those discussed above. The mathematical challenge is here the treatment of the ball as the domain. One could get the idea that it suffices to use one of the well-known approximation methods on $\mathbb{R}^3$ for this problem, that is, the use of a 1D method for each cartesian coordinate x_1, x_2, x_3. However, this would not correspond to the physical or medical reality. The structures inside the Earth as well as the human head consist of layers with approximately spherical boundaries, which is why the polar coordinates appear to be an intuitive choice. For this reason, we will combine the gained knowledge on functions on an interval (for the radial part) with the knowledge on those on a sphere (for the angular part) in the chapter on the approximation on a ball.

Something should be clarified here. Despite the disadvantages connected to polynomial approximation, we need to learn how it works—for two reasons: First, these methods still have their legitimation and are appropriate and preferred in many applications. Second, we will need the polynomial bases to construct more elaborate methods.

The methods taught here can be subdivided into two classes with respect to algorithmic aspects. We will refer to this distinction as Principles 1 and 2:

Principle 1. *A basis system $\{B_n\}_n$ is given, find coefficients $\{a_n\}_n$ for the representation*

$$F(x) = \sum_n a_n B_n(x) . \tag{1.3}$$

In practice, we need a finite basis system which means that the relevant infinite-dimensional function space (like the set of all continuous functions) has to be reduced to a finite-dimensional subspace.

If we know F on a point grid, the coefficients $\{a_n\}_n$ can be determined from a linear system

$$\sum_{n=0}^{N} a_n B_n(x_j) = F(x_j) \quad \text{for all } j = 1, \ldots, G . \tag{1.4}$$

Usually, the associated matrix

$$A = (B_n(x_j))_{\substack{j=1,\ldots,G \\ n=0,\ldots,N}} \tag{1.5}$$

is not quadratic. Whereas the case $G < N+1$ is not desirable (not enough data and, thus, infinitely many solutions), the case $G > N+1$ is not only a practical reality—it is also recommendable. Of course, an overdetermined system forces us to give up the hope for the existence of a solution (i.e., an interpolating function, viz., a function which exactly has the given values). However, exactness is not desirable, if the values $F(x_j)$ contain measuring errors as it will always be the case. An overdetermined system can much better compensate a few strongly perturbed data. By the way, remember that such systems $Ax = b$ are treated by solving the normal equation $A^{\mathrm{T}}Ax = A^{\mathrm{T}}b$.

The practical implementation is connected to a couple of other problems. One of them is the condition number of the matrix. This quantity is a measure for the sensitivity of a matrix with respect to rounding errors on a computer during the inversion process (if you want to know more about it, please consult a book of your choice on numerical linear algebra, e.g., [102, p. 80], [147, p.85], and [153, p. 36]). Depending on the choice of the basis system and its size,[2] the matrix can be very ill-conditioned, that is, the numerically calculated solution of (1.4) can be far away from the real one. In the worst case, the matrix can even be numerically singular, that is, no solution is obtained. For this purpose, so-called regularization methods are used which only give an approximate solution of $A^{\mathrm{T}}Ax = A^{\mathrm{T}}b$ but a stable one, that is, a solution which shows a low sensitivity with respect to perturbances.

By the way, if the basis is orthonormal, the coefficients are also obtainable as inner products $a_n = \langle F, B_n \rangle$ (if you do not feel confident with this theory, please consult Sect. 2.2). Although orthonormal bases are often used, this procedure of determining the coefficients is less common (e.g., for reasons of numerical efficiency). Note that the pointwise data of F can also be used here, if the inner product (such as the L^2-inner product) is discretized by

$$\langle F, B_n \rangle \approx \sum_{k=0}^{K} w_k F(x_k) B(x_k) \tag{1.6}$$

with given weights $w_0, \ldots, w_K$.

Principle 2. *The function F is convolved with a so-called Approximate Identity $\{\Phi_J\}_J$, that is, a family of approximations*

$$F_J(y) := \int_D \Phi_J(y,x) F(x) \, \mathrm{d}x, \quad y \in D, \tag{1.7}$$

is calculated which converges to F in a certain limit (such as $J \to \infty$). Here, D is the domain of F.

Again, a numerical integration method is used to discretize the integral:

$$\int_D \Phi_J(y,x) F(x) \, \mathrm{d}x \approx \sum_{k=0}^{K} w_k \Phi_J(y, x_k) \, F(x_k), \tag{1.8}$$

where w_k are certain (given) weights. There are many motivations for this approach. One of them is based on a common mistake: Some physicists and engineers (and, unfortunately, also some mathematicians) occasionally use a phantom called the "Dirac delta function" δ. This mysterious object is claimed to have the following property:

---

[2]For instance, the Earth gravity model EGM96 (see [110, 203]) uses more than 130,000 basis functions, and EGM 2008 (see [146, 204]) even has more then 4,000,000 basis functions.

$$F(y) = \int_{\mathbb{R}} \delta(y-x)\, F(x)\, \mathrm{d}x \tag{1.9}$$

for all $y \in \mathbb{R}$ and all F, which are, say, continuous and have a compact support, briefly $F \in \mathrm{C}_0(\mathbb{R})$. However, there exists a well-known indirect mathematical proof with a clear message: Such a function δ can never exist. We will prove this in Remark 3.28 on p. 63.

Certainly, it would be great to have δ. If we knew F on the point grid and (1.8) were exact, then we would be able to calculate F outside the point grid. Since this is apparently impossible, can we at least replace "=" in (1.9) by "≈"?

Mathematicians found a way to rescue the theories which use δ. δ is not a function but a functional, that is, a mapping from a normed space to its corresponding field: For each $y \in \mathbb{R}$, we define the mapping

$$\begin{aligned} \delta_y : \mathrm{C}_0(\mathbb{R}) &\to \mathbb{R} \\ f &\mapsto f(y). \end{aligned} \tag{1.10}$$

δ_y is linear and continuous. As you might know, there is the Riesz representation theorem which says that if $\mathscr{F} : H \to \mathbb{R}$ is linear and continuous and H is a Hilbert space then there exists $g \in H$ such that $\mathscr{F} f = \langle f, g \rangle$ for all $f \in H$. In terms of the L^2-inner product, this would mean

$$f(y) = \delta_y f = \int_{\mathbb{R}} f(x)\, g(x)\, \mathrm{d}x. \tag{1.11}$$

So, isn't $\delta(y-x) := g(x), x \in \mathbb{R}$, the desired Dirac delta function? No, it isn't, because $(\mathrm{C}_0(\mathbb{R}), \langle\cdot,\cdot\rangle_{\mathrm{L}^2(\mathbb{R})})$ is not a Hilbert space! It is not complete. The space $\mathrm{L}^2(\mathbb{R})$ is a Hilbert space, but the functional δ_y cannot be extended to it in a sensible manner, because "$f(y)$" is not uniquely defined in $\mathrm{L}^2(\mathbb{R})$ (if you don't know what $\mathrm{L}^2(\mathbb{R})$ is, please consult Sect. 2.2).

This might be a reason for this common mistake. Enough about this! Here is the way out: In the language of functional analysis, δ would be associated to the identity operator

$$\mathrm{Id} : \mathrm{C}_0(\mathbb{R}) \to \mathrm{C}_0(\mathbb{R}). \tag{1.12}$$

This identity operator certainly exists, but we know that it is not representable by (1.9). However, we can find a sequence of operators $T_J : \mathrm{C}_0(\mathbb{R}) \to \mathrm{C}_0(\mathbb{R})$ which converges to Id in some sense, that is, $T_J F \to F$ pointwise on $\mathbb{R}$ for every function $F \in \mathrm{C}_0(\mathbb{R})$. Moreover, these operators can be associated to functions Φ_J such that

$$(T_J F)(y) = \int_{\mathbb{R}} \Phi_J(y,x)\, F(x)\, \mathrm{d}x. \tag{1.13}$$

This motivates the name "Approximate Identity" for $\{T_J\}_J$ or, alternatively, $\{\Phi_J\}_J$. Whereas this example refers to functions on $\mathbb{R}$, similar techniques can be established on the sphere and the ball.

A particular feature of this approach is that we get the above mentioned multiscale analysis. The functions Φ_J will be called scaling functions and the sequence $(T_J F)_J$ allows us to look at F at different resolutions. The details from one resolution step to the next can be represented by integrals of the type

$$\int_{\mathbb{R}} \Psi_J(y,x)\, F(x)\, \mathrm{d}x, \tag{1.14}$$

where Ψ_J is called a wavelet.

Part I
Basics

Chapter 2
Basic Fundamentals: What You Need to Know

If you want to learn how to cook the perfect soufflé, you first have to know how to beat egg white. It is not easy but easier than cooking the whole soufflé. However, if your beaten egg white does not have the right consistency, your soufflé won't become satisfactory.

Now, don't wonder. This is not a cooking book. The allegory of the soufflé means here: This textbook will give you a small glimpse of a topic in today's research. You will only reach this point if you are versed in the use of some basic mathematical theories. You will need the mathematics of the first two, or at least the first one and a half, years of undergraduate courses. That's why I won't explain here, for example, what a partial derivative, an integral on a volume in $\mathbb{R}^3$, a vector space, or a determinant are. You should know that or recapitulate that if you do not feel confident with it.

Due to my experience in teaching this course, I will give you a short introduction to some mathematical basics which are essential for the presented subjects but are usually not known to all students attending this course. These are functional analysis and the theory of integrals on curves and surfaces in $\mathbb{R}^3$.

2.1 Preliminaries

The set of all positive integers is denoted by $\mathbb{N}$, where $\mathbb{N}_0 := \mathbb{N} \cup \{0\}$, and $\mathbb{Z}$ is the set of all integers. Moreover, the set of all real numbers is, as usual, represented by $\mathbb{R}$, and the set of all rational numbers is called $\mathbb{Q}$.

Furthermore, the set of all polynomials on the real line, that is, all polynomials in one variable, is called $\mathrm{Pol}(\mathbb{R})$ or, briefly, Pol. More general, $\mathrm{Pol}(\mathbb{R}^n)$ represents the set of all polynomials in n variables. In addition, the restrictions of polynomials to a domain D are collected in $\mathrm{Pol}(D)$. For instance,

$$\mathrm{Pol}[a,b] := \mathrm{Pol}([a,b]) := \left\{ P\big|_{[a,b]} \,\middle|\, P \in \mathrm{Pol} \right\}. \tag{2.1}$$

V. Michel, *Lectures on Constructive Approximation*, Applied and Numerical Harmonic Analysis, DOI 10.1007/978-0-8176-8403-7_2,

Finally, if we only consider polynomials of degree $\leq m$, then the sets $\mathrm{Pol}_{0\ldots m}$, $\mathrm{Pol}_{0\ldots m}(\mathbb{R}^n)$, and $\mathrm{Pol}_{0\ldots m}(D)$, respectively, are used.

For two univariate functions F and G and for $x_0 \in \mathbb{R} \cup \{-\infty, +\infty\}$, the **Landau symbols** $\mathscr{O}$ and o are defined as follows:

- $F(x) = \mathscr{O}(G(x))$ as $x \to x_0 :\Leftrightarrow \frac{F(x)}{G(x)}$ is bounded as $x \to x_0$.
- $F(x) = o(G(x))$ as $x \to x_0 :\Leftrightarrow \lim_{x \to x_0} \frac{F(x)}{G(x)} = 0$.

2.2 Basics of Functional Analysis

This textbook is about the approximation of functions. Some of the questions will be: Which functions can be approximated? In which sense does the sequence of approximations converge to the function? The answers require the knowledge of spaces of functions and their properties and a concept of a convergence of functions. Such structures are provided by functional analysis. We will need here pre-Hilbert spaces and normed spaces. For further details, please consult any standard textbook on functional analysis such as [93, 207].

Definition 2.1. A real vector space X is called a **pre-Hilbert space** if it is equipped with an **inner product**, that is, a mapping $\langle \cdot, \cdot \rangle : X \times X \to \mathbb{R}$ with the following properties:

(IP1) $\langle x, x \rangle \geq 0$ for all $x \in X$, where $\langle x, x \rangle = 0$ if and only if $x = 0$ (positive definiteness).
(IP2) $\langle x, y \rangle = \langle y, x \rangle$ for all $x, y \in X$ (symmetry).
(IP3) $\langle \lambda x + \mu y, z \rangle = \lambda \langle x, z \rangle + \mu \langle y, z \rangle$ for all $x, y, z \in X$ and all $\lambda, \mu \in \mathbb{R}$ (bilinearity).

Note that (IP2) in combination with (IP3) yields

$$\langle x, \lambda y + \mu z \rangle = \langle \lambda y + \mu z, x \rangle = \lambda \langle y, x \rangle + \mu \langle z, x \rangle = \lambda \langle x, y \rangle + \mu \langle x, z \rangle$$

for all $x, y, z \in X$ and all $\lambda, \mu \in \mathbb{R}$.

Definition 2.2. Let X be a real vector space. If X is equipped with a mapping $\|\cdot\| : X \to \mathbb{R}$ which satisfies

(N1) $\|x\| \geq 0$ for all $x \in X$, where $\|x\| = 0$ if and only if $x = 0$ (positive definiteness),
(N2) $\|\lambda x\| = |\lambda| \, \|x\|$ for all $x \in X$ and all $\lambda \in \mathbb{R}$ (homogeneity),
(N3) $\|x + y\| \leq \|x\| + \|y\|$ for all $x, y \in X$ (triangle inequality),

then X (or, more precisely, $(X, \|\cdot\|)$) is called a **normed space** and $\|\cdot\|$ is called a **norm**.

Note that, since all vector spaces here are real vector spaces, the expression "real" will be omitted from here on.

There are some essential relations between normed spaces and pre-Hilbert spaces.

Theorem 2.3. *Let X be a vector space.*

(a) Every pre-Hilbert space is a normed space. More precisely, if $\langle \cdot, \cdot \rangle$ is an inner product on X, then the mapping $\|\cdot\| : X \to \mathbb{R}$ defined by

$$\|x\| := \sqrt{\langle x, x \rangle} \quad \textit{for all } x \in X \tag{2.2}$$

*is always a norm. It is called the **induced norm**.*

*(b) A norm $\|\cdot\|$ on X is an induced norm, that is, it originates from an inner product, if and only if the **parallelogram identity***

$$\|x+y\|^2 + \|x-y\|^2 = 2\|x\|^2 + 2\|y\|^2 \quad \textit{for all } x, y \in X \tag{2.3}$$

holds.

*(c) Let $\langle \cdot, \cdot \rangle$ be an inner product on X with the corresponding induced norm $\|\cdot\|$. Then the **Cauchy–Schwarz inequality***

$$|\langle x, y \rangle| \leq \|x\| \cdot \|y\| \quad \textit{for all } x, y \in X \tag{2.4}$$

holds.

Unless anything else is stated, $\|\cdot\|$ in the context of a pre-Hilbert space always represents the corresponding induced norm.

Let us have a look at some examples of pre-Hilbert spaces and normed spaces which will be important for our further work.

Example 2.4. Of course, $\mathbb{R}^n$ with the Euclidean dot product

$$\langle x, y \rangle := \sum_{i=1}^{n} x_i y_i =: x \cdot y; \quad x = (x_1, \ldots, x_n)^{\mathrm{T}}, y = (y_1, \ldots, y_n)^{\mathrm{T}} \in \mathbb{R}^n; \tag{2.5}$$

and its induced norm

$$\|x\| := \sqrt{\sum_{i=1}^{n} x_i^2} =: |x| \tag{2.6}$$

is a pre-Hilbert space, but this is not really exciting. Let us look at some more interesting examples.

1. The space $\mathrm{C}(D, \mathbb{R}^m)$, where $D \subset \mathbb{R}^n$ is compact, for example, $D = [a, b] \subset \mathbb{R}$, of all continuous functions $f : D \to \mathbb{R}^m$ can be equipped with different norms: For $p \in [1, +\infty[$, we set

$$\|f\|_p := \|f\|_{\mathrm{L}^p(D, \mathbb{R}^m)} := \left(\int_D |f(x)|^p \, \mathrm{d}x \right)^{1/p}, \tag{2.7}$$

where $|\cdot|$ refers to the Euclidean norm, and, formally, for $p = +\infty$, we set

$$\|f\|_\infty := \|f\|_{\mathrm{C}(D, \mathbb{R}^m)} := \max_{x \in D} |f(x)|. \tag{2.8}$$

Note that the longer indices are used if the choice of the domain is not clear. For instance, for $f : \mathbb{R} \to \mathbb{R}, x \mapsto x^2$, we have

$$\begin{aligned} \|f\|_{\mathrm{C}([-1,0],\mathbb{R})} &= 1, \\ \|f\|_{\mathrm{C}([0,\sqrt{2}],\mathbb{R})} &= 2. \end{aligned} \tag{2.9}$$

Moreover, the case $p = 2$ is special because then and only then we have the parallelogram identity

$$\begin{aligned} &\int_D (f+g)^2(x)\,\mathrm{d}x + \int_D (f-g)^2(x)\,\mathrm{d}x \\ &= \int_D \left(f^2(x) + 2f(x)g(x) + g^2(x) + f^2(x) - 2f(x)g(x) + g^2(x)\right)\,\mathrm{d}x \\ &= 2\int_D (f(x))^2\,\mathrm{d}x + 2\int_D (g(x))^2\,\mathrm{d}x. \end{aligned} \tag{2.10}$$

Hence, this norm is an induced norm. The corresponding inner product is

$$\langle f,g\rangle_2 := \langle f,g\rangle_{\mathrm{L}^2(D,\mathbb{R}^m)} := \int_D f(x)\cdot g(x)\,\mathrm{d}x, \quad f,g \in \mathrm{C}(D,\mathbb{R}^m), \tag{2.11}$$

where "·" refers to the Euclidean dot product.

2. In part (1), the norms for finite p do not require continuity but merely p-integrability, that is, the integrability of $|f|^p$. By the way, all integrals in this book are Lebesgue integrals. If you only know the Riemann integral, don't worry. The differences are minor, although they are sometimes important. Let us have a look at the space $\mathscr{L}^p(D,\mathbb{R}^m)$, $1 \le p < +\infty$, which consists of all (Lebesgue) measurable functions $f : D \to \mathbb{R}^m$ (the "usual" functions are measurable) with

$$\int_D |f(x)|^p\,\mathrm{d}x < +\infty, \tag{2.12}$$

where the set $D \subset \mathbb{R}^n$ only has to be a (Lebesgue) measurable set (e.g., an open or closed set or nearly every other set that occurs in practice). The formula (2.7) can obviously also be used for $f \in \mathscr{L}^p(D,\mathbb{R}^m)$, but $\|\cdot\|_p$ is not a norm on $\mathscr{L}^p(D,\mathbb{R}^m)$. The reason is that a Lebesgue integral of a nonnegative function vanishes if and only if the function vanishes almost everywhere, that is, if the set where the integrand is not zero has the Lebesgue measure zero.[1] For instance, countable sets have a vanishing Lebesgue measure. Thus, for

$$F(x) := \begin{cases} 1, & x \in [0,1] \cap \mathbb{Q} \\ 0, & x \in [0,1] \cap (\mathbb{R} \setminus \mathbb{Q}), \end{cases} \tag{2.13}$$

---

[1] Nonmathematicians might want to ignore this problem and the following considerations of part (2) and just accept that there is a particular difference between $\mathscr{L}^p$ and L^p. The rest of this book will, nevertheless, be comprehensible.

we have $\|F\|_{\mathrm{L}^p([0,1],\mathbb{R})} = 0$ for all p. This problem is solved by considering functions to be identical, if they are almost everywhere equal, that is, if they only differ on a set of measure zero. This has the consequence that exactly these functions f with $\|f\|_p = 0$ are identified with the zero function. For a rigorous mathematical formalism, one uses the algebraic technique of equivalence classes: Let $\mathscr{N}^p(D,\mathbb{R}^m)$ denote the set of all Lebesgue measurable functions $f : D \to \mathbb{R}^m$ with

$$\int_D |f(x)|^p \,\mathrm{d}x = 0. \tag{2.14}$$

Then the normed space $(\mathrm{L}^p(D,\mathbb{R}^m), \|\cdot\|_p)$ is defined by

$$\mathrm{L}^p(D,\mathbb{R}^m) := \mathscr{L}^p(D,\mathbb{R}^m) / _{\mathscr{N}^p(D,\mathbb{R}^m)}. \tag{2.15}$$

Note that $\mathrm{L}^2(D,\mathbb{R}^m)$ is a pre-Hilbert space with the inner product $\langle\cdot,\cdot\rangle_2$.

For scalar functions, we will use the shorter notations $\mathrm{C}(D) := \mathrm{C}(D,\mathbb{R})$ and $\mathrm{L}^p(D) := \mathrm{L}^p(D,\mathbb{R})$. Moreover, $\mathrm{C}^{(k)}(D,\mathbb{R}^m)$ denotes the set of all k-times ($k \in \mathbb{N}_0 \cup \{\infty\}$) continuously differentiable functions $f : D \to \mathbb{R}^m$, where $\mathrm{C}^{(k)}(D) := \mathrm{C}^{(k)}(D,\mathbb{R})$.

A fundamental property of the L^p-spaces is the Hölder inequality (see, e.g., [99, p. 224]).

Theorem 2.5 (Hölder's Inequality). *Let $D \subset \mathbb{R}^n$ be a measurable set and $F \in \mathrm{L}^p(D)$, $G \in \mathrm{L}^q(D)$, where $p,q \in]1,+\infty[$ with $\frac{1}{p}+\frac{1}{q} = 1$. Then $F \cdot G \in \mathrm{L}^1(D)$ and*

$$\|F \cdot G\|_{\mathrm{L}^1(D)} \le \|F\|_{\mathrm{L}^p(D)} \|G\|_{\mathrm{L}^q(D)}. \tag{2.16}$$

Moreover, for bounded domains D, the $\mathrm{L}^p(D)$-spaces are nested.

Theorem 2.6. *Let $D \subset \mathbb{R}^n$ be measurable and bounded. If $1 \le q \le p < +\infty$, then*

$$\mathrm{L}^p(D) \subset \mathrm{L}^q(D). \tag{2.17}$$

Proof. Let $f \in \mathrm{L}^p(D)$. Then we can subdivide D into two parts:

$$A := \left\{x \in D \,\middle|\, |f(x)| \le 1\right\}, \quad B := D \setminus A \tag{2.18}$$

(which are well defined up to a set of measure zero and are measurable). Then

$$\begin{aligned}\int_D |f(x)|^q \,\mathrm{d}x &= \int_A |f(x)|^q \,\mathrm{d}x + \int_B |f(x)|^q \,\mathrm{d}x \\ &\le \int_A 1 \,\mathrm{d}x + \int_B |f(x)|^p \,\mathrm{d}x \\ &\le \lambda(A) + \|f\|^p_{\mathrm{L}^p(D)} \end{aligned} \tag{2.19}$$

is obviously finite, where $\lambda(A)$ is the Lebesgue measure of A. □

For $L^1[a,b]$, an analogue of the fundamental theorem of calculus is known (see, e.g., [99, p. 550]). For formulating this theorem, we need the concept of absolutely continuous functions.

Definition 2.7. A function $f : [a,b] \to \mathbb{R}$ is called **absolutely continuous** (or briefly **AC**) if the following holds true: For every $\varepsilon > 0$, there exists $\delta > 0$ such that, for every finite set of points

$$a \le x_1 < y_1 \le x_2 < y_2 \le \cdots \le x_m < y_m \le b \tag{2.20}$$

with

$$\sum_{j=1}^{m} (y_j - x_j) < \delta, \tag{2.21}$$

we get

$$\sum_{j=1}^{m} \left| f(y_j) - f(x_j) \right| < \varepsilon. \tag{2.22}$$

The case $m = 1$ immediately yields the following embedding.

Theorem 2.8. *Every absolutely continuous function is continuous.*

With a lot of more work, one can prove the announced analogue of the fundamental theorem of calculus.

Theorem 2.9. *Let $f : [a,b] \to \mathbb{R}$ be given. Then f is absolutely continuous if and only if there exists $g \in L^1[a,b]$ with*

$$f(x) = f(a) + \int_a^x g(t)\,\mathrm{d}t \quad \text{for all } x \in [a,b]. \tag{2.23}$$

In this case, f is almost everywhere differentiable on $[a,b]$, where $f'(x) = g(x)$ for almost every $x \in [a,b]$.

Remember that one of the reasons for the use of functional analysis in approximation theory is the concept of the convergence of functions which is provided just by the choice of the norm as the following definition shows.

Definition 2.10. Let $(X, \|\cdot\|)$ be a normed space, (x_n) be a sequence in X, and $\xi \in X$ be an element.

(a) We say that (x_n) **converges** to ξ if

$$\lim_{n\to\infty} \|x_n - \xi\| = 0. \tag{2.24}$$

In this case, we write $\xi = \lim_{n\to\infty} x_n$. If the choice of the normed space is not clear, notations such as $\xi =^X \lim_{n\to\infty} x_n$ or $\xi =^{\|\cdot\|} \lim_{n\to\infty} x_n$ are used.

(b) (x_n) is called a **Cauchy sequence** if the following holds true: For every $\varepsilon > 0$, there exists an index $n_0 \in \mathbb{N}$ such that, for all $n, m \geq n_0$, the inequality

$$\|x_n - x_m\| < \varepsilon \tag{2.25}$$

holds.

As you might know, there are more general spaces, the so-called metric spaces. They are equipped with a metric d which measures distances. A norm always induces a metric by $d(x,y) := \|x-y\|$. Thus, convergence in a normed space simply means that the distance (measured by the norm) between the sequence and the limit tends to zero (in the ordinary sense of a convergence in $\mathbb{R}$ which you learned in analysis). The concept of a Cauchy sequence can correspondingly be interpreted. In analogy to finite-dimensional spaces, we have that every convergent sequence is a Cauchy sequence, where the opposite conclusion is false in general.

Definition 2.11. A normed space is called **complete** if every Cauchy sequence in this space converges to an element of this space. A complete normed space is also called a **Banach space**, whereas a complete pre-Hilbert space is called a **Hilbert space**.

Example 2.12. Again, the boring examples from your first year of studies are as follows: $\mathbb{R}^n$, $n \in \mathbb{N}$, in combination with the Euclidean dot product is a Hilbert space, where $\mathbb{Q}^n$, $n \in \mathbb{N}$, with the same inner product is not complete.

The function spaces which are relevant for this course have the following properties:

$(\mathrm{C}(D,\mathbb{R}^m), \|\cdot\|_\infty)$, where $D \subset \mathbb{R}^n$ is compact, is a Banach space.
$(\mathrm{C}(D,\mathbb{R}^m), \|\cdot\|_p)$, where $D \subset \mathbb{R}^n$ is compact and $1 \leq p < +\infty$, is in general not complete.
$(\mathrm{L}^p(D,\mathbb{R}^m), \|\cdot\|_p)$, where $D \subset \mathbb{R}^n$ is Lebesgue measurable and $1 \leq p < +\infty$, is a Banach space. In particular, $\left(\mathrm{L}^2(D,\mathbb{R}^m), \langle\cdot,\cdot\rangle_2\right)$ is a Hilbert space.

There exist two different concepts of convergence in pre-Hilbert spaces.[2] The relation between both is an immediate consequence of the Cauchy–Schwarz inequality.

Theorem 2.13. *Let $(X, \langle\cdot,\cdot\rangle)$ be a pre-Hilbert space. If (y_k) is a sequence in X and $y \in X$ is an element such that $\lim_{k\to\infty} \|y_k - y\| = 0$ (i.e., (y_k) **converges (strongly)** to y), then*

$$\lim_{k\to\infty} \langle y_k, x\rangle = \langle y, x\rangle \quad \textit{for all } x \in X \tag{2.26}$$

*(i.e., (y_k) **converges weakly** to y).*

In the context of convergence, the following relation is helpful.

[2] Actually, this is a particular case of a more general concept on normed spaces.

Theorem 2.14. *Let $D \subset \mathbb{R}^n$ be a compact set, and let (F_n) be a sequence in $C(D,\mathbb{R}^m)$ which converges with respect to $\|\cdot\|_\infty$ to $F : D \to \mathbb{R}^m$. Then F is continuous and (F_n) also converges to F with respect to $\|\cdot\|_p$ for every $p \in [1,+\infty[$.*

Proof. Note that the convergence in the sense of $\|\cdot\|_\infty$ is the same as the uniform convergence, since we have (by using the quantors $\forall$ for "for all" and $\exists$ for "there exists") in this case

$$\begin{aligned} &\forall \varepsilon > 0 \exists n_0 \forall n \geq n_0 \quad \|F_n - F\|_\infty < \varepsilon, \\ \Leftrightarrow &\forall \varepsilon > 0 \exists n_0 \forall n \geq n_0 \quad \max_{x\in D} |F_n(x) - F(x)| < \varepsilon, \\ \Leftrightarrow &\forall \varepsilon > 0 \exists n_0 \forall n \geq n_0 \quad |F_n(x) - F(x)| < \varepsilon \quad \forall x \in D. \end{aligned} \tag{2.27}$$

The first line corresponds to the $\|\cdot\|_\infty$-convergence, whereas the last line defines the uniform convergence. Now, remember your lectures on analysis. The uniform limit of a sequence of continuous functions is continuous. Thus, F is continuous.

The convergence in the $\|\cdot\|_p$-sense is a consequence of the following considerations: For each $G \in C(D,\mathbb{R}^n)$, we have

$$\|G\|_p = \left(\int_D |G(x)|^p \,\mathrm{d}x\right)^{1/p} \leq \left(\int_D \underbrace{\left(\max_{y\in D} |G(y)|\right)}_{=\|G\|_\infty}^p \mathrm{d}x\right)^{1/p} = \|G\|_\infty \left(\int_D \mathrm{d}x\right)^{1/p}. \tag{2.28}$$

Since D is compact and, consequently, bounded, the last integral is always finite. Hence, we obtain from the convergence in the $\|\cdot\|_\infty$-sense, that is, from (2.27),

$$\forall \varepsilon > 0 \exists n_0 \forall n \geq n_0 \quad \|F_n - F\|_p < \left(\int_D \mathrm{d}x\right)^{1/p} \cdot \varepsilon. \tag{2.29}$$

□

In analogy to the concepts in $\mathbb{R}^n$, topological expressions such as "closed" and "open" can be transferred to normed spaces.

Definition 2.15. Let $(X, \|\cdot\|)$ be a normed space, $M \subset X$ be a subset, and $x \in X$ be an element.

(a) The **open ball** with center x and radius $r > 0$ is denoted by

$$B_r(x) := \{y \in X \mid \|y - x\| < r\}. \tag{2.30}$$

(b) The element x is called an **inner point** of M if there exists $\varepsilon > 0$ such that $B_\varepsilon(x) \subset M$. The set $\operatorname{int} M$ consists of all inner points of M and is called the **interior** of M.

(c) The set M is called an **open set** if every element of M is an inner point, that is, if $M = \operatorname{int} M$.
(d) The element x is called an **accumulation point** of M if the following holds true: For every $\varepsilon > 0$, there exists an element $y \in B_\varepsilon(x)$ with $y \neq x$ and $y \in M$. The set $\overline{M}$ denotes the union of M and all of its accumulation points and is called the **closure** of M. If different norms may be chosen, the notation $\overline{M}^{\|\cdot\|}$ avoids confusions.
(e) The set M is called **closed**[3] if it contains all of its accumulation points, that is, if $M = \overline{M}$.
(f) The set M is called **dense** in X if $\overline{M} = X$.

Hilbert spaces have a very helpful property, which is of particular interest in constructive approximation: the existence of complete orthonormal systems.

Definition 2.16. Let $(X, \langle\cdot,\cdot\rangle)$ be a pre-Hilbert space and $\{x_\alpha\}_{\alpha \in A}$ be a subset of X, where A is an index set (which need not be countable).

(a) $\{x_\alpha\}_{\alpha \in A}$ is called an **orthogonal system** if $\langle x_\alpha, x_\beta \rangle = 0$ whenever $\alpha \neq \beta$ (for $\alpha, \beta \in A$).
(b) $\{x_\alpha\}_{\alpha \in A}$ is called an **orthonormal system (ons)** if

$$\langle x_\alpha, x_\beta \rangle = \delta_{\alpha\beta} := \begin{cases} 1, & \text{if } \alpha = \beta \\ 0, & \text{if } \alpha \neq \beta \end{cases}, \tag{2.31}$$

where $\delta_{\alpha\beta}$ is called the **Kronecker delta**.
(c) $\{x_\alpha\}_{\alpha \in A}$ is called **complete** if the only element which is orthogonal to $\{x_\alpha\}_{\alpha \in A}$ is the zero element, that is, if $f \in X$ with $\langle f, x_\alpha \rangle = 0$ for all $\alpha \in A$, then $f = 0$.
(d) A complete ons is called an **orthonormal basis (onb)**. A complete orthogonal system is called an **orthogonal basis**.

A preliminary result for Theorem 2.18 is the following lemma.

Lemma 2.17 (Bessel Inequality). *Let $(X, \langle\cdot,\cdot\rangle)$ be a pre-Hilbert space and let $\{x_n\}_{n \in \mathbb{N}_0}$ be a countable orthonormal system in X. Then the inequality*

$$\sum_{n=0}^{\infty} \langle f, x_n \rangle^2 \leq \|f\|^2 \tag{2.32}$$

is valid for all $f \in X$.

There are several equivalent criteria for the completeness of an orthonormal system. They can be formulated for uncountable sets as well, but we will need the countable case only.

---

[3] More precisely, this is the concept of a closed set in the topology. Later, we will also get to know the concept of a closed set in the sense of the approximation theory.

Theorem 2.18. *Let $(X,\langle\cdot,\cdot\rangle)$ be a Hilbert space and let $\{x_n\}_{n\in\mathbb{N}_0}$ be a countable orthonormal system in X. Then the following properties are equivalent:*

1. *$\{x_n\}_{n\in\mathbb{N}_0}$ is complete.*
2. *Every element of the space can be expanded in a (generalized)* ***Fourier series****: If $f\in X$, then*

$$\lim_{N\to\infty}\left\|f-\sum_{n=0}^{N}\langle f,x_n\rangle x_n\right\|=0, \tag{2.33}$$

   where $\|\cdot\|$ is the induced norm. Briefly,

$$\text{“}f=\sum_{n=0}^{\infty}\langle f,x_n\rangle x_n \quad \textit{in the sense of } (X,\langle\cdot,\cdot\rangle)\text{.”} \tag{2.34}$$

3. *The* ***Parseval identities*** *hold: If $f,g\in X$, then*

$$\langle f,g\rangle=\sum_{n=0}^{\infty}\langle f,x_n\rangle\,\langle g,x_n\rangle \tag{2.35}$$

   and, in particular,

$$\|f\|^2=\sum_{n=0}^{\infty}\langle f,x_n\rangle^2. \tag{2.36}$$

4. *$\{x_n\}_{n\in\mathbb{N}_0}$ is a basis, that is,*

$$\overline{\operatorname{span}\{x_n|n\in\mathbb{N}_0\}}^{\|\cdot\|}=X. \tag{2.37}$$

Now, this is an essential key to constructive approximation. Consider, for example, $\mathrm{L}^2(D,\mathbb{R}^m)$ (we will get to know further Hilbert spaces, but this one suffices here). If (!) we know a complete orthonormal system $\{g_n\}_{n\in\mathbb{N}_0}$ in $\mathrm{L}^2(D,\mathbb{R}^m)$, then every $f\in\mathrm{L}^2(D,\mathbb{R}^m)$ may be represented as

$$f=\sum_{n=0}^{\infty}\langle f,g_n\rangle_2\,g_n \tag{2.38}$$

in the sense of $\mathrm{L}^2(D,\mathbb{R}^m)$.

This is a fundamental result for two reasons:

1. If we want to analyze f, we need to be able to decompose f. The contribution by g_1 might have a different meaning than the contribution by g_{10}. We can compare f in this respect with other functions. We can detect similarities or differences which we might not observe otherwise.
2. In practice, we do not know f explicitly like $f(x)=x^2\sin x$. We only have discrete data, that is, data $\{f(x_j)\}_{j=1,\dots,G}$ on a finite point grid. Nevertheless,

we want to find out as much about f as possible. Equation (2.38) allows us to calculate an approximation to f as

$$f \approx \sum_{n=0}^{N} \langle f, g_n \rangle_2 \, g_n, \tag{2.39}$$

if N is chosen sufficiently large. The inner products $\langle f, g_n \rangle_2$, $n = 0, \ldots, N$, can be calculated approximately out of the discrete data of f by numerical means. Remember Principle 1 in Chap. 1.

We discussed this in the introduction in further detail.

Let me draw your attention to (2.38) from another point of view. Remember the definition of the L^p-spaces and the trick that is needed to obtain a norm. This means that we cannot be sure that the left-hand side and the right-hand side of (2.38) are equal at every $x \in D$. They are merely equal "in the sense of $\mathrm{L}^2(D, \mathbb{R}^m)$," which means that they may differ on a set of measure zero. Never forget this!

Remark 2.19. Actually, properties 2–4 in Theorem 2.18 are also equivalent for (countable) orthonormal systems in pre-Hilbert spaces, that is, the completeness of the space is not required, if we skip property 1 (see, for example, [93, pp. 176–178]). Moreover, property 4 in the case of a *non*-orthogonal system does *not* imply property 2, e.g., $(1,0)^{\mathrm{T}}$ and $\frac{1}{\sqrt{2}}(1,1)^{\mathrm{T}}$ constitute a (normed) basis of $\mathbb{R}^2$, but

$$\left\langle (x,y)^{\mathrm{T}}, (1,0)^{\mathrm{T}} \right\rangle (1,0)^{\mathrm{T}} + \frac{1}{2} \left\langle (x,y)^{\mathrm{T}}, (1,1)^{\mathrm{T}} \right\rangle (1,1)^{\mathrm{T}} = \frac{1}{2}(3x+y, x+y)^{\mathrm{T}} \neq (x,y)^{\mathrm{T}} \tag{2.40}$$

(except $(x,y)^{\mathrm{T}} = (0,0)^{\mathrm{T}}$).

We will also need the concept of an orthogonal complement and a direct sum, which can be explained as follows.

Definition 2.20. Let V be a vector space and $U, W \subset V$ be linear subspaces. Then we define the sum $U + W$ by

$$U + W := \{u + w \,|\, u \in U, w \in W\}. \tag{2.41}$$

If every $x \in U + W$ is uniquely representable by $x = u + w$, $u \in U$, $w \in W$, that is, if u and w are uniquely determined by x, then we write $U \oplus W$ and call this a **direct sum**. If $V = U \oplus W$, then W is called the **algebraic complement** of U (in V).

Example 2.21. We study a finite-dimensional and an infinite-dimensional example.

(a) Let $V = \mathbb{R}^2$. If

$$U_1 := \{(x,0) \,|\, x \in \mathbb{R}\}, \quad W_1 := \{(0,y) \,|\, y \in \mathbb{R}\}, \tag{2.42}$$

then, obviously, $V = U_1 \oplus W_1$, since every $(x,y) \in \mathbb{R}^2$ is representable as $(x,y) = (x,0) + (0,y)$, where x and y are unique. The same holds true for

$$U_2 := \{(x,-x)\,|\,x \in \mathbb{R}\}, \quad W_2 := \{(y,y)\,|\,y \in \mathbb{R}\}, \tag{2.43}$$

since

$$(\xi,\eta) = (x,-x) + (y,y) \tag{2.44}$$

is uniquely solvable by

$$x = \frac{1}{2}(\xi - \eta), \quad y = \frac{1}{2}(\xi + \eta). \tag{2.45}$$

Hence, $V = U_2 \oplus W_2$.

(b) Let $V = \mathrm{C}(\mathbb{R})$ be the space of all continuous functions on $\mathbb{R}$. Then

$$U_1 := \{\lambda \exp\,|\,\lambda \in \mathbb{R}\}, \quad W_1 := \left\{x \mapsto \mu x^2 \,\middle|\, \mu \in \mathbb{R}\right\} \tag{2.46}$$

generate a direct sum $U_1 \oplus W_1$, which is a proper subspace of V. Moreover, the sum of

$$U_2 := \mathrm{Pol}\,(\mathbb{R}), \quad W_2 := \left\{x \mapsto \lambda \exp(x) + \mu x^2 \,\middle|\, \lambda,\mu \in \mathbb{R}\right\} \tag{2.47}$$

is not direct, since, for example, $f(x) := x^2$ is representable as

$$f(x) = \underbrace{x^2}_{\in U_2} + \underbrace{0\exp(x) + 0x^2}_{\in W_2} \tag{2.48}$$

and as

$$f(x) = \underbrace{0}_{\in U_2} + \underbrace{0\exp(x) + 1x^2}_{\in W_2}. \tag{2.49}$$

Theorem 2.22. *Let V be a vector space and $U,W \subset V$ be linear subspaces. Then the sum $U+W$ is direct if and only if $U \cap W = \{0\}$.*

Proof.
"$\Rightarrow$":
Let $x \in U \cap W$. Then

$$x = \underbrace{x}_{\in U} + \underbrace{0}_{\in W} = \underbrace{0}_{\in U} + \underbrace{x}_{\in W}. \tag{2.50}$$

Since the sum is direct, we conclude that $x = 0$.
"$\Leftarrow$":
Let $x = u_1 + w_1 = u_2 + w_2 \in V$ with $u_1,u_2 \in U$ and $w_1,w_2 \in W$. As a consequence,

$$U \ni u_1 - u_2 = w_2 - w_1 \in W. \tag{2.51}$$

Since $U \cap W = \{0\}$, we conclude that

$$u_1 - u_2 = 0 = w_2 - w_1. \qquad (2.52)$$

□

Theorem 2.23 (Orthogonal Decomposition). *Let $(X, \langle \cdot, \cdot \rangle)$ be an inner product space and $Y \subset X$ be a complete linear subspace. Then*

$$X = Y \oplus Y^{\perp}, \qquad (2.53)$$

where

$$Y^{\perp} := \{z \in X \mid \langle z, y \rangle = 0 \textit{ for all } y \in Y\} \qquad (2.54)$$

is called the ***orthogonal complement*** *of Y (in X). Moreover, the orthogonal complement $Y^{\perp}$ is a closed linear subspace of X and $Y^{\perp\perp} = Y$.*

Note that Y has to be a complete[4] subspace. In the case of an infinite-dimensional X, one could otherwise find a proper linear subspace $Y \subsetneq X$ which is nevertheless dense in X, that is, $\overline{Y} = X$. In this case, $Y^{\perp} = \{0\}$ since $\overline{Y} = X$ implies that every $z \in Y^{\perp} \subset X$ is the limit of a sequence (y_n) in Y, where, consequently,

$$0 = \langle z, y_n \rangle \qquad (2.55)$$

for all $n \in \mathbb{N}_0$, because $z \in Y^{\perp}$, and

$$\langle z, z \rangle = \lim_{n \to \infty} \langle z, y_n \rangle = 0 \qquad (2.56)$$

due to Theorem 2.13. Hence, $Y \oplus Y^{\perp} \neq X$ in this case.

The last topic in this section is about linear and bounded operators. By the way, if X and Y are normed spaces, then mappings $T : X \to Y$ are called **operators** and mappings $T : X \to \mathbb{R}$ (which is a particular case, since $\mathbb{R}$ has a norm, too) are called **functionals**.

Definition 2.24. Let $(X, \|\cdot\|_X)$ and $(Y, \|\cdot\|_Y)$ be normed spaces and $T : X \to Y$ be an operator.

(a) T is called a **linear operator** if

$$T(\lambda x + \mu y) = \lambda Tx + \mu Ty \qquad (2.57)$$

for all $x, y \in X$ and all $\lambda, \mu \in \mathbb{R}$ (note that $Tx = T(x)$, etc.).

(b) T is called a **bounded operator** if

$$\|T\|_{\mathscr{L}(X,Y)} := \sup_{\substack{x \in X, \\ x \neq 0}} \frac{\|Tx\|_Y}{\|x\|_X} < +\infty. \qquad (2.58)$$

The vector space $\mathscr{L}(X, Y)$ consists of all linear bounded operators $T : X \to Y$. Moreover, $X^* := \mathscr{L}(X, \mathbb{R})$ is called the **dual space** of X.

[4]If X is complete, it suffices to know that Y is closed.

(c) T is called a **continuous operator** if the following holds true for each $x \in X$: For every $\varepsilon > 0$, there exists $\delta(=\delta(\varepsilon,x)) > 0$ such that $\|y-x\| < \delta$, $y \in X$, always implies $\|Tx - Ty\| < \varepsilon$.

Theorem 2.25. *Let $(X, \|\cdot\|_X)$ and $(Y, \|\cdot\|_Y)$ be normed spaces. Then the following holds true:*

(a) $\left(\mathscr{L}(X,Y), \|\cdot\|_{\mathscr{L}(X,Y)}\right)$ is a normed space. It is a Banach space, if Y is complete.
(b) If $T : X \to Y$ is linear, then T is bounded if and only if T is continuous. Moreover, there exist the following equivalent notations for the operator norm:

$$\|T\|_{\mathscr{L}(X,Y)} = \sup_{\substack{x\in X,\\ \|x\|_X=1}} \|Tx\|_Y = \sup_{\substack{x\in X,\\ \|x\|_X\leq 1}} \|Tx\|_Y. \tag{2.59}$$

Theorem 2.26 (Riesz Representation Theorem). *Let $(X, \langle\cdot,\cdot\rangle)$ be a Hilbert space with the dual space X^*. Then the following holds true:*

(a) Every $x \in X$ defines a linear and bounded functional $f_x \in X^*$ by

$$f_x(y) := \langle y,x\rangle \quad \forall y \in X. \tag{2.60}$$

(b) Every $f \in X^*$ can (vice versa) be represented by a unique element $x_f \in X$ as

$$f(y) = \langle y,x_f\rangle \quad \forall y \in X. \tag{2.61}$$

Moreover, $\|f\|_{X^*} = \|x_f\|$.

The following theorem is a consequence of the Riesz representation theorem.

Theorem 2.27. *Let $(X, \langle\cdot,\cdot\rangle_X)$ and $(Y, \langle\cdot,\cdot\rangle_Y)$ be Hilbert spaces and $T : X \to Y$ be a linear and bounded operator. Then there exists one and only one linear and bounded operator $T^* : Y \to X$ which satisfies*

$$\langle Tx,y\rangle_Y = \langle x,T^*y\rangle_X \qquad \forall x \in X \quad \forall y \in Y. \tag{2.62}$$

Moreover, $\|T\|_{\mathscr{L}(X,Y)} = \|T^*\|_{\mathscr{L}(Y,X)}$. The operator T^* is called the ***adjoint operator*** *of T.*

Definition 2.28. Let $(X, \langle\cdot,\cdot\rangle)$ be a Hilbert space and $T : X \to X$ be a linear and bounded operator. T is called **self-adjoint** if $T = T^*$.

2.3 Curves and Surfaces

The surfaces in this textbook will usually be spheres and the curves are usually also rather harmless. For this reason, this section is not kept more complicated than necessary. For further details see [108, Chaps. XV and XX], [150, Chap. 16], [158, Chap. 11], [185, Chap. 15], and [191, Chaps. 5 and 6].

Definition 2.29. Let $I := [a,b]$ be an interval in $\mathbb{R}$ and $R := [a,b] \times [c,d]$ be a rectangle in $\mathbb{R}^2$, where certainly $a < b$ and $c < d$.

(a) A mapping $\varphi : I \to \mathbb{R}^n$, $n \geq 2$, is called a **simple arc** and $\varphi(I)$ is called a **curve** if φ is continuously differentiable on $[a,b]$ and $\varphi'(x) \neq 0$ for all $x \in]a,b[$. If $\varphi(a) = \varphi(b)$, then the curve is called **closed**.
(b) A mapping $\Phi \in \mathrm{C}^{(1)}(R, \mathbb{R}^3)$ which is injective on $\operatorname{int} R$ and has a Jacobian matrix with a full rank, that is, $\operatorname{rk} \Phi' = 2$ on $\operatorname{int} R$, is called a **parametrization** of the **surface** $\Phi(R)$ with **parameter range** R.

Example 2.30. The following are some simple examples of a curve and a surface:

(a) A circle with the radius $r > 0$ and the center x in $\mathbb{R}^2$ can be parametrized by

$$\varphi(t) := x + \begin{pmatrix} r\cos t \\ r\sin t \end{pmatrix}, \quad t \in [0, 2\pi]. \tag{2.63}$$

(b) A sphere with the radius $r > 0$ and the center $x \in \mathbb{R}^3$ can be parametrized by

$$\Phi(\varphi, \vartheta) = x + \begin{pmatrix} r\sin\vartheta\cos\varphi \\ r\sin\vartheta\sin\varphi \\ r\cos\vartheta \end{pmatrix}, \quad \varphi \in [0, 2\pi],\ \vartheta \in [0, \pi]. \tag{2.64}$$

Definition 2.31. Let $\varphi(I)$ be a curve and $\Phi(R)$ be a surface. Moreover, let $f : \varphi(I) \to \mathbb{R}$ and $g : \Phi(R) \to \mathbb{R}$ be continuous functions.

(a) The **line integral** of f along $\varphi(I)$ is defined by

$$\int_{\varphi(I)} f(x)\, \mathrm{d}l(x) := \int_I f(\varphi(t))\, |\varphi'(t)|\, \mathrm{d}t. \tag{2.65}$$

(b) The **surface integral** of g across $\Phi(R)$ is defined by

$$\int_{\Phi(R)} g(x)\, \mathrm{d}\omega(x) := \int_R g(\Phi(u,v)) \left| \left(\frac{\partial \Phi}{\partial u} \wedge \frac{\partial \Phi}{\partial v} \right)(u,v) \right| \mathrm{d}(u,v), \tag{2.66}$$

where the **vector product** (**cross product**) of $y, z \in \mathbb{R}^3$ is (as usual) defined by

$$y \wedge z := \begin{pmatrix} y_2 z_3 - y_3 z_2 \\ y_3 z_1 - y_1 z_3 \\ y_1 z_2 - y_2 z_1 \end{pmatrix}. \tag{2.67}$$

In physics, the symbol $\times$ is used for the vector product. Here, we prefer the symbol $\wedge$ to avoid confusion between the symbol $\times$ and the letter x.
(c) The length of $\varphi(I)$ and the surface area of $\Phi(R)$ are obtained by choosing $f \equiv 1$ and $g \equiv 1$, respectively.

Example 2.32. Let us calculate the length of a circle line and the area of a sphere. For the circle, we use the parametrization (2.63) and get the length

$$\begin{aligned}\int_{\varphi([0,2\pi])} 1\,\mathrm{dl}(x) &= \int_0^{2\pi} \left| \begin{pmatrix} -r\sin t \\ r\cos t \end{pmatrix} \right| \mathrm{d}t \\ &= \int_0^{2\pi} \sqrt{r^2\sin^2 t + r^2\cos^2 t}\,\mathrm{d}t \\ &= \int_0^{2\pi} r\,\mathrm{d}t \\ &= 2\pi r. \end{aligned} \tag{2.68}$$

Finally, the area of a sphere which is parametrized by (2.64) is given by

$$\begin{aligned} &\int_{\Phi([0,2\pi]\times[0,\pi])} 1\,\mathrm{d}\omega(x) \\ &= \int_0^{\pi}\int_0^{2\pi} \left| \begin{pmatrix} -r\sin\vartheta\sin\varphi \\ r\sin\vartheta\cos\varphi \\ 0 \end{pmatrix} \wedge \begin{pmatrix} r\cos\vartheta\cos\varphi \\ r\cos\vartheta\sin\varphi \\ -r\sin\vartheta \end{pmatrix} \right| \mathrm{d}\varphi\,\mathrm{d}\vartheta \\ &= \int_0^{\pi}\int_0^{2\pi} \left| \begin{pmatrix} -r^2\sin^2\vartheta\cos\varphi \\ -r^2\sin^2\vartheta\sin\varphi \\ -r^2\sin\vartheta\cos\vartheta\left(\sin^2\varphi+\cos^2\varphi\right) \end{pmatrix} \right| \mathrm{d}\varphi\,\mathrm{d}\vartheta \\ &= \int_0^{\pi}\int_0^{2\pi} \sqrt{r^4\sin^4\vartheta\left(\cos^2\varphi+\sin^2\varphi\right)+r^4\sin^2\vartheta\cos^2\vartheta}\,\mathrm{d}\varphi\,\mathrm{d}\vartheta \\ &= \int_0^{\pi}\int_0^{2\pi} r^2\sqrt{\sin^2\vartheta\left(\sin^2\vartheta+\cos^2\vartheta\right)}\,\mathrm{d}\varphi\,\mathrm{d}\vartheta \\ &= \int_0^{\pi}\int_0^{2\pi} r^2|\sin\vartheta|\,\mathrm{d}\varphi\,\mathrm{d}\vartheta. \end{aligned} \tag{2.69}$$

We observe that $\sin\vartheta \geq 0$ for $\vartheta \in [0,\pi]$ and get

$$\int_{\Phi([0,2\pi]\times[0,\pi])} 1\,\mathrm{d}\omega(x) = 2\pi r^2 \int_0^{\pi} \sin\vartheta\,\mathrm{d}\vartheta = 4\pi r^2. \tag{2.70}$$

You should know what a volume integral is. However, to allow you a recapitulation, let us calculate the volume of a ball $\overline{B_R(x)}$, that is, a ball with the radius R and the center x. For simplicity, let us take the $\mathbb{R}^3$. The parametrization of $\overline{B_R(x)}$ is the same as in (2.64) where additionally r runs from 0 to R. Hence, the volume of the ball is

$$\begin{aligned} &\int_{\overline{B_R(x)}} 1\,\mathrm{d}y \\ &= \int_0^R\int_0^{\pi}\int_0^{2\pi} \left| \det \begin{pmatrix} \sin\vartheta\cos\varphi & r\cos\vartheta\cos\varphi & -r\sin\vartheta\sin\varphi \\ \sin\vartheta\sin\varphi & r\cos\vartheta\sin\varphi & r\sin\vartheta\cos\varphi \\ \cos\vartheta & -r\sin\vartheta & 0 \end{pmatrix} \right| \mathrm{d}\varphi\,\mathrm{d}\vartheta\,\mathrm{d}r \end{aligned}$$

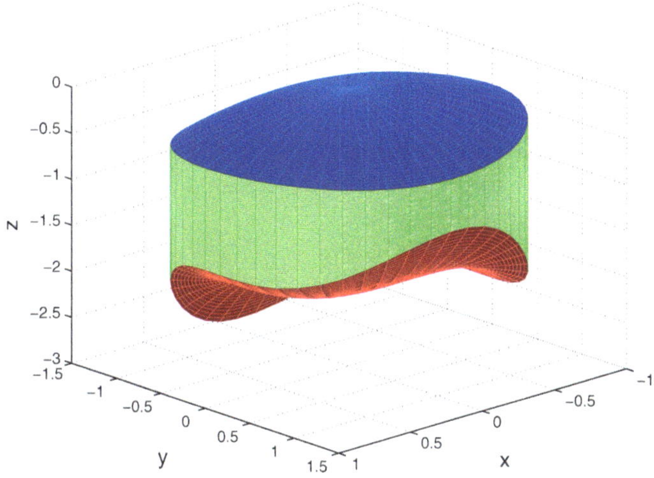

Fig. 2.1 An example for the set V and the graphs of g and h (*red and blue*) in (2.72)

$$
\begin{aligned}
&= \int_0^R \int_0^\pi \int_0^{2\pi} r^2 \sin\vartheta \,\mathrm{d}\varphi\,\mathrm{d}\vartheta\,\mathrm{d}r \\
&= 2\pi \int_0^R r^2 \int_0^\pi \sin\vartheta\,\mathrm{d}\vartheta\,\mathrm{d}r \\
&= 2\pi \cdot \frac{1}{3} R^3 \cdot 2 \\
&= \frac{4}{3}\pi R^3. \qquad (2.71)
\end{aligned}
$$

There is a couple of fundamental theorems that connect volume and surface integrals in a particular way.

Theorem 2.33. *Let $G \subset \mathbb{R}^2$ be open and bounded and let $g, h : G \to \mathbb{R}$ be given continuous functions (with $g(x_1,x_2) < h(x_1,x_2)$ for all $(x_1,x_2) \in G$) that define the set*

$$V := \left\{ x \in \mathbb{R}^3 \,\middle|\, (x_1,x_2) \in G,\ g(x_1,x_2) < x_3 < h(x_1,x_2) \right\}, \qquad (2.72)$$

see Fig. 2.1. Moreover, let ∂V be the surface of V and let n be the outer unit normal field on ∂V.

(a) ***(Gauß's Law)*** *If $f : \overline{V} \to \mathbb{R}^3$ is continuously differentiable on V with bounded derivatives on V and continuous on $\overline{V}$, then*

$$\int_V \operatorname{div} f(x)\,\mathrm{d}x = \int_{\partial V} f(x) \cdot n(x)\,\mathrm{d}\omega(x), \qquad (2.73)$$

where

$$\operatorname{div} f(x) := \sum_{j=1}^{3} \frac{\partial f_j}{\partial x_j}(x). \tag{2.74}$$

(b) ***(First and Second Green's Identity)*** *If $u, v : \overline{V} \to \mathbb{R}$ are twice continuously differentiable in V with bounded second-order derivatives and continuously differentiable in $\overline{V}$, then*

$$\int_V u(x)\Delta v(x) + (\nabla u(x)) \cdot (\nabla v(x))\,\mathrm{d}x = \int_{\partial V} u(x)\frac{\partial v}{\partial n}(x)\,\mathrm{d}\omega(x) \tag{2.75}$$

and

$$\int_V u(x)\Delta v(x) - v(x)\Delta u(x)\,\mathrm{d}x = \int_{\partial V} \left(u\frac{\partial v}{\partial n} - v\frac{\partial u}{\partial n}\right)(x)\,\mathrm{d}\omega(x), \tag{2.76}$$

where ∇ represents the gradient (nabla) operator, Δ denotes the ***Laplace operator (Laplacian)***

$$\Delta := \frac{\partial^2}{\partial x_1^2} + \frac{\partial^2}{\partial x_2^2} + \frac{\partial^2}{\partial x_3^2}, \tag{2.77}$$

that is, the trace of the Hessian, and $\frac{\partial v}{\partial n}$ stands for the directional derivative of v in the direction n, that is, $\frac{\partial v}{\partial n} = n \cdot \nabla v$.

The set V can, for instance, be an open ball $B_R(0)$. In this case, we can choose

$$G := \left\{ (x_1, x_2) \in \mathbb{R}^2 \middle| \, x_1^2 + x_2^2 < R^2 \right\} \tag{2.78}$$

and

$$\begin{aligned} g(x_1, x_2) &:= -\sqrt{R^2 - x_1^2 - x_2^2}, \\ h(x_1, x_2) &:= +\sqrt{R^2 - x_1^2 - x_2^2}. \end{aligned} \tag{2.79}$$

Gauß's Law is sometimes very helpful for the calculation of integrals. For instance,

$$\int_{\partial B_R(0)} \begin{pmatrix} x - \tan\frac{1}{1+y^2} + \ln\sqrt{z^2+2} \\ \sinh\left(xe^z\right) + y \\ z + \frac{yx^2 - \ln\left(y^2+x^4+5\right)}{\cosh y} \end{pmatrix} \cdot n(x,y,z)\,\mathrm{d}\omega(x,y,z) \tag{2.80}$$

is very hard to calculate if one tries to do this directly. However, due to Gauß's Law, this integral equals

$$\int_{B_R(0)} \operatorname{div} \begin{pmatrix} x - \tan\frac{1}{1+y^2} + \ln\sqrt{z^2+2} \\ \sinh\left(xe^z\right) + y \\ z + \frac{yx^2 - \ln\left(y^2+x^4+5\right)}{\cosh y} \end{pmatrix} \mathrm{d}(x,y,z) = \int_{B_R(0)} 3\,\mathrm{d}(x,y,z) = 4\pi R^3. \tag{2.81}$$

Chapter 3
Approximation of Functions on the Real Line

3.1 Orthogonal Basis Systems on Intervals

As we know from functional analysis, if we have a Hilbert space $(H, \langle\cdot,\cdot\rangle)$ and a corresponding orthonormal basis $\{b_n\}_{n\in\mathbb{N}_0}$, then every element $f \in H$ can be represented by a (generalized) Fourier series

$$f = \sum_{n=0}^{\infty} \langle f, b_n\rangle\, b_n \tag{3.1}$$

and can, consequently, be approximated by a truncation

$$f \approx \sum_{n=0}^{N} \langle f, b_n\rangle\, b_n, \tag{3.2}$$

provided that N is sufficiently large. More general, in the case of an orthogonal basis $\{b_n\}_{n\in\mathbb{N}_0}$, the series has the form

$$f = \sum_{n=0}^{\infty} \left\langle f, \frac{1}{\|b_n\|} b_n \right\rangle \frac{1}{\|b_n\|} b_n = \sum_{n=0}^{\infty} \frac{\langle f, b_n\rangle}{\|b_n\|^2} b_n. \tag{3.3}$$

In practice, observations of 1D-functions usually have a bounded domain. Thus, we will restrict our attention here to functions on intervals $[a,b]$. The intuitive choice of a corresponding Hilbert space is $\mathrm{L}^2([a,b])$, which is usually abbreviated as $\mathrm{L}^2[a,b]$. We will consider a generalization of this space as follows.

Definition 3.1. Let $w\colon [a,b] \to \mathbb{R}$ (with $a, b \in \mathbb{R}$ and $a < b$) be a continuous function on $[a,b]$ which is positive on $]a,b[$. For all measurable $F, G\colon [a,b] \to \mathbb{R}$ with

$$\int_a^b F(x)^2 w(x)\,\mathrm{d}x < +\infty, \qquad \int_a^b G(x)^2 w(x)\,\mathrm{d}x < +\infty, \tag{3.4}$$

V. Michel, *Lectures on Constructive Approximation*, Applied and Numerical Harmonic Analysis, DOI 10.1007/978-0-8176-8403-7_3,

we define

$$\langle F,G\rangle_{\mathrm{L}^2_w[a,b]} := \int_a^b F(x)G(x)w(x)\,\mathrm{d}x. \tag{3.5}$$

In analogy to the definition of $\mathrm{L}^2[a,b]$, the Hilbert space $(\mathrm{L}^2_w[a,b], \langle\cdot,\cdot\rangle_{\mathrm{L}^2_w[a,b]})$ is defined by (see also Example 2.4)

$$\mathrm{L}^2_w[a,b] := \left\{ F:[a,b]\to\mathbb{R} \;\middle|\; F \text{ measurable and } \int_a^b F(x)^2 w(x)\,\mathrm{d}x < +\infty \right\} \Big/ {\mathscr{N}^2_w[a,b]}, \tag{3.6}$$

where

$$\mathscr{N}^2_w[a,b] := \left\{ F:[a,b]\to\mathbb{R} \;\middle|\; F \text{ measurable and } \int_a^b F(x)^2\, w(x)\,\mathrm{d}x = 0 \right\}. \tag{3.7}$$

The proof that $\mathrm{L}^2_w[a,b]$ is a Hilbert space is similar to the proof in the case of $\mathrm{L}^2[a,b]$. Alternatively, the fact that $\mathrm{L}^2[a,b]$ is a Hilbert space can be used to prove the completeness of $\mathrm{L}^2_w[a,b]$.

We will now consider the question how a complete orthonormal system in $\mathrm{L}^2_w[a,b]$ can be constructed. Let us start with the construction of an orthonormal system. If we simply considered polynomials on $[a,b]$, it would be quite clear that the monomials $\{x^n\}_{n\in\mathbb{N}_0}$ provide a (certainly non-orthogonal) basis of the polynomials. We will see later that it really suffices to use polynomials. From linear algebra, we know Schmidt's orthonormalization procedure. If we apply this algorithm to the monomials and use $\langle\cdot,\cdot\rangle_{\mathrm{L}^2_w[a,b]}$ as the inner product, we obtain a system of orthonormal polynomials (see also [26, 180] for further details on orthogonal polynomials).

Theorem 3.2 (Orthonormal Polynomials). *Let $\mathrm{L}^2_w[a,b]$ be given as defined above. Then there exists a system of polynomials $\{P_{w,n}\}_{n\in\mathbb{N}_0}$ such that:*

1. *Each $P_{w,n}$, $n\in\mathbb{N}_0$, is a polynomial of degree n.*
2. *$\langle P_{w,n},P_{w,m}\rangle_{\mathrm{L}^2_w[a,b]} = \delta_{nm}$ for all $n,m\in\mathbb{N}_0$.*

If $\{Q_{w,n}\}_{n\in\mathbb{N}_0}$ is another system with this property, then either $Q_{w,n}=P_{w,n}$ or $Q_{w,n} = -P_{w,n}$ for each $n\in\mathbb{N}_0$.

Proof. As we observed above, the existence of such a system is a consequence of Schmidt's orthonormalization procedure. The uniqueness up to $\pm$ can be seen from the following iterative construction method (which also gives the existence of the system):

$P_{w,0}$ is a constant. Since $\|P_{w,0}\|_{\mathrm{L}^2_w[a,b]} = 1$ is required, only the sign of the constant is free to choose. Now, let $P_{w,0},\ldots,P_{w,n}$ be constructed. For the polynomial

$$P_{w,n+1}(x) = \sum_{k=0}^{n+1} a_k x^k \tag{3.8}$$

and its $n+2$ coefficients $a_0,\dots,a_{n+1}$, we have the following conditions:

$$\int_a^b \left(\sum_{k=0}^{n+1} a_k x^k\right) P_{w,m}(x)\, w(x)\, \mathrm{d}x = 0 \quad \text{for all } m = 0,\dots,n, \tag{3.9}$$

$$\int_a^b \left(\sum_{k=0}^{n+1} a_k x^k\right)^2 w(x)\, \mathrm{d}x = 1. \tag{3.10}$$

Note that (3.9) is invariant with respect to the selected signs of $P_{w,0},\dots,P_{w,n}$. The system (3.9) is linear and corresponds to the $(n+1)\times(n+2)$-matrix

$$\left(\int_a^b x^k P_{w,m}(x)\, w(x)\, \mathrm{d}x\right)_{\substack{m=0,\dots,n\\ k=0,\dots,n+1}}. \tag{3.11}$$

Now remember the properties of the polynomials already constructed: They are orthogonal. Hence, they are, in particular, linearly independent. In combination with the requirement that $\deg P_{w,j} = j$ for all j, we can conclude that for each $k(\le n)$, $P_{w,0},\dots,P_{w,k}$ is a basis of $\mathrm{Pol}_{0\dots k}$. Consequently, every x^k is a linear combination of $P_{w,0},\dots,P_{w,k}$. Thus, the monomial x^k is L^2_w-orthogonal to $P_{w,m}$ for each $m > k$. This corresponds to the entries below the main diagonal. Thus, the determinant of the upper left $(n+1)\times(n+1)$-block is

$$\prod_{k=0}^{n} \int_a^b x^k P_{w,k}(x)\, w(x)\, \mathrm{d}x. \tag{3.12}$$

None of the integrals $\int_a^b x^k P_{w,k}(x)\, w(x)\, \mathrm{d}x$ vanishes. Otherwise, the representation of x^k by $P_{w,0},\dots,P_{w,k}$ would be reduced to a representation by $P_{w,0},\dots,P_{w,k-1}$, which contradicts the degree of x^k. Therefore, the matrix (3.11) has the full rank $n+1$, and the unknown coefficient vector $(a_0,\dots,a_{n+1})$ is uniquely determined up to a constant factor. Due to condition (3.10), this constant is uniquely determined up to a factor ± 1 and nonzero.

Wait! There's no reason to celebrate yet. The proof is not finished. We have to explain why $a_{n+1} \neq 0$, since $P_{w,n+1}$ must have the degree $n+1$. However, this is an immediate consequence of the fact that $P_{w,0},\dots,P_{w,n}$ provide a basis for $\mathrm{Pol}_{0\dots n}$ and $P_{w,n+1}$ is orthogonal to this basis and nonzero. □

This proof also shows us how we can determine such orthogonal polynomials. However, this appears to be only practicable for low degrees. We will see that there is a more useful technique for the calculation of orthogonal polynomials. Before we study this, we have to take care of the fact that we only have an orthonormal system yet but not necessarily a basis. In order to investigate this further, we need a couple of theorems. The first one is a very famous theorem of approximation theory (see, e.g., [43, p. 107], [51, p. 1], [159, p. 11]).

Theorem 3.3 (Weierstraß Approximation Theorem). *Every continuous function on an interval $[a,b]$ can be uniformly approximated by a polynomial. More precisely, for every continuous function $f : [a,b] \to \mathbb{R}$ and every $\varepsilon > 0$, there exists a polynomial p such that*

$$\|f - p\|_{\mathrm{C}[a,b]} < \varepsilon. \tag{3.13}$$

In other words, $\mathrm{Pol}\,[a,b]$ *is dense in* $\left(\mathrm{C}\,[a,b], \|\cdot\|_{\mathrm{C}[a,b]}\right)$*:*

$$\overline{\mathrm{Pol}\,[a,b]}^{\|\cdot\|_{\mathrm{C}[a,b]}} = \mathrm{C}\,[a,b]. \tag{3.14}$$

Proof. First of all, it suffices to consider the interval $[0,1]$. For general intervals $[a,b]$, just use the transformation $g(t) = f\left(\frac{t-a}{b-a}\right)$. For a better understanding of the proof, the "divide and conquer" principle can help.

(1) The Bernstein operator and the idea of the proof:
For every $n \in \mathbb{N}$, we define the operator $B_n : \mathrm{C}\,[0,1] \to \mathrm{Pol}_{0\ldots n}[0,1]$ by

$$(B_n f)(x) := \sum_{k=0}^{n} f\left(\frac{k}{n}\right) \binom{n}{k} x^k (1-x)^{n-k}, \tag{3.15}$$

where $x \in [0,1]$, $f \in \mathrm{C}\,[0,1]$, and $\binom{n}{k}$ denotes the usual binomial coefficient "n choose k." We will show in this proof that $\|B_n f - f\|_{\mathrm{C}[0,1]} \to 0$ as $n \to \infty$ for every $f \in \mathrm{C}[0,1]$. This will complete the proof since this implies that for every $\varepsilon > 0$, there exists n_0 such that for $n = n_0$ (actually, for all $n \geq n_0$) $\|B_n f - f\|_{\mathrm{C}[0,1]} < \varepsilon$, where $B_n f$ is a polynomial. Let us see how we get there.

(2) The determination of $B_n f$ for some simple f:
Assume that g_0 is the constant 1, when we determine $B_n g_0$. Did you notice it? Right, it is the binomial theorem which gives us

$$(B_n 1)(x) = \sum_{k=0}^{n} \binom{n}{k} x^k (1-x)^{n-k} = (x + (1-x))^n = 1 \tag{3.16}$$

for all $x \in [0,1]$ (and elsewhere, of course). Note that this result can formally be extended (since B_n has only been defined for $n \geq 1$ so far) by

$$B_0 1 = \binom{0}{0} x^0 (1-x)^{0-0} = 1. \tag{3.17}$$

Suppose that $g_1(x) = x$, $x \in [0,1]$, then

$$(B_n g_1)(x) = \sum_{k=0}^{n} \frac{k}{n} \binom{n}{k} x^k (1-x)^{n-k}, \quad x \in [0,1], \tag{3.18}$$

where the summand for $k = 0$ vanishes. Since

$$\frac{k}{n} \binom{n}{k} = \frac{k}{n} \cdot \frac{n!}{(n-k)!k!} = \frac{(n-1)!}{(n-1-(k-1))!(k-1)!} = \binom{n-1}{k-1}, \tag{3.19}$$

we get (by setting $j := k-1$)

$$\begin{aligned}(B_n g_1)(x) &= \sum_{k=1}^{n} \binom{n-1}{k-1} x^k (1-x)^{n-k} \\ &= x \cdot \underbrace{\sum_{j=0}^{n-1} \binom{n-1}{j} x^j (1-x)^{n-1-j}}_{=(B_{n-1}1)(x)} \\ &= x. \end{aligned} \tag{3.20}$$

If $g_2(x) = x^2$ and $n \geq 2$, we obtain

$$\begin{aligned}(B_n g_2)(x) &= \sum_{k=0}^{n} \frac{k^2}{n^2} \binom{n}{k} x^k (1-x)^{n-k} \\ &= \sum_{k=0}^{n} \frac{k}{n} \left(\frac{k}{n} - \frac{1}{n}\right) \binom{n}{k} x^k (1-x)^{n-k} + \frac{1}{n} (B_n g_1)(x), \quad x \in [0,1], \end{aligned} \tag{3.21}$$

where in the last sum the summands vanish for $k = 0$ and $k = 1$. Now we use

$$\begin{aligned}\frac{k}{n} \cdot \frac{k-1}{n} \cdot \binom{n}{k} &= \frac{k(k-1)}{n^2} \cdot \frac{n!}{(n-k)!k!} \\ &= \frac{n-1}{n} \cdot \frac{(n-2)!}{(n-2-(k-2))!(k-2)!} \\ &= \left(1 - \frac{1}{n}\right) \binom{n-2}{k-2} \end{aligned} \tag{3.22}$$

for the derivation ($j := k-2$)

$$\begin{aligned}(B_n g_2)(x) &= \left(1 - \frac{1}{n}\right) x^2 \underbrace{\sum_{j=0}^{n-2} \binom{n-2}{j} x^j (1-x)^{n-2-j}}_{=(B_{n-2}1)(x)} + \frac{1}{n} \cdot x \\ &= x^2 + \frac{1}{n} \left(x - x^2\right). \end{aligned} \tag{3.23}$$

Note that

$$\begin{aligned}(B_1 g_2)(x) &= 0 + x^1 (1-x)^{1-1} \\ &= x \\ &= x^2 + \frac{1}{1} \left(x - x^2\right) \end{aligned} \tag{3.24}$$

shows that the obtained formula is also valid for $n = 1$.

(3) A partition of unity:
Equation (3.16) is called a partition of unity. Such structures are extremely helpful in constructive approximation, since they allow the following trick:

$$f(x) = \sum_{k=0}^{n} f(x) \binom{n}{k} x^k (1-x)^{n-k}. \tag{3.25}$$

You might want to ask, "What's so special about this?" Well, we want to show that $\|B_n f - f\|_{C[0,1]} \to 0$ as $n \to \infty$ for every $f \in C[0,1]$, right? So, we have to look at $\max_{x\in[0,1]} |(B_n f)(x) - f(x)|$. And here's the whole trick:

$$\begin{aligned} |(B_n f)(x) - f(x)| &= \left| \sum_{k=0}^{n} f\left(\frac{k}{n}\right) \binom{n}{k} x^k (1-x)^{n-k} - \sum_{k=0}^{n} f(x) \binom{n}{k} x^k (1-x)^{n-k} \right| \\ &\le \sum_{k=0}^{n} \left| f\left(\frac{k}{n}\right) - f(x) \right| \binom{n}{k} x^k (1-x)^{n-k} \end{aligned} \tag{3.26}$$

for all $x \in [0,1]$.
This estimate is a step forward. It is easier to show that the right-hand side uniformly converges to 0. For this purpose, we subdivide the sum into two parts.

(4) A subdivision of the sum I:
For the purpose of the convergence proof, let us keep $f \in C[0,1]$ and $\varepsilon > 0$ fixed (whereas the choice is certainly arbitrary). f is continuous, and $[0,1]$ is a compact set. Hence, f is uniformly continuous, i.e., there exists $\delta = \delta(\varepsilon) > 0$ such that $|x-y| < \delta$, $x,y \in [0,1]$, always implies $|f(x) - f(y)| < \varepsilon$. Thus, those $k \in \{0,\ldots,n\}$ with $\left|\frac{k}{n} - x\right| < \delta$ yield in (3.26) the estimate, which is true *for all* $x \in [0,1]$,

$$\begin{aligned} \sum_{\substack{k=0 \\ \left|\frac{k}{n}-x\right|<\delta}}^{n} \left| f\left(\frac{k}{n}\right) - f(x) \right| \binom{n}{k} x^k (1-x)^{n-k} &\le \varepsilon \sum_{\substack{k=0 \\ \left|\frac{k}{n}-x\right|<\delta}}^{n} \binom{n}{k} x^k (1-x)^{n-k} \\ &\le \varepsilon \sum_{k=0}^{n} \binom{n}{k} x^k (1-x)^{n-k} \\ &= \varepsilon. \end{aligned} \tag{3.27}$$

(5) A subdivision of the sum II:
We consider those $k \in \{0,\ldots,n\}$ with $\left|\frac{k}{n} - x\right| \ge \delta$. They satisfy obviously the inequality $\frac{1}{\delta^2}\left(\frac{k}{n} - x\right)^2 \ge 1$. Consequently, the associated part of the sum in (3.26) yields the estimate

$$
\begin{aligned}
&\sum_{\substack{k=0\\ \left|\frac{k}{n}-x\right|\geq\delta}}^{n} \left|f\left(\frac{k}{n}\right)-f(x)\right|\binom{n}{k}x^k(1-x)^{n-k}\\
&\leq 2\|f\|_{C[0,1]}\sum_{\substack{k=0\\ \left|\frac{k}{n}-x\right|\geq\delta}}^{n}\binom{n}{k}x^k(1-x)^{n-k}\\
&\leq \frac{2\|f\|_{C[0,1]}}{\delta^2}\sum_{\substack{k=0\\ \left|\frac{k}{n}-x\right|\geq\delta}}^{n}\left(\frac{k}{n}-x\right)^2\binom{n}{k}x^k(1-x)^{n-k}\\
&\leq \frac{2\|f\|_{C[0,1]}}{\delta^2}\sum_{k=0}^{n}\left(\frac{k}{n}-x\right)^2\binom{n}{k}x^k(1-x)^{n-k}\\
&= \frac{2\|f\|_{C[0,1]}}{\delta^2}\cdot\left[\sum_{k=0}^{n}\frac{k^2}{n^2}\binom{n}{k}x^k(1-x)^{n-k}-2x\sum_{k=0}^{n}\frac{k}{n}\binom{n}{k}x^k(1-x)^{n-k}\right.\\
&\qquad\left.+x^2\sum_{k=0}^{n}\binom{n}{k}x^k(1-x)^{n-k}\right]\\
&= \frac{2\|f\|_{C[0,1]}}{\delta^2}\cdot\left[(B_ng_2)(x)-2x(B_ng_1)(x)+x^2(B_ng_0)(x)\right]\\
&= \frac{2\|f\|_{C[0,1]}}{\delta^2}\left[x^2+\frac{1}{n}(x-x^2)-2x\cdot x+x^2\cdot 1\right]\\
&= \frac{2\|f\|_{C[0,1]}}{\delta^2}\cdot\frac{1}{n}(x-x^2),\quad x\in[0,1].
\end{aligned}
\tag{3.28}
$$

Since $0\leq x-x^2\leq\frac{1}{4}$ on $[0,1]$, we can find an integer n_0 (which is independent of x) such that for all $n\geq n_0$ *and all* $x\in[0,1]$,

$$
\sum_{\substack{k=0\\ \left|\frac{k}{n}-x\right|\geq\delta}}^{n}\left|f\left(\frac{k}{n}\right)-f(x)\right|\binom{n}{k}x^k(1-x)^{n-k}\leq\varepsilon. \tag{3.29}
$$

(6) The final step:
Combining parts 4 and 5 in (3.26), we obtain that for all $\varepsilon>0$ and all $f\in C[0,1]$, there exists n_0 such that, for every $n\geq n_0$,

$$
\max_{x\in[0,1]}|(B_nf)(x)-f(x)|\leq 2\varepsilon. \tag{3.30}
$$

□

This result leads us to the concept of a closed subset in the approximation theory, which clearly has to be distinguished from the concept of a closed set in the topology (see, e.g., [43, p. 188]).

Definition 3.4. Let $(X, \|\cdot\|_X)$ be a normed space and let $S \subset X$ be a subset. The set S is called **closed** in $(X, \|\cdot\|_X)$ (in the sense of the approximation theory), if the following holds true: For every $f \in X$ and every $\varepsilon > 0$, there exists a finite linear combination $\sum_{n=0}^N a_n g_n$ of elements g_n of S such that

$$\left\| f - \sum_{n=0}^{N} a_n g_n \right\|_X < \varepsilon. \tag{3.31}$$

In other words,

$$\overline{\operatorname{span} S}^{\|\cdot\|_X} = X. \tag{3.32}$$

The name "closed" is a bit confusing here. It is *not* the "closed" which one knows from topology (see Definition 2.15). One could say that S is closed in X (in the sense of the approximation theory) if $\operatorname{span} S$ is dense in X (in the sense of the topology).

Note that, in Hilbert spaces, an orthogonal system is closed if and only if it is complete (see Theorem 2.18). In fact, this can be generalized, that is, the orthogonality is unnecessary for the equivalence.

Theorem 3.5. *In every inner product space, a system $\{x_\alpha\}_{\alpha \in A}$ is closed if and only if it is complete.*

The proof uses basic arguments of functional analysis (see, e.g., [43, p. 263]).

We know, based on the previous considerations, the following closed systems (where the closure with respect to $\|\cdot\|_{\mathrm{L}^2_w[a,b]}$ is a consequence of $\|f\|_{\mathrm{L}^2_w[a,b]} \leq \|f\|_\infty \sqrt{\|w\|_\infty (b-a)}\, \forall f \in \mathrm{C}[a,b]$).

Theorem 3.6. *The following systems are closed in $(\mathrm{C}[a,b], \|\cdot\|_\infty)$ as well as in $(\mathrm{C}[a,b], \|\cdot\|_{\mathrm{L}^2_w[a,b]})$ (for $a, b \in \mathbb{R}$ with $a < b$):*

(a) All monomials: $\{x^k\}_{k=0,1,\ldots}$
(b) The sequence of orthonormal polynomials $\{P_{w,n}\}$ on $[a,b]$ for an arbitrary but fixed choice of w based on Theorem 3.2.

We can extend this result to $\mathrm{L}^2_w[a,b]$.

Theorem 3.7. *Every system of orthonormal polynomials $\{P_{w,n}\}_{n \in \mathbb{N}_0}$ as introduced in Theorem 3.2 is complete in $\mathrm{L}^2_w[a,b]$, i.e., it is an orthonormal basis of the Hilbert space $(\mathrm{L}^2_w[a,b], \langle \cdot,\cdot \rangle_{\mathrm{L}^2_w[a,b]})$.*

Proof. Due to Theorem 3.5, this statement is equivalent to the completeness of the monomials in $\mathrm{L}^2_w[a,b]$. Thus, suppose that $f \in \mathrm{L}^2_w[a,b]$ with

$$\int_a^b x^n f(x) w(x)\, \mathrm{d}x = 0 \quad \text{for all } n \in \mathbb{N}_0. \tag{3.33}$$

Integration by parts yields, with $F(x) := \int_a^x f(t)w(t)\,\mathrm{d}t$, $x \in [a,b]$, for all $n \in \mathbb{N}_0$ the identity

$$0 = \int_a^b x^n f(x)w(x)\,\mathrm{d}x = x^n F(x)\bigg|_a^b - \int_a^b n x^{n-1} F(x)\,\mathrm{d}x. \tag{3.34}$$

By the construction of F we get

$$F(a) = \int_a^a f(x)w(x)\,\mathrm{d}x = 0,$$
$$F(b) = \int_a^b f(x)w(x)\,\mathrm{d}x = \int_a^b x^0 f(x)w(x)\,\mathrm{d}x = 0. \tag{3.35}$$

Hence, (3.34) implies

$$\int_a^b x^{n-1} F(x)\,\mathrm{d}x = 0 \qquad \text{for all } n \in \mathbb{N}. \tag{3.36}$$

Note that the Cauchy–Schwarz inequality yields

$$\int_a^b |f(t)w(t)|\,\mathrm{d}t \le \left(\int_a^b (f(t))^2 w(t)\,\mathrm{d}t\right)^{1/2} \left(\int_a^b w(t)\,\mathrm{d}t\right)^{1/2} \tag{3.37}$$

such that $fw \in \mathrm{L}^1[a,b]$. Hence, F is absolutely continuous due to Theorem 2.9 and, in particular, continuous.

Since F is continuous and the monomials are complete in $(\mathrm{C}[a,b], \|\cdot\|_{\mathrm{L}^2[a,b]})$ (see also Definition 2.16) due to Theorems 3.5 and 3.6, it follows that $F = 0$ and, consequently, $f = \frac{1}{w}F' = 0$ in the sense of $\mathrm{L}^2_w[a,b]$. □

Hence, every $f \in \mathrm{L}^2_w[a,b]$ can be represented by

$$f = \sum_{n=0}^{\infty} \langle f, P_{w,n}\rangle_{\mathrm{L}^2_w[a,b]} P_{w,n} \tag{3.38}$$

in the sense of $\mathrm{L}^2_w[a,b]$. Remember what this type of convergence means. The series (3.38) does not necessarily converge pointwise. In particular, we should avoid to write "$f(x) = \sum_{n=0}^{\infty} \langle f, P_{w,n}\rangle_{\mathrm{L}^2_w[a,b]} P_{w,n}(x)$."

We will now investigate some general properties of orthogonal polynomials before we study some particular examples.

Theorem 3.8 (Zeros of Orthonormal Polynomials). *Let $\{P_{w,n}\}_{n\in\mathbb{N}_0}$ be a system of $\mathrm{L}^2_w[a,b]$-orthonormal polynomials based on Theorem 3.2. Then each $P_{w,n}$ has n real and pairwise distinct zeros in $]a,b[$.*

Proof. Since $P_{w,0}$ is constant, we have

$$0 = \langle P_{w,n}, P_{w,0}\rangle_{\mathrm{L}^2_w[a,b]} = P_{w,0} \int_a^b P_{w,n}(x)w(x)\,\mathrm{d}x \tag{3.39}$$

for each $n \geq 1$. Since $w > 0$ on $]a,b[$ and $P_{w,0} \neq 0$, $P_{w,n}$ must, consequently, have positive and negative values in $[a,b]$. Due to the continuity of $P_{w,n}$, there must be at least one zero of $P_{w,n}$ in $]a,b[$ where $P_{w,n}$ changes sign due to the intermediate value theorem.

Let $x_1 < \cdots < x_k$ be all zeros of $P_{w,n}$ where the polynomial changes sign (i.e., all zeros with odd multiplicity) in $]a,b[$. Obviously, $k \leq n$. We now generate a new polynomial where all zeros in $]a,b[$ have an even multiplicity:

$$Q_n(x) := P_{w,n}(x) \cdot \prod_{j=1}^{k} (x - x_j) . \tag{3.40}$$

Let us collect what we can find out about Q_n:

1. Q_n is either nonnegative on the whole interval $[a,b]$ or nonpositive on $[a,b]$ (with a finite number of roots only).
2. $\prod_{j=1}^{k} (x - x_j)$ is a polynomial of degree k and can, therefore, be represented in the basis $P_{w,0}, \ldots, P_{w,k}$ of $\mathrm{Pol}_{0\ldots k}[a,b]$.

Property (1) in combination with the positivity of w on $]a,b[$ implies that

$$\int_a^b Q_n(x) w(x) \, \mathrm{d}x \neq 0, \tag{3.41}$$

whereas property (2) yields, by definition of the orthogonal polynomials, that

$$\int_a^b Q_n(x) w(x) \, \mathrm{d}x = \int_a^b P_{w,n}(x) \prod_{j=1}^{k} (x - x_j) \, w(x) \, \mathrm{d}x = 0, \qquad \text{if } k < n. \tag{3.42}$$

Hence, $k = n$, that is, *all* zeros of $P_{w,n}$ are in $]a,b[$ and all have the multiplicity 1. □

This is a surprising property. In particular, there are no imaginary zeros. Moreover, there is one particular consequence: $P_{w,n}(b) \neq 0$ for all w and all n. This opens the door to an alternative definition of orthogonal polynomials. We skip the requirement of normalized polynomials. This yields unique polynomials up to a constant factor $(\neq 0)$ as we learned above. This factor can be fixed by defining the values at b. Note that it is important for this procedure that $P_{w,n}(b)$ indeed does not vanish.

Theorem 3.9 (Orthogonal Polynomials). *Let $w \in \mathrm{C}[a,b]$ with $w > 0$ on $]a,b[$ and let $(\gamma_n)_{n \in \mathbb{N}_0} \subset \mathbb{R} \setminus \{0\}$ be a given sequence. Then there exists one and only one sequence of polynomials $\{\tilde{P}_{w,n}\}_{n \in \mathbb{N}_0}$ with the following properties:*

1. *Each $\tilde{P}_{w,n}$ has the degree n, that is, $\deg \tilde{P}_{w,n} = n$.*
2. *If $n \neq m$,*

$$\int_a^b \tilde{P}_{w,n}(x) \tilde{P}_{w,m}(x) w(x) \, \mathrm{d}x = 0.$$

3. $\tilde{P}_{w,n}(b) = \gamma_n$ *for every* $n \in \mathbb{N}_0$.

This sequence of polynomials is an orthogonal basis of $\mathrm{L}^2_w[a,b]$.

No further proving is necessary. This theorem is an immediate consequence of the preliminary considerations.

Of course, a vast selection of orthogonal polynomials based on the choice of w, $[a,b]$, and (γ_n) is possible. In practice, a few particular systems are relevant. In the following, we will only take care of the Jacobi polynomials and some particular examples of them. In some cases, different values γ_n are common.

Definition 3.10. The following choices of w, $[a,b]$, and (γ_n) correspond to the named orthogonal polynomials:

(a) The **Jacobi polynomials** $P_n^{(\alpha,\beta)} := \tilde{P}_{w,n}$, $n \in \mathbb{N}_0$, are given by

$$w(x) := (1-x)^\alpha (1+x)^\beta$$

for $\alpha, \beta > -1$,

$$[a,b] := [-1,1],$$

and

$$\gamma_n := \binom{n+\alpha}{n},$$

where $\binom{c}{d} := \frac{\Gamma(c+1)}{\Gamma(c-d+1)\Gamma(d+1)}$ and

$$\Gamma(x) := \int_0^\infty \mathrm{e}^{-t} t^{x-1}\,\mathrm{d}t, \quad x > 0,$$

denotes the Gamma function[1] with $\Gamma(n+1) = n!$ for all $n \in \mathbb{N}_0$ and $\Gamma(x+1) = x\Gamma(x)$ for all $x > 0$.

(b) The **ultraspherical polynomials** (or **Gegenbauer polynomials**) are Jacobi polynomials with $\alpha = \beta$ up to constant factors. More precisely, one writes

$$P_n^{(\lambda)} := \frac{\Gamma\left(\lambda + \frac{1}{2}\right)}{\Gamma(2\lambda)} \frac{\Gamma(n+2\lambda)}{\Gamma\left(n+\lambda+\frac{1}{2}\right)} P_n^{(\lambda-1/2,\lambda-1/2)}, \quad \lambda \neq 0, \quad \lambda > -\frac{1}{2}. \tag{3.43}$$

(c) The **Legendre polynomials** P_n, $n \in \mathbb{N}_0$, represent the particular case $P_n := P_n^{(0,0)}$.

[1] Though the integral representation of the Gamma function is valid on the domain $\mathbb{R}^+$, the function itself can be analytically continued to a meromorphic function on the complex plane $\mathbb{C}$. In particular, the Gamma function is also declared on the real interval $]-1,0[$. This is important for the definition of the ultraspherical polynomials below.

(d) Moreover, for $\alpha = \beta = -\frac{1}{2}$, we have the **Chebyshev polynomials of the first kind**

$$T_n := 2^{2n} \binom{2n}{n}^{-1} P_n^{(-1/2,-1/2)}. \tag{3.44}$$

(e) Finally, for $\alpha = \beta = \frac{1}{2}$, we have the **Chebyshev polynomials of the second kind**

$$U_n := 2^{2n} \binom{2n+1}{n+1}^{-1} P_n^{(1/2,1/2)}. \tag{3.45}$$

Note that the cases $-1 < \alpha < 0$ and $-1 < \beta < 0$ are not covered by our general approach, since in these cases w is not defined on $[-1,1]$ but merely on $]-1,1[$. However, the theory can be extended accordingly, since the requirement $\alpha, \beta > -1$ leaves the singularity "weak enough" (for instance, $\int_{-1}^{1} (1-x)^{\alpha}\, \mathrm{d}x$ is finite for $\alpha > -1$).

It should be noted that general ultraspherical polynomials will not be discussed further here. Whenever you read $P_n^{(k)}$, it will refer to the k-th derivative of the Legendre polynomial.

For the Chebyshev polynomials of the first and the second kind, explicit representations can be derived. This can be seen from the orthogonality conditions

$$\int_{-1}^{1} T_n(x)T_m(x)(1-x)^{-1/2}(1+x)^{-1/2}\,\mathrm{d}x = 0 \;\text{ for } n \neq m\,,$$

$$\int_{-1}^{1} U_n(x)U_m(x)(1-x)^{1/2}(1+x)^{1/2}\,\mathrm{d}x = 0 \;\text{ for } n \neq m, \tag{3.46}$$

which can be simplified as follows:

$$\int_{-1}^{1} T_n(x)T_m(x)\left(1-x^2\right)^{-1/2}\mathrm{d}x = 0 \;\text{ for } n \neq m\,,$$

$$\int_{-1}^{1} U_n(x)U_m(x)\left(1-x^2\right)^{1/2}\mathrm{d}x = 0 \;\text{ for } n \neq m. \tag{3.47}$$

The substitution $x = \cos\vartheta$, $\vartheta \in [0,\pi]$, yields

$$\int_{0}^{\pi} T_n(\cos\vartheta)T_m(\cos\vartheta)\frac{1}{\sin\vartheta}\cdot\sin\vartheta\,\mathrm{d}\vartheta = 0 \;\text{ for } n \neq m\,,$$

$$\int_{0}^{\pi} U_n(\cos\vartheta)U_m(\cos\vartheta)\sin\vartheta\cdot\sin\vartheta\,\mathrm{d}\vartheta = 0 \;\text{ for } n \neq m$$

$$\Leftrightarrow \int_{0}^{\pi} T_n(\cos\vartheta)T_m(\cos\vartheta)\,\mathrm{d}\vartheta = 0 \;\text{ for } n \neq m\,,$$

$$\int_{0}^{\pi} U_n(\cos\vartheta)U_m(\cos\vartheta)\sin^2\vartheta\,\mathrm{d}\vartheta = 0 \;\text{ for } n \neq m. \tag{3.48}$$

With experience in integration and addition theorems, one obtains the following result.

Theorem 3.11. *The Chebyshev polynomials can be represented as follows:*

$$T_n(\cos\vartheta) = \cos(n\vartheta), \quad U_n(\cos\vartheta) = \frac{\sin[(n+1)\vartheta]}{\sin\vartheta}, \quad n \in \mathbb{N}_0,\ \vartheta \in [0,\pi]. \tag{3.49}$$

Note that for U_n, the values at $\vartheta = 0$ and $\vartheta = \pi$ are limits.

Let us also have a closer look at the Legendre polynomials. They are defined by the following requirements:

1. Every P_n is a polynomial of degree n.
2. If $n \neq m$,
$$\int_{-1}^{1} P_n(t)P_m(t)\,\mathrm{d}t = 0.$$
3. $P_n(1) = 1$ for every $n \in \mathbb{N}_0$.

Orthogonal polynomials represent a part of the mathematical area of "special functions." There exist many books with detailed studies of special functions (such as [1, 113]), orthogonal polynomials in particular (e.g., [26, 180]), and even more specific topics such as Legendre polynomials (see [160–162]). We will only summarize here some selected and most commonly used properties of the Jacobi polynomials. For further details and the proofs, please consult the corresponding references.

Theorem 3.12 (Rodriguez Formula). *The Jacobi polynomials $P_n^{(\alpha,\beta)}$, $n \in \mathbb{N}_0$, where $\alpha, \beta > -1$, satisfy*

$$(1-x)^\alpha(1+x)^\beta P_n^{(\alpha,\beta)}(x) = \frac{(-1)^n}{2^n n!}\left(\frac{\mathrm{d}}{\mathrm{d}x}\right)^n \left[(1-x)^{n+\alpha}(1+x)^{n+\beta}\right] \tag{3.50}$$

for all $x \in [-1,1]$. In particular, the Rodriguez formula for the Legendre polynomials is

$$P_n(x) = \frac{(-1)^n}{2^n n!}\left(\frac{\mathrm{d}}{\mathrm{d}x}\right)^n (1-x^2)^n, \quad x \in [-1,1],\ n \in \mathbb{N}_0. \tag{3.51}$$

In the proof of Theorem 3.2, we already learned how orthogonal polynomials can be calculated iteratively by solving systems of equations (where here the last condition $\|P_{w,n}\| = 1$ has to be replaced by $\tilde{P}_{w,n}(b) = \gamma_n$). The Rodriguez formula provides us with a direct way. However, both ways are not practicable for the determination of $P_n^{(\alpha,\beta)}$ for very large n. The Rodriguez formula is more important for proving further properties of the Jacobi polynomials. The most elegant way of calculating Jacobi polynomials is the following three-term recurrence relation.

Theorem 3.13 (Recurrence Formula for Jacobi Polynomials). *For every $\alpha, \beta > -1$, the Jacobi polynomials are given by*

$$P_0^{(\alpha,\beta)}(x) = 1,$$
$$P_1^{(\alpha,\beta)}(x) = \frac{1}{2}(\alpha+\beta+2)x + \frac{1}{2}(\alpha-\beta), \tag{3.52}$$

$$\begin{aligned} &2n(n+\alpha+\beta)(2n+\alpha+\beta-2)P_n^{(\alpha,\beta)}(x) \\ &\quad = (2n+\alpha+\beta-1)\left[(2n+\alpha+\beta)(2n+\alpha+\beta-2)x+\alpha^2-\beta^2\right]P_{n-1}^{(\alpha,\beta)}(x) \\ &\qquad -2(n+\alpha-1)(n+\beta-1)(2n+\alpha+\beta)P_{n-2}^{(\alpha,\beta)}(x), \quad n \geq 2, \end{aligned} \tag{3.53}$$

$x \in [-1,1]$. *In particular,*

$$P_0(x) = 1,$$
$$P_1(x) = x,$$
$$P_n(x) = \frac{2n-1}{n} x P_{n-1}(x) - \frac{n-1}{n} P_{n-2}(x), \quad n \geq 2, \tag{3.54}$$

$x \in [-1,1]$.

Figures 3.1 and 3.2 show the graphs of some Jacobi polynomials.

The Legendre polynomials and the Jacobi polynomials of type $P_n^{(0,2)}$, $P_n^{(0,m+1/2)}$ are also important for the analysis on the sphere and the ball, respectively.

What do we have now? In the spaces $\mathrm{L}_w^2[a,b]$, we can represent elements F in terms of Fourier series. For instance, for $F \in \mathrm{L}_w^2[-1,1]$ with $w(x) = (1-x)^\alpha(1+x)^\beta$, $\alpha, \beta > -1$, we have

$$F = \sum_{n=0}^{\infty} \left\langle F, P_n^{(\alpha,\beta)} \right\rangle_{\mathrm{L}_w^2[-1,1]} \left\| P_n^{(\alpha,\beta)} \right\|_{\mathrm{L}_w^2[-1,1]}^{-2} P_n^{(\alpha,\beta)} \tag{3.55}$$

in the sense of $\mathrm{L}_w^2[-1,1]$. The required norms can be taken from the following theorem.

Theorem 3.14. *For every $\alpha, \beta > -1$, the norm of $P_n^{(\alpha,\beta)}$, $n \in \mathbb{N}_0$, is given by*

$$\left\| P_n^{(\alpha,\beta)} \right\|_{\mathrm{L}_w^2[-1,1]}^2 = \frac{2^{\alpha+\beta+1}}{2n+\alpha+\beta+1} \cdot \frac{\Gamma(n+\alpha+1)\Gamma(n+\beta+1)}{n!\,\Gamma(n+\alpha+\beta+1)}. \tag{3.56}$$

In particular,

$$\|P_n\|_{\mathrm{L}^2[-1,1]}^2 = \frac{2}{2n+1}. \tag{3.57}$$

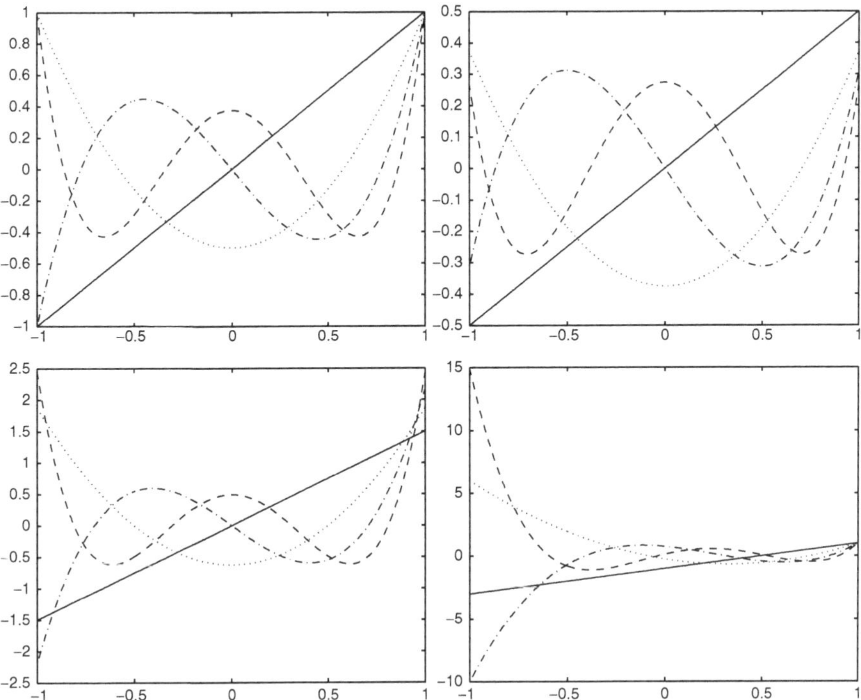

Fig. 3.1 Graphs of Jacobi polynomials of degrees 1 (*solid line*), 2 (*dotted line*), 3 (*dash-dotted line*), and 4 (*dashed line*); *left-hand top*: $P_n = P_n^{(0,0)}$, *right-hand top*: $P_n^{(-1/2,-1/2)}$, *left-hand bottom*: $P_n^{(1/2,1/2)}$, *right-hand bottom*: $P_n^{(0,2)}$

Therefore, the expansion of $F \in \mathrm{L}^2[-1,1]$ in terms of Legendre polynomials is given by

$$F = \sum_{n=0}^{\infty} \langle F, P_n \rangle_{\mathrm{L}^2[-1,1]} \frac{2n+1}{2} P_n \tag{3.58}$$

in the sense of $\mathrm{L}^2[-1,1]$.

Definition 3.15. For every $F \in \mathrm{L}^2[-1,1]$, the values

$$F^{\wedge}(n) := 2\pi \langle F, P_n \rangle_{\mathrm{L}^2[-1,1]}, \quad n \in \mathbb{N}_0, \tag{3.59}$$

are called the **Legendre coefficients** of F.

In this notation,

$$F = \sum_{n=0}^{\infty} F^{\wedge}(n) \frac{2n+1}{4\pi} P_n \tag{3.60}$$

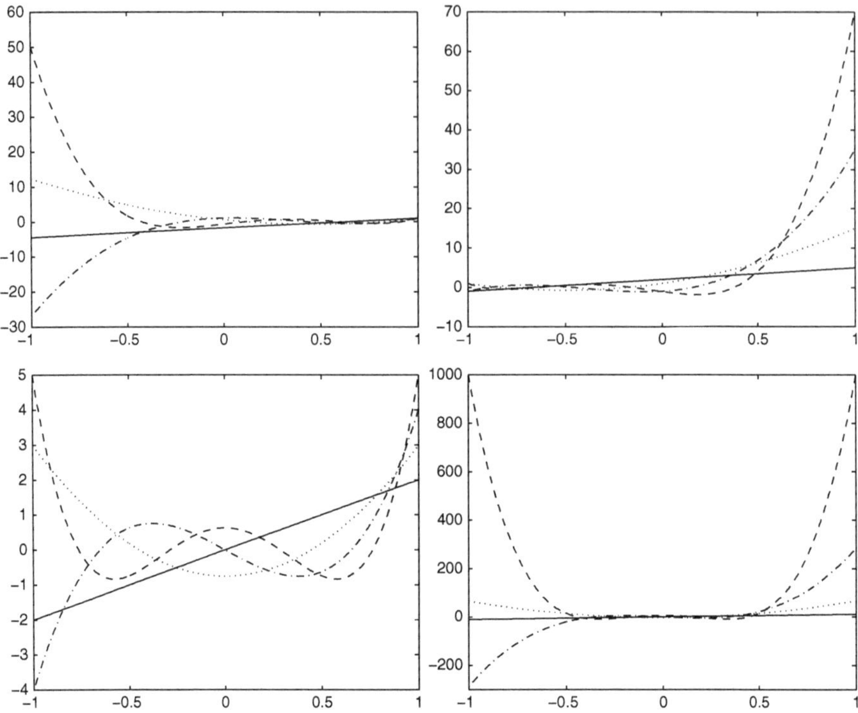

Fig. 3.2 Graphs of Jacobi polynomials of degrees 1 (*solid line*), 2 (*dotted line*), 3 (*dash-dotted line*), and 4 (*dashed line*); *left-hand top*: $P_n = P_n^{(0,3.5)}$, *right-hand top*: $P_n^{(4,0)}$, *left-hand bottom*: $P_n^{(1,1)}$, *right-hand bottom*: $P_n^{(10,10)}$

for each $F \in \mathrm{L}^2[-1,1]$. The Legendre series (3.60) will also play an important role when we study the spherical analysis.

Here are two examples of Legendre series.

Theorem 3.16. *For all* $h \in]-1,1[$ *and all* $t \in [-1,1]$,

$$\frac{1}{\sqrt{1+h^2-2ht}} = \sum_{n=0}^{\infty} h^n P_n(t),$$

$$\frac{1-h^2}{(1+h^2-2ht)^{3/2}} = \sum_{n=0}^{\infty} (2n+1) h^n P_n(t). \tag{3.61}$$

The proofs are easy exercises. For the first identity, show that both sides satisfy the same ordinary differential equation (where h is the variable!) with equal initial value, and use the Picard–Lindelöf theorem to prove that this initial value problem is uniquely solvable. For the second identity, differentiate the first one with respect to h and combine the differentiated series with the non-differentiated one in an appropriate way. Try it!

We have already learned in the introduction how to determine the coefficients in a truncated form of the expansion (3.55). After this procedure, the approximation of F, which is represented by the corresponding truncated expansion, requires the calculation of a sum of the form

$$S_N(x) = \sum_{k=0}^{N} A_k T_k(x), \tag{3.62}$$

where the coefficients A_k are given and $(T_k)_k$ is a system (of orthogonal polynomials) which can be calculated via a recurrence relation. A very efficient way of implementing the recursion in this summation is provided by the Clenshaw algorithm (see [44]). Since later the coefficients A_k may depend on x (in a different context), we allow this here, too.

Theorem 3.17 (Clenshaw Algorithm). *Let a system of 1D-functions T_k satisfy the linear second-order recurrence relation*

$$T_k(x) - a_k(x)\, T_{k-1}(x) - b_k(x)\, T_{k-2}(x) = 0; \quad k = 2,\ldots,N; \tag{3.63}$$

where the coefficients $a_k(x)$ and $b_k(x)$ as well as the first two functions $T_0(x) \neq 0$ and $T_1(x)$ are given. Then the sum

$$S_N(x) := \sum_{k=0}^{N} A_k(x)\, T_k(x) \tag{3.64}$$

can be calculated by

$$\begin{aligned} U_{N+1} &:= U_{N+2} := 0 \\ \text{for } k = N,\ldots,1: U_k &:= a_{k+1}U_{k+1} + b_{k+2}U_{k+2} + A_k \\ S_N &= (A_0 + b_2 U_2)\, T_0 + U_1 T_1. \end{aligned} \tag{3.65}$$

Note that the dependence on x was not mentioned in (3.65) for reasons of readability. We will also omit it in the following proof.

Proof. We define the additional coefficient $a_1 := \frac{T_1}{T_0}$ such that $T_1 - a_1 T_0 = 0$. Then the recurrence relation (3.63) can be represented as a linear system as follows:

$$\underbrace{\begin{pmatrix} 1 & 0 & 0 & \cdots & \cdots & 0 \\ -a_1 & 1 & 0 & \cdots & \cdots & \vdots \\ -b_2 & -a_2 & 1 & \ddots & \ddots & \vdots \\ 0 & \ddots & \ddots & \ddots & \ddots & \vdots \\ \vdots & \ddots & \ddots & \ddots & \ddots & 0 \\ 0 & \cdots & 0 & -b_N & -a_N & 1 \end{pmatrix}}_{=:M} \underbrace{\begin{pmatrix} T_0 \\ T_1 \\ T_2 \\ \vdots \\ T_N \end{pmatrix}}_{=:t} = \underbrace{\begin{pmatrix} T_0 \\ 0 \\ 0 \\ \vdots \\ 0 \end{pmatrix}}_{=:r}. \tag{3.66}$$

Obviously, $\det M = 1$. Hence, M^{-1} exists.

Let $a := (A_0, \ldots, A_N)^{\mathrm{T}}$, $u := (U_0, \ldots, U_N)^{\mathrm{T}}$, where $U_0 := a_1 U_1 + b_2 U_2 + A_0$. With the introduced notations, the sum that has to be calculated can be represented by

$$S_N = a^{\mathrm{T}} t = a^{\mathrm{T}} M^{-1} r = \left(\left(M^{-1} \right)^{\mathrm{T}} a \right)^{\mathrm{T}} r = \left(\left(M^{\mathrm{T}} \right)^{-1} a \right)^{\mathrm{T}} r. \tag{3.67}$$

The algorithmic iteration in (3.65) is equivalent to

$$\underbrace{\begin{pmatrix} 1 & -a_1 & -b_2 & 0 & \ldots & 0 \\ 0 & 1 & -a_2 & -b_3 & & \vdots \\ \vdots & 0 & \ddots & \ddots & \ddots & 0 \\ \vdots & & \ddots & 1 & -a_{N-1} & -b_N \\ \vdots & & & \ddots & 1 & -a_N \\ 0 & \ldots & \ldots & \ldots & 0 & 1 \end{pmatrix}}_{=M^{\mathrm{T}}} \begin{pmatrix} U_0 \\ U_1 \\ U_2 \\ \vdots \\ U_N \end{pmatrix} = \begin{pmatrix} A_0 \\ A_1 \\ \vdots \\ \vdots \\ A_N \end{pmatrix}. \tag{3.68}$$

Consequently, the auxiliary variables $U_0, \ldots, U_N$ satisfy

$$u = \left(M^{\mathrm{T}} \right)^{-1} a. \tag{3.69}$$

By inserting (3.69) in (3.67), we obtain

$$\begin{aligned} S_N &= u^{\mathrm{T}} r = U_0 T_0 = a_1 U_1 T_0 + b_2 U_2 T_0 + A_0 T_0 \\ &= U_1 T_1 + (b_2 U_2 + A_0) T_0. \end{aligned} \tag{3.70}$$

□

For instance, for the Legendre polynomials, the algorithm is, due to (3.54), as follows:

1. $U_{N+1}(x) := U_{N+2}(x) := 0$.
2. For $k = N, \ldots, 1$:

$$U_k(x) := \frac{2k+1}{k+1} x U_{k+1}(x) - \frac{k+1}{k+2} U_{k+2}(x) + A_k(x). \tag{3.71}$$

3. Finally,

$$\sum_{k=0}^{N} A_k(x) P_k(x) = A_0(x) - \frac{1}{2} U_2(x) + U_1(x) x. \tag{3.72}$$

Two further properties of Jacobi polynomials are added here. The first theorem is about the absolute maximum of these polynomials (see, e.g., [180, p. 168]). We will find a simple way of proving this result in the case of the Legendre polynomials when we study functions on the sphere.

Theorem 3.18. *For all $\alpha, \beta > -1$ with $q := \max(\alpha, \beta) \geq -\frac{1}{2}$ and every $n \in \mathbb{N}_0$, the following holds true:*

$$\max_{x \in [-1,1]} \left| P_n^{(\alpha,\beta)}(x) \right| = \binom{n+q}{n} = \max\left(\left| P_n^{(\alpha,\beta)}(-1) \right|, P_n^{(\alpha,\beta)}(1) \right) = \mathcal{O}\left(n^q\right) \quad (3.73)$$

(as $n \to \infty$), where $\mathcal{O}$ is the usual Landau symbol (see Sect. 2.1). In particular,

$$\max_{x \in [-1,1]} |P_n(x)| = 1 = P_n(1). \quad (3.74)$$

Note that one can prove (see, e.g., [180, p. 59]) that

$$P_n^{(\alpha,\beta)}(-1) = (-1)^n \binom{n+\beta}{n}, \quad (3.75)$$

whereas by Definition 3.10, we have

$$P_n^{(\alpha,\beta)}(1) = \binom{n+\alpha}{n}, \quad (3.76)$$

for all $n \in \mathbb{N}_0$ and all $\alpha, \beta \in \mathbb{R}$ with $\alpha, \beta > -1$.

An estimate for the derivatives of the Legendre polynomials can be derived, too (see also [138]).

Theorem 3.19. *For every $n \in \mathbb{N}_0$ and every $t \in [-1, 1]$,*

$$\left| P_n'(t) \right| \leq P_n'(1) = \frac{n(n+1)}{2}. \quad (3.77)$$

Proof.

(1) The Legendre coefficients of P_n':
Since we know that $P_n' \in \mathrm{L}^2[-1,1]$, we can expand this function with respect to the Legendre polynomials P_j. We get the coefficients via integration by parts:

$$\begin{aligned} \frac{1}{2\pi} \left(P_n'\right)^\wedge(j) &= \int_{-1}^{1} P_n'(t) P_j(t)\,\mathrm{d}t \\ &= P_n(t)P_j(t)\Big|_{-1}^{1} - \int_{-1}^{1} P_n(t)P_j'(t)\,\mathrm{d}t \\ &= 1 - P_n(-1)P_j(-1) - \frac{1}{2\pi}\left(P_j'\right)^\wedge(n). \end{aligned} \quad (3.78)$$

Since P_k' is a polynomial of degree $k-1$ ($k \geq 1$), the orthogonality of the Legendre polynomials implies

$$\left(P_n'\right)^{\wedge}(j) = 0, \quad \text{if } j \geq n \tag{3.79}$$

and

$$\frac{1}{2\pi}\left(P_n'\right)^{\wedge}(j) = 1 - P_n(-1)P_j(-1), \quad \text{if } j < n. \tag{3.80}$$

(2) We calculate $P_n(-1)$:
From the recurrence formula (see Theorem 3.13), we get

$$P_n(-1) = -\frac{2n-1}{n}P_{n-1}(-1) - \frac{n-1}{n}P_{n-2}(-1) \tag{3.81}$$

such that we can easily prove by induction that $P_n(-1) = (-1)^n$ for all $n \in \mathbb{N}_0$:

$$\begin{aligned} P_0(-1) &= 1 = (-1)^0, \\ P_1(-1) &= -1 = (-1)^1, \\ P_n(-1) &= -\frac{2n-1}{n}(-1)^{n-1} - \frac{n-1}{n}(-1)^{n-2} \\ &= (-1)^{n-2}\left(\frac{2n-1}{n} - \frac{n-1}{n}\right) \\ &= (-1)^{n-2} \\ &= (-1)^n. \end{aligned} \tag{3.82}$$

(3) The absolute maximum of P_n':
As a consequence of (3.79), (3.80), and (3.82), the Legendre series (3.60) for $F = P_n'$ is

$$P_n'(t) = \sum_{j=0}^{n-1}\frac{2j+1}{2}\left[1-(-1)^{n+j}\right]P_j(t) \quad \forall t \in [-1,1]. \tag{3.83}$$

Using Theorem 3.18, we get

$$\left|P_n'(t)\right| \leq \sum_{j=0}^{n-1}\frac{2j+1}{2}\left[1-(-1)^{n+j}\right] = P_n'(1). \tag{3.84}$$

(4) We calculate $P_n'(1)$:

$P_n'(1)$ can be calculated from (3.84) as follows[2]: If n is even, then only the summands for odd j (i.e., $j = 2k+1$) are different from zero. Hence, we get, for even n,

$$\begin{aligned} P_n'(1) &= \sum_{\substack{j=1 \\ j \text{ odd}}}^{n-1} \frac{2j+1}{2} \cdot 2 \\ &= \sum_{k=0}^{(n-2)/2} [2(2k+1)+1] \\ &= \sum_{k=0}^{(n-2)/2} (4k+3) \end{aligned}$$

---

[2] Alternatively, the recurrence formula (Theorem 3.13)

$$P_n(t) = \frac{2n-1}{n} t P_{n-1}(t) - \frac{n-1}{n} P_{n-2}(t) \quad \forall t \in [-1,1] \quad \forall n \geq 2 \tag{3.85}$$

can be differentiated such that (for all $n \geq 2$)

$$P_n'(t) = \frac{2n-1}{n} P_{n-1}(t) + \frac{2n-1}{n} t P_{n-1}'(t) - \frac{n-1}{n} P_{n-2}'(t) \quad \forall t \in [-1,1]. \tag{3.86}$$

By induction, we obtain $P_n'(1) = \frac{n(n+1)}{2}$ as follows:

$$\begin{aligned} P_0'(1) &= 0 = \frac{0 \cdot (0+1)}{2}, \\ P_1'(1) &= 1 = \frac{1 \cdot (1+1)}{2}, \\ P_n'(1) &= \frac{2n-1}{n} + \frac{2n-1}{n} \frac{(n-1)n}{2} - \frac{n-1}{n} \frac{(n-2)(n-1)}{2} \\ &= \frac{1}{2n} \left(4n - 2 - (n-1)^2(n-2)\right) + \frac{(2n-1)(n-1)}{2} \\ &= \frac{1}{2n} \left(4n - 2 - n^3 + 2n^2 + 2n^2 - 4n - n + 2\right) + \frac{2n^2 - 3n + 1}{2} \\ &= \frac{1}{2n} \left(-n^3 + 4n^2 - n\right) + \frac{2n^2 - 3n + 1}{2} \\ &= \frac{1}{2} \left(-n^2 + 4n - 1 + 2n^2 - 3n + 1\right) \\ &= \frac{1}{2} \left(n^2 + n\right) \\ &= \frac{1}{2} n(n+1). \end{aligned} \tag{3.87}$$

$$\begin{aligned}
&= 4 \cdot \frac{1}{2} \cdot \frac{n-2}{2} \cdot \frac{n}{2} + 3 \cdot \frac{n}{2} \\
&= \frac{1}{2}\left(n^2 - 2n + 3n\right) \\
&= \frac{1}{2} n\,(n+1). \qquad (3.88)
\end{aligned}$$

Finally, odd numbers n yield

$$\begin{aligned}
P_n'(1) &= \sum_{\substack{j=0 \\ j \text{ even}}}^{n-1} \frac{2j+1}{2} \cdot 2 \\
&= \sum_{k=0}^{(n-1)/2} (4k+1) \\
&= 4 \cdot \frac{1}{2} \cdot \frac{n-1}{2} \cdot \frac{n+1}{2} + \frac{n+1}{2} \\
&= \frac{1}{2}\left(n^2 - 1 + n + 1\right) \\
&= \frac{1}{2} n\,(n+1). \qquad (3.89)
\end{aligned}$$

□

Corollary 3.20. *For all* $n \in \mathbb{N}_0$, $P_n(-1) = (-1)^n$.

By differentiating (3.83) $k-1$ times, we get the following generalization of Theorem 3.19 by induction.

Theorem 3.21. *For every* $n, k \in \mathbb{N}_0$ *and every* $t \in [-1,1]$,

$$\left|P_n^{(k)}(t)\right| \leq P_n^{(k)}(1) = \mathcal{O}\left(n^{2k}\right) \text{ as } n \to \infty. \qquad (3.90)$$

Note that the theorem is certainly trivial in the case $k > n$.

Proof. If $|P_n^{(k-1)}(t)| \leq P_n^{(k-1)}(1)$ for all $n \in \mathbb{N}_0$, then

$$\begin{aligned}
\left|P_n^{(k)}(t)\right| &\leq \sum_{j=k-1}^{n-1} \frac{2j+1}{2}\left[1 - (-1)^{n+j}\right]\left|P_j^{(k-1)}(t)\right| \\
&\leq \sum_{j=k-1}^{n-1} \frac{2j+1}{2}\left[1 - (-1)^{n+j}\right] P_j^{(k-1)}(1) \\
&= P_n^{(k)}(1) \qquad (3.91)
\end{aligned}$$

for all $t \in [-1,1]$ and $n \in \mathbb{N}_0$. Moreover, the definition of the Legendre polynomials yields $P_n^{(0)}(1) = 1$ for all $n \in \mathbb{N}_0$. Provided that there exists a constant $C \in \mathbb{R}$ such that $P_n^{(k-1)}(1) \leq C n^{2k-2}$ for all $n \in \mathbb{N}_0$ and a given $k \in \mathbb{N}_0$, we get

$$\begin{aligned} P_n^{(k)}(1) &= \sum_{j=k-1}^{n-1} \frac{2j+1}{2} \left[1-(-1)^{n+j}\right] P_j^{(k-1)}(1) \\ &\leq \sum_{\substack{j=k-1 \\ n+j \text{ odd}}}^{n-1} (2j+1) C j^{2k-2} \\ &\leq \frac{n+1}{2} (2n-1) C n^{2k-2} \\ &= \mathcal{O}\left(n^{2k}\right) \quad \text{as } n \to \infty. \end{aligned} \tag{3.92}$$

□

Moreover, each Jacobi polynomial is a solution of a corresponding differential equation (see, e.g., [26, p. 148], [180, p. 60]). In other words, these functions are eigenfunctions of a particular differential operator. We will also find a way of proving this property for the Legendre polynomials in the context of the spherical analysis (see Theorem 5.29).

Theorem 3.22. *For* $\alpha, \beta > -1$; $w(x) := (1-x)^\alpha (1+x)^\beta$, $x \in [-1,1]$; $\lambda_n := -n(n+\alpha+\beta+1)$, $n \in \mathbb{N}_0$; *the Jacobi polynomial* $y = P_n^{(\alpha,\beta)}$ *is a solution of the ordinary differential equation*

$$\frac{\mathrm{d}}{\mathrm{d}x}\left[\left(1-x^2\right) w(x) y'(x)\right] - \lambda_n w(x) y(x) = 0, \tag{3.93}$$

$x \in [-1,1]$.

Note that (3.93) is equivalent to

$$\left(1-x^2\right) y''(x) + [\beta - \alpha - (\alpha+\beta+2)x] y'(x) + n(n+\alpha+\beta+1) y(x) = 0, \tag{3.94}$$

$x \in [-1,1]$.

There are more orthogonal basis systems in $\mathrm{L}^2[a,b]$ than only the translates of the Legendre polynomials (i.e., $Q_n(t) := P_n(2\frac{t-a}{b-a} - 1)$, $t \in [a,b]$, $n \in \mathbb{N}_0$, provides an orthogonal basis of $\mathrm{L}^2[a,b]$).

For instance, instead of orthogonal algebraic polynomials, as we discussed up to now, one can take orthogonal trigonometric polynomials. The details are explained in the following theorem.

Theorem 3.23 (Trigonometric ONB of $\mathbf{L}^2[0,2\pi]$). *An orthonormal basis of the Hilbert space* $\mathrm{L}^2[0,2\pi]$ *is given by the union of the following functions:*

$$
\begin{aligned}
F_{0,1}(t) &:= \frac{1}{\sqrt{2\pi}}, \quad t \in [0,2\pi], \\
F_{n,1}(t) &:= \frac{1}{\sqrt{\pi}} \cos(nt), \quad t \in [0,2\pi], \quad n \in \mathbb{N}, \\
F_{n,2}(t) &:= \frac{1}{\sqrt{\pi}} \sin(nt), \quad t \in [0,2\pi], \quad n \in \mathbb{N}. \qquad (3.95)
\end{aligned}
$$

The proof of the orthonormality is an easy exercise in integration by parts. For a proof of the completeness, see [27, pp. 36–43].

A decomposition of $G \in \mathrm{L}^2[0,2\pi]$ in terms of

$$
G = \left\langle G, \frac{1}{\sqrt{2\pi}} \right\rangle_{\mathrm{L}^2[0,2\pi]} \frac{1}{\sqrt{2\pi}} + \sum_{n=1}^{\infty} \sum_{j=1}^{2} \langle G, F_{n,j} \rangle_{\mathrm{L}^2[0,2\pi]} F_{n,j} \qquad (3.96)
$$

(with respect to $\mathrm{L}^2[0,2\pi]$) corresponds to the representation of a signal as a superposition of oscillations of different frequencies. The Fourier coefficients $\langle G, F_{n,j} \rangle$ represent the amplitudes where n is a kind of a frequency. Note that there exist fast methods for calculating the Fourier coefficients with respect to this onb, see, for example, [32], [153, Sect. 10.9.2], [195, Sect. 3.2.3].

As we have already discussed in the introduction, the problem of such a representation is the complete loss of a temporal reference. The sequence of amplitudes (which can be called here the Fourier transform of G) only represents the mean contribution of each frequency over the whole time interval. We cannot distinguish, for example, if the high-frequency parts equally occurred over the whole observation period or if they dominated the initial or the final part only.

A helpful result in this context is the construction of localized analogues of the trigonometric onb, which provide an onb of $\mathrm{L}^2(\mathbb{R})$.

Theorem 3.24 (Localized Trigonometric ONB of $\mathrm{L}^2(\mathbb{R})$). *Let $\varepsilon > 0$ and $r \in]0,\varepsilon]$ be given numbers and let $(a_j)_{j\in\mathbb{Z}}$ be a sequence satisfying $\lim_{j\to\pm\infty} a_j = \pm\infty$ and $a_{j+1} - a_j \geq \varepsilon$ for all $j \in \mathbb{Z}$. Moreover, let*

$$
\beta(t) := \sin\left[\frac{\pi}{4}\left(1 + \sin\left(\frac{\pi}{2}t\right)\right)\right], \quad t \in \mathbb{R}, \qquad (3.97)
$$

and the sequence of functions $(b_j)_{j\in\mathbb{Z}}$ with

$$
b_j(t) := \begin{cases} \beta\left(\frac{t-a_j}{r}\right), & \text{if } a_j - r \leq t < a_j + r \\ 1, & \text{if } a_j + r \leq t < a_{j+1} - r \\ \beta\left(\frac{a_{j+1}-t}{r}\right), & \text{if } a_{j+1} - r \leq t < a_{j+1} + r \\ 0 & \text{else} \end{cases} \qquad (3.98)
$$

be given. Then the system $\{\Psi_{k,j}\}_{k\in\mathbb{N}_0,\, j\in\mathbb{Z}}$ defined by

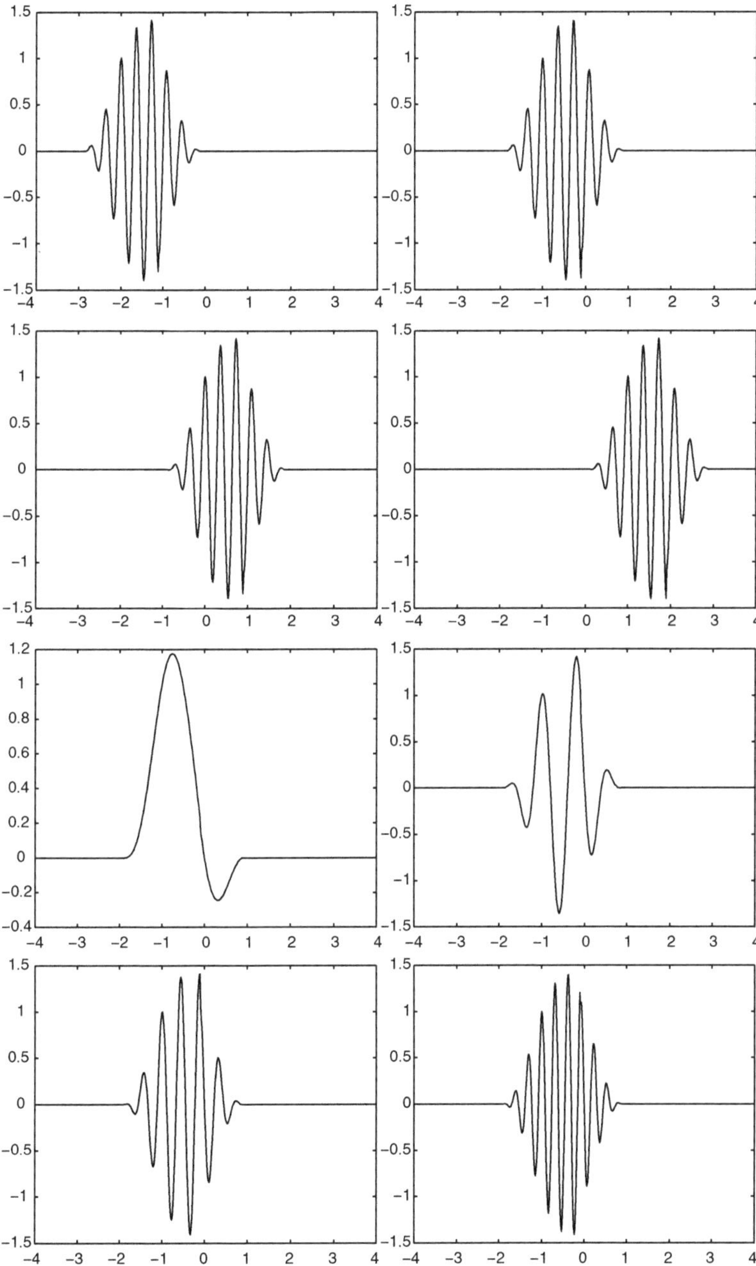

Fig. 3.3 Localized trigonometric basis functions $\Psi_{k,j}$: the functions concentrate on different subintervals of $\mathbb{R}$. This translation is controlled by the index j. Moreover, the index k provides functions at different "frequencies." In this example, the parameters $a_j := j - 3$ and $r := 0.9$ were chosen. The images show (in reading direction) the functions $\Psi_{5,1}$, $\Psi_{5,2}$, $\Psi_{5,3}$, $\Psi_{5,4}$, $\Psi_{0,2}$, $\Psi_{2,2}$, $\Psi_{4,2}$, and $\Psi_{6,2}$

$$\Psi_{k,j}(t) := b_j(t)\sqrt{\frac{2}{a_{j+1}-a_j}}\cos\left[\frac{\pi}{a_{j+1}-a_j}\left(k+\frac{1}{2}\right)(t-a_j)\right], \quad t\in\mathbb{R}, \tag{3.99}$$

is an orthonormal basis of $\mathrm{L}^2(\mathbb{R})$.

Elaborated numerical methods such as a best-basis algorithm exist for the efficient choice of a finite subsystem for practical purposes. For further details, we refer to [30, 196, 197]. Figure 3.3 shows a selection of some basis functions.

3.2 A Brief Introduction to Cubic Splines

Polynomial interpolation has some drawbacks. One of them is connected to the fact that the graphs of polynomials oscillate. Although there are polynomials such as x^{2n+1}, $n\in\mathbb{N}_0$, which are monotonically increasing, too many grid points or a few extraordinary values can produce ugly, badly approximating polynomials.

Remember that the data $\{F(x_j)\}_{j=0,\ldots,n}$ and the ansatz

$$P(x) = \sum_{k=0}^{n} a_k B_k(x), \tag{3.100}$$

where $B_0,\ldots,B_n$ are chosen basis functions, yield the linear equations

$$\sum_{k=0}^{n} a_k B_k(x_j) = F(x_j) \quad \forall j=0,\ldots,n. \tag{3.101}$$

In the case of monomials $B_k(x)=x^k$, the corresponding matrix is the Vandermonde matrix, which is known to be regular but very ill-conditioned (see Fig. 3.4). Even for moderate values of n, the matrix becomes numerically singular. On the other hand, the analytical regularity of the Vandermonde matrix (for each $n\geq 1$) implies the uniqueness of a polynomial $P\in\mathrm{Pol}_{0\ldots n}(\mathbb{R})$ satisfying $P(x_j)=F(x_j)$, $j=0,\ldots,n$. Figure 3.5 shows these interpolating polynomials for different n, where the Legendre polynomials were used as basis functions. As it was mentioned above, the single "false value" causes heavy oscillations in this case.

Due to this undesired behavior, cubic splines were developed. Details on this approach can be learned in every textbook on basic numerical analysis (such as [153, Sect. 8.7], [195, Sect. 3.3]). We will only discuss some essentials of this method.

Definition 3.25. Let $G_n := \{x_0,\ldots,x_n\}$ be a grid of $n+1$ points in $[a,b]$, where $a=x_0<x_1<\cdots<x_{n-1}<x_n=b$. A function $S:[a,b]\to\mathbb{R}$ is called a **cubic spline** if it satisfies the following conditions:

(S1) For each $j=0,\ldots,n-1$, the restriction $S\big|_{[x_j,x_{j+1}]}$ is a polynomial of degree 3 or lower.

(S2) $S\in\mathrm{C}^{(2)}[a,b]$.

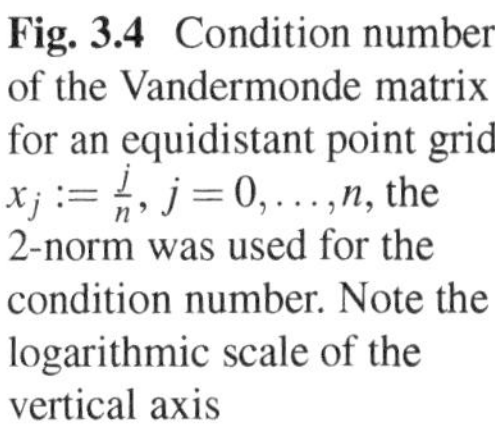

Fig. 3.4 Condition number of the Vandermonde matrix for an equidistant point grid $x_j := \frac{j}{n}$, $j = 0, \ldots, n$, the 2-norm was used for the condition number. Note the logarithmic scale of the vertical axis

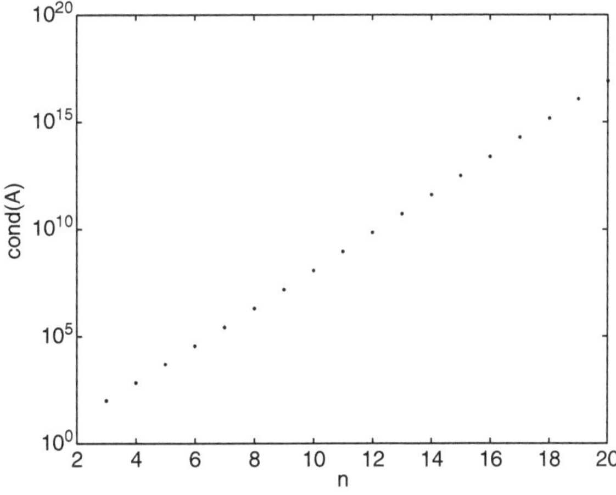

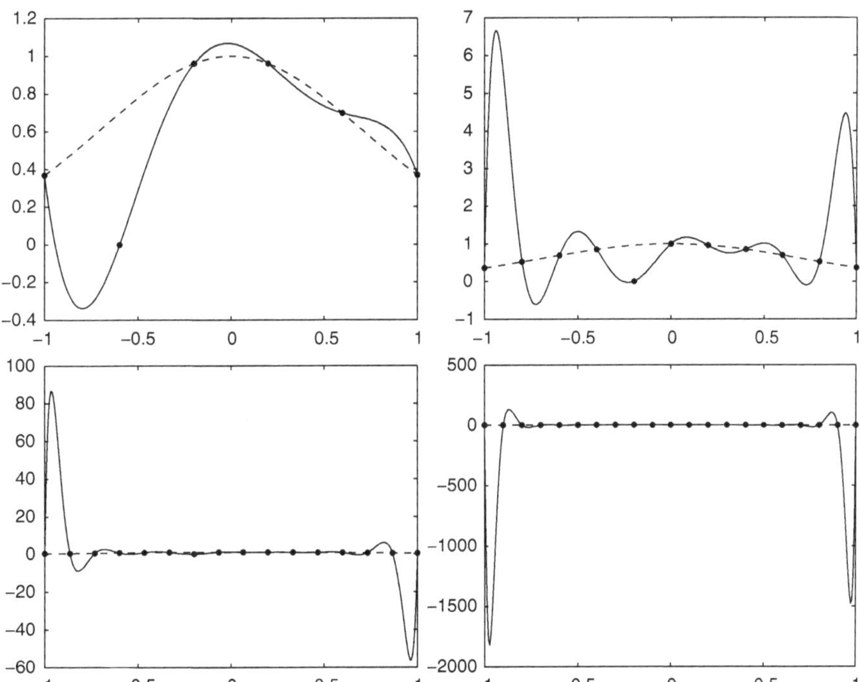

Fig. 3.5 Different equidistant point grids $x_0, \ldots, x_n$ on $[-1, 1]$ with $x_0 = -1$ and $x_n = 1$ were used to generate data $y_j = \mathrm{e}^{-x_j^2}$, $j = 0, \ldots, n$, where one value was replaced by 0. In each plot, the unique interpolating polynomial of degree $\leq n$ and the associated points (x_j, y_j), $j = 0, \ldots, n$, are shown. The *dashed line* corresponds to $F(t) = \mathrm{e}^{-t^2}$. The chosen values for n are: 5 (*top left*), 10 (*top right*), 15 (*bottom left*), and 20 (*bottom right*)

Hence, a cubic spline can be represented as

$$S(x) = \sum_{j=0}^{n-1} \left[a_j + b_j (x - x_j) + c_j (x - x_j)^2 + d_j (x - x_j)^3 \right] \chi_{[x_j, x_{j+1}[}(x), \qquad (3.102)$$

where χ_D is the **characteristic function (indicator function)** of the set D with $\chi_D(x) := 0$ for $x \notin D$ and $\chi_D(x) := 1$ for $x \in D$. Note that the value at x_n is automatically given due to the continuity of S.

In the combination with the interpolation conditions $S(x_j) = y_j$, $j = 0, \ldots, n$ where $(y_0, \ldots, y_n)$ is given, most of the coefficients of S are uniquely determined. More precisely, $S(x_j) = y_j$, $j = 0, \ldots, n-1$, yields in (3.102) immediately $a_j = y_j$ for all $j = 0, \ldots, n-1$. Furthermore, the $\mathrm{C}^{(2)}[a,b]$-requirement in combination with $S(x_n) = y_n$ implies that

$$y_j + b_j \left(x_{j+1} - x_j\right) + c_j \left(x_{j+1} - x_j\right)^2 + d_j \left(x_{j+1} - x_j\right)^3 = y_{j+1} \ \forall j = 0, \ldots, n-1, \qquad (3.103)$$

$$b_j + 2c_j \left(x_{j+1} - x_j\right) + 3d_j \left(x_{j+1} - x_j\right)^2 = b_{j+1} \ \forall j = 0, \ldots, n-2, \qquad (3.104)$$

$$2c_j + 6d_j \left(x_{j+1} - x_j\right) = 2c_{j+1} \ \forall j = 0, \ldots, n-2. \qquad (3.105)$$

With the abbreviation $h_j := x_{j+1} - x_j$, (3.105) yields

$$d_j = \frac{c_{j+1} - c_j}{3h_j} \quad \forall j = 0, \ldots, n-2. \qquad (3.106)$$

Inserting (3.106) in (3.103) gives

$$\begin{aligned} b_j &= \frac{y_{j+1} - y_j}{h_j} - c_j h_j - \frac{c_{j+1} - c_j}{3h_j} h_j^2 \\ &= \frac{y_{j+1} - y_j}{h_j} - \frac{1}{3}\left(c_{j+1} + 2c_j\right) h_j \quad \forall j = 0, \ldots, n-2. \end{aligned} \qquad (3.107)$$

Finally, inserting (3.106) and (3.107) in (3.104) yields

$$\begin{aligned} &\frac{y_{j+1} - y_j}{h_j} - \frac{1}{3}\left(c_{j+1} + 2c_j\right) h_j + 2c_j h_j + \left(c_{j+1} - c_j\right) h_j \\ &\quad = \frac{y_{j+2} - y_{j+1}}{h_{j+1}} - \frac{1}{3}\left(c_{j+2} + 2c_{j+1}\right) h_{j+1} \quad \forall j = 0, \ldots, n-3, \end{aligned} \qquad (3.108)$$

which is equivalent to

$$\begin{aligned} &\frac{1}{3} h_j c_j + \frac{2}{3}\left(h_{j+1} + h_j\right) c_{j+1} + \frac{1}{3} h_{j+1} c_{j+2} \\ &\quad = \frac{y_{j+2} - y_{j+1}}{h_{j+1}} - \frac{y_{j+1} - y_j}{h_j} \quad \forall j = 0, \ldots, n-3. \end{aligned} \qquad (3.109)$$

Equation (3.109) is a linear system with a right-hand side which is given by the data and the point grid. The corresponding unknown vector is $(c_0,\dots,c_{n-1})^{\mathrm{T}}$, and the associated matrix is

$$\begin{pmatrix} \frac{1}{3}h_0 & \frac{2}{3}(h_1+h_0) & \frac{1}{3}h_1 & 0 & & & & \\ 0 & \frac{1}{3}h_1 & \frac{2}{3}(h_2+h_1) & \frac{1}{3}h_2 & 0 & & & \\ & \ddots & \ddots & \ddots & \ddots & \ddots & & \\ & & \ddots & \ddots & \ddots & \ddots & \ddots & \\ & & & \ddots & \ddots & \ddots & \ddots & \ddots \\ & & & 0 & \frac{1}{3}h_{n-4} & \frac{2}{3}(h_{n-3}+h_{n-4}) & \frac{1}{3}h_{n-3} & 0 \\ & & & & 0 & \frac{1}{3}h_{n-3} & \frac{2}{3}(h_{n-2}+h_{n-3}) & \frac{1}{3}h_{n-2} \end{pmatrix}. \tag{3.110}$$

This is a $(n-2)\times n$-matrix which obviously has the maximal rank $n-2$, since $h_j \neq 0$ for all j. Thus, two further conditions are needed to obtain $(c_0,\dots,c_{n-1})^{\mathrm{T}}$ uniquely (and, automatically, also the other coefficients via (3.106) and (3.107) for $j=0,\dots,n-2$). Three possibilities of adding two independent conditions are common, where we will study the first one in further detail.

(N) The **natural cubic spline** has to satisfy $S''(x_0)=S''(x_n)=0$ which yields $c_0 = 0$ (which cuts out the first column of the matrix above) and the additional last equation

$$c_{n-1}+3d_{n-1}h_{n-1}=0 \tag{3.111}$$

which is equivalent to

$$d_{n-1}=-\frac{c_{n-1}}{3h_{n-1}}. \tag{3.112}$$

Inserting (3.112) in (3.103) for $j=n-1$ yields

$$\begin{aligned} b_{n-1} &= \frac{y_n-y_{n-1}}{h_{n-1}} - c_{n-1}h_{n-1} + \frac{1}{3}c_{n-1}h_{n-1} \\ &= \frac{y_n-y_{n-1}}{h_{n-1}} - \frac{2}{3}c_{n-1}h_{n-1}. \end{aligned} \tag{3.113}$$

We observe now that (3.107) is also valid for $j=n-1$ if we define $c_n := 0$. Hence, (3.109) can analogously also be derived for $j=n-2$. We get

$$\frac{1}{3}h_{n-2}c_{n-2}+\frac{2}{3}(h_{n-1}+h_{n-2})\,c_{n-1} = \frac{y_n-y_{n-1}}{h_{n-1}} - \frac{y_{n-1}-y_{n-2}}{h_{n-2}}. \tag{3.114}$$

This adds the last line

$$\left(0,\dots,0,\frac{1}{3}h_{n-2},\frac{2}{3}(h_{n-1}+h_{n-2})\right) \tag{3.115}$$

to the matrix above which is now tridiagonal (with size $(n-1)\times(n-1)$). This matrix is now, indeed, regular. It is even positive definite. This can be seen as follows (see, e.g., [195, Lemma 3.6]): We first observe that the matrix (let us call it $A = (a_{ij})_{i,j=1,\dots,n-1}$) satisfies[3]

$$\sum_{\substack{j=1\\ j\neq i}}^{n-1} |a_{ij}| < a_{ii} \qquad \forall i = 1,\dots,n-1, \tag{3.116}$$

for the very trivial reason that $\frac{1}{3} < \frac{2}{3}$. Now let (λ, x) be an arbitrary eigenvalue–eigenvector pair, where we assume—without loss of generality—that $\|x\|_\infty := \max_{1\le j\le n-1} |x_j| = 1$. We choose an index $i \in \{1,\dots,n-1\}$ with $|x_i| = \|x\|_\infty$. Then

$$\lambda x_i = \sum_{j=1}^{n-1} a_{ij}x_j = a_{ii}x_i + \sum_{\substack{j=1\\ j\neq i}}^{n-1} a_{ij}x_j \tag{3.117}$$

and, consequently,

$$|\lambda - a_{ii}| \underbrace{|x_i|}_{=1} = \left|\sum_{\substack{j=1\\ j\neq i}}^{n-1} a_{ij}x_j\right| \le \sum_{\substack{j=1\\ j\neq i}}^{n-1} |a_{ij}| \underbrace{|x_j|}_{\le\|x\|_\infty=1} < a_{ii}. \tag{3.118}$$

Due to the absolute value on the left-hand side, we get

$$-a_{ii} < \lambda - a_{ii} < a_{ii}. \tag{3.119}$$

Hence,

$$0 < \lambda < 2a_{ii}. \tag{3.120}$$

Thus, λ is positive.

(H) The **Hermitian cubic spline** has given derivatives at the boundary: $S'(x_0)$ and $S'(x_n)$ are given.

(P) The **periodic cubic spline** is periodic in the zeroth to second derivatives, that is, $y_0 = y_n$, $S'(x_0) = S'(x_n)$, and $S''(x_0) = S''(x_n)$.

Cubic splines and, in particular, the natural cubic spline are much smoother than an interpolating polynomial (see Fig. 3.6), where "smooth" refers here to a low curvature of the graph. This property is demonstrated by the following theorem.

Theorem 3.26 (Holladay's Theorem). *Let $a = x_0 < x_1 < \dots < x_{n-1} < x_n = b$ and $y = (y_0,\dots,y_n)^{\mathrm{T}} \in \mathbb{R}^{n+1}$ be given. Then the natural cubic spline given by $S(x_j) = y_j$, $j = 0,\dots,n$, minimizes the functional*

[3] This property is called the strict diagonal dominance.

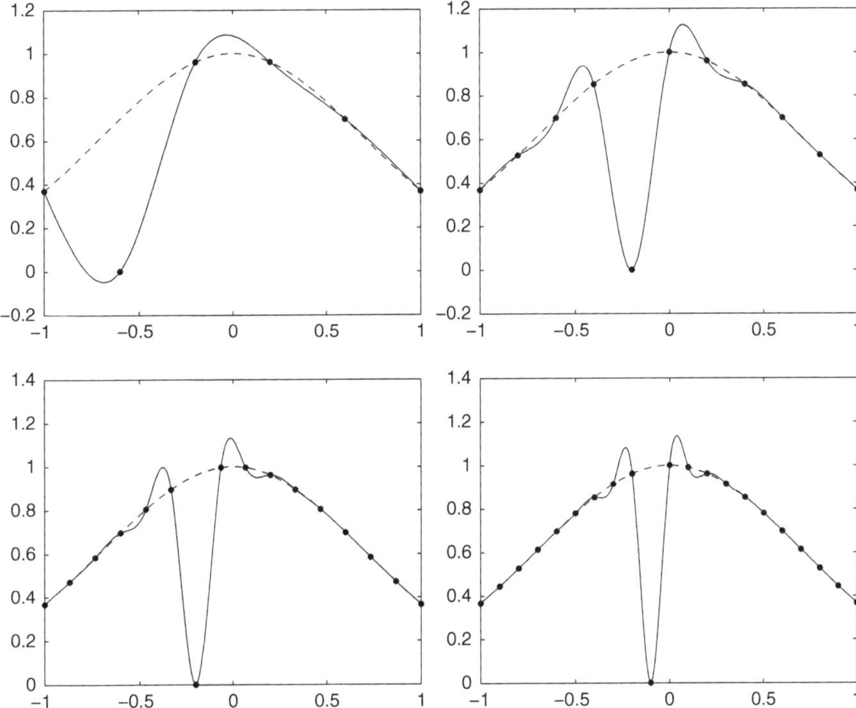

Fig. 3.6 In comparison to Fig. 3.5, the same interpolation problems were solved by natural cubic splines. The "false value" has a smaller effect on the curvature (and, in particular, only a local effect!) in the case of the cubic spline

$$G \mapsto \int_a^b \left(G''(x)\right)^2 \mathrm{d}x \quad \left(= \|G''\|^2_{\mathrm{L}^2[a,b]}\right) \tag{3.121}$$

among all functions $G \in \mathrm{C}^{(2)}[a,b]$ which satisfy $G(x_j) = y_j$, $j = 0,\ldots,n$.

Note that $\|G''\|_{\mathrm{L}^2[a,b]}$ is the linearized curvature of the curve defined by the graph of G.

Furthermore, splines react less sensitive to gaps in the point grid [remember the data set (D1)] than polynomials, see Fig. 3.7.

Remark 3.27. Natural cubic splines can also be used for numerical integration and differentiation. Following [84, 163, 167], one defines the linear functional $\mathscr{F} : \mathrm{C}^{(1)}[a,b] \to \mathbb{R}$ by

$$\mathscr{F}F := \int_a^b a_0(x)\, F(x) + a_1(x)\, F'(x)\, \mathrm{d}x + \sum_{j=0}^{1} \sum_{i=1}^{N_j} b_{ij}\, F^{(j)}\left(\xi_{ij}\right), \quad F \in \mathrm{C}^{(1)}[a,b], \tag{3.122}$$

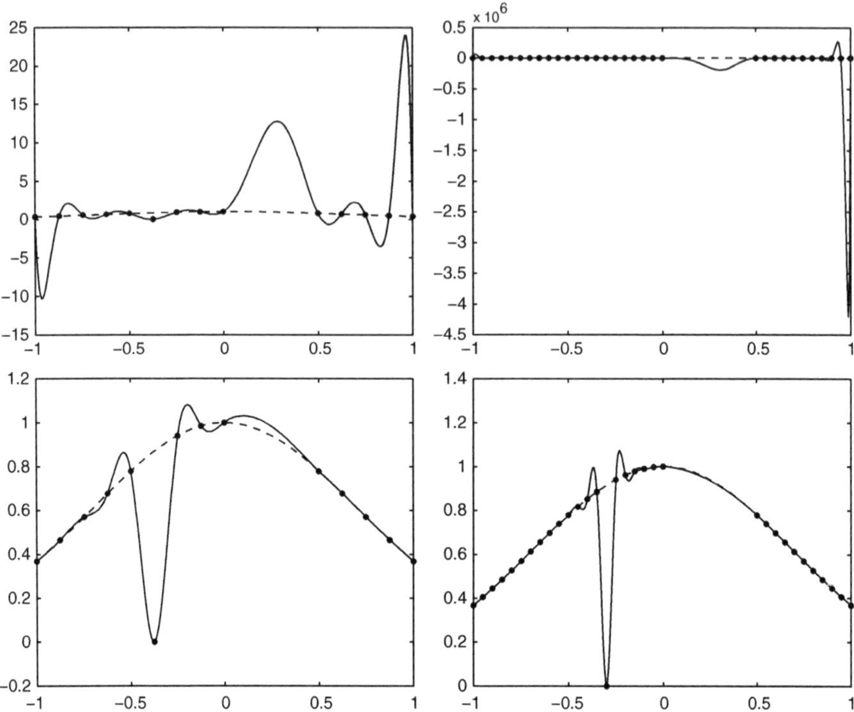

Fig. 3.7 A new point grid with a gap at $]0,0.5[$ was used in the example of Figs. 3.5 and 3.6. The *first row* shows the interpolating polynomial in the case of 14 (*left*) and 32 (*right*) grid points, whereas the *second row* shows the corresponding natural cubic spline which behaves less extremely at the gap

where each $a_j : [a,b] \to \mathbb{R}$ is piecewise continuous and all ξ_{ij} are points in $[a,b]$. This covers, in particular, the cases where $\mathscr{F}$ yields the integral over (subintervals of) $[a,b]$ or the value of F' at a given point in $[a,b]$. Since usually F is only given on a point grid, one commonly tries to approximate $\mathscr{F}$ by a functional $\mathscr{G}$ of the form

$$\mathscr{G}F := \sum_{j=0}^{n} c_j F(x_j), \tag{3.123}$$

where $a = x_0 < \cdots < x_n = b$. Often, one requires some polynomial exactness, that is, we want that $(\mathscr{F} - \mathscr{G})P = 0$ for all $P \in \mathrm{Pol}_{0\ldots r}(\mathbb{R})$ for a given $r \in \mathbb{N}$ (we say that $\mathscr{G}$ is exact for the degree r). Peano's theorem says that, in this case of exactness for the degree r, the approximation error can be represented as

$$(\mathscr{F} - \mathscr{G})\,F = \int_a^b K(t)\,F^{(r+1)}(t)\,\mathrm{d}t \tag{3.124}$$

for all $F \in \mathrm{C}^{(r+1)}[a,b]$, where the function K can be represented by a particular formula, which is not of further interest here. We only need to know that K depends on $\mathscr{F}$, $\mathscr{G}$, and r only and is independent of F. According to Sard, a functional $\mathscr{G}$ which is exact for the degree r is called a **best approximation** if

$$J(\mathscr{G}) := \int_a^b (K(t))^2 \,\mathrm{d}t \tag{3.125}$$

is minimized by $\mathscr{G}$ in comparison to all other functionals of the form (3.123) which are exact for the degree r. Finally, Schoenberg's theorem says the following: If S_F^* denotes the natural cubic spline given by $S_F^*(x_j) = F(x_j) \forall j = 0,\ldots,n$, and $\mathscr{F}$ is given by (3.122), then there is one and only one functional $\mathscr{G}$ of the form (3.123) which satisfies $\mathscr{G}F = \mathscr{F}S_F^*$ for all $F \in \mathrm{C}^{(2)}[a,b]$. This functional $\mathscr{G}$ is also exact for the degree 1 and is *the* best approximation to $\mathscr{F}$, that is,

$$J(\mathscr{G}) < J\left(\tilde{\mathscr{G}}\right) \tag{3.126}$$

for all $\tilde{\mathscr{G}} \neq \mathscr{G}$ of the form (3.123) which approximate $\mathscr{F}$ exactly for the degree 1.

What does this mean? We need not know what the functional $\mathscr{G}$ looks like. If $\mathscr{G}$ is the best approximation to $\mathscr{F}$, that is, $\mathscr{G}F = \mathscr{F}S_F^*$ for all $F \in \mathrm{C}^{(2)}[a,b]$, we simply have to interpolate F by the natural cubic spline S_F^* and to apply $\mathscr{F}$ to S_F^*. We can expect $\mathscr{F}S_F^*$ to be an appropriate approximation to $\mathscr{F}F$.

3.3 An Approximate Identity for $\mathbb{R}$

It might be worth reading the remarks on Principle 2 in the introduction once again. We already discussed the (in-)famous Dirac delta function here. Let us prove its nonexistence.

Remark 3.28. Let us denote again the set of all continuous functions $F : \mathbb{R} \to \mathbb{R}$ with a compact support by $\mathrm{C}_0(\mathbb{R})$. We will show now that there is no function $\delta : \mathbb{R} \to \mathbb{R}$ such that

$$F(y) = \int_{\mathbb{R}} \delta(y-x)\, F(x)\,\mathrm{d}x \quad \forall y \in \mathbb{R} \quad \forall F \in \mathrm{C}_0(\mathbb{R}). \tag{3.127}$$

The proof is as follows: Let us assume that there exists such a function δ, which satisfies, say, the local absolute integrability,[4] that is,

$$\int_a^b |\delta(x)|\,\mathrm{d}x < +\infty \tag{3.128}$$

[4]Otherwise, the integral itself in (3.127) would not make sense.

for all $a,b \in \mathbb{R}$ with $a < b$. Then (3.127) must also be valid for the family of functions $\{F_c\}_{c>0} \subset \mathrm{C}_0(\mathbb{R})$ which is given by

$$F_c(x) := \begin{cases} \exp\left(\frac{c^2}{x^2-c^2}\right), & |x| < c \\ 0 & \text{else} \end{cases}. \tag{3.129}$$

Note that $F_c(x) = F_c(-x)$. Hence, we get

$$\begin{aligned}
\mathrm{e}^{-1} &= F_c(0) && (3.130)\\
&= \int_{-\infty}^{+\infty} \delta(-x) F_c(x)\, \mathrm{d}x && (3.131)\\
&= \int_{-\infty}^{+\infty} \delta(x) F_c(x)\, \mathrm{d}x && (3.132)\\
&= \int_{-c}^{c} \delta(x) F_c(x)\, \mathrm{d}x && (3.133)\\
&\leq \int_{-c}^{c} |\delta(x)| \underbrace{|F_c(x)|}_{\leq F_c(0) = \mathrm{e}^{-1}} \mathrm{d}x && (3.134)\\
&\leq \mathrm{e}^{-1} \int_{-c}^{c} |\delta(x)|\, \mathrm{d}x && (3.135)
\end{aligned}$$

for all $c > 0$. Consequently,

$$\int_{-c}^{c} |\delta(x)|\, \mathrm{d}x \geq 1 \tag{3.136}$$

for all $c > 0$, in particular, for $c \to 0+$, which is a contradiction. This completes the proof.

Following Principle 2 of the introduction, we are seeking now a sequence of kernels $\Phi_J : \mathbb{R} \times \mathbb{R} \to \mathbb{R}$ such that

$$\int_{\mathbb{R}} \Phi_J(x,y) F(y)\, \mathrm{d}y \tag{3.137}$$

converges to $F(x)$ for certain F. It turns out that, for example, the Gaussian functions

$$\Phi_J(x,y) := \frac{J}{2\sqrt{\pi}} \mathrm{e}^{-(J(x-y))^2/4}, \quad x,y \in \mathbb{R}, J \in \mathbb{R}^+, \tag{3.138}$$

can be chosen for this purpose. We, consequently, have a family $(T_J)_J$ of operators

$$T_J : F \mapsto \int_{\mathbb{R}} \Phi_J(\cdot, y) F(y)\, \mathrm{d}y \tag{3.139}$$

which converges to the identity operator (but only in the sense that each function $T_J F$ converges *pointwise* to F). This family of operators $(T_J)_J$ (or, alternatively, the family of kernels $(\Phi_J)_J$) is, therefore, called an Approximate Identity.

Definition 3.29. The space $\mathrm{C}_1(\mathbb{R})$ is defined to consist of all continuous functions $F : \mathbb{R} \to \mathbb{R}$ satisfying

$$\int_{\mathbb{R}} |F(x)| \, \mathrm{d}x < +\infty. \tag{3.140}$$

Theorem 3.30 (Approximate Identity). *For each $J \in \mathbb{R}^+$, the kernels $\Phi_J : \mathbb{R} \times \mathbb{R} \to \mathbb{R}$ are defined by*

$$\Phi_J(x,y) := \frac{J}{2\sqrt{\pi}} \mathrm{e}^{-(J(x-y))^2/4}, \quad x,y \in \mathbb{R}. \tag{3.141}$$

Then the following (so-called convolution identity[5]) holds true for all $F \in \mathrm{C}_1(\mathbb{R})$:

$$\lim_{J\to\infty} \int_{\mathbb{R}} \Phi_J(x,y) F(y) \, \mathrm{d}y = F(x) \quad \textit{for all } x \in \mathbb{R}. \tag{3.142}$$

Proof. The proof is subdivided into several parts. First of all, we keep $x \in \mathbb{R}$ arbitrary but fixed.

(1) Rewriting the integral:
A simple substitution $y := x - \tilde{y}$ yields

$$\begin{aligned} \int_{\mathbb{R}} \Phi_J(x,\tilde{y}) \, F(\tilde{y}) \, \mathrm{d}\tilde{y} &= \frac{J}{2\sqrt{\pi}} \int_{\mathbb{R}} \mathrm{e}^{-(J(x-\tilde{y}))^2/4} F(\tilde{y}) \, \mathrm{d}\tilde{y} \\ &= \frac{J}{2\sqrt{\pi}} \int_{\mathbb{R}} \mathrm{e}^{-(Jy)^2/4} F(x-y) \, \mathrm{d}y. \end{aligned} \tag{3.143}$$

We set

$$G_J(y) := \frac{J}{2\sqrt{\pi}} \mathrm{e}^{-(Jy)^2/4}, \quad y \in \mathbb{R}, J \in \mathbb{R}^+, \tag{3.144}$$

and observe that we have to show that

$$\lim_{J\to\infty} \int_{\mathbb{R}} G_J(y) F(x-y) \, \mathrm{d}y = F(x) \quad \forall x \in \mathbb{R}. \tag{3.145}$$

(2) A partition of unity:
It is an easy exercise in analysis to calculate the following integral ($z := \frac{Jy}{2}$):

$$\int_{\mathbb{R}} \mathrm{e}^{-(Jy)^2/4} \, \mathrm{d}y = \int_{\mathbb{R}} \mathrm{e}^{-z^2} \, \mathrm{d}z \cdot \frac{2}{J} = \sqrt{\pi} \cdot \frac{2}{J}. \tag{3.146}$$

---

[5] See also Definition 3.31 below.

Hence,

$$\int_{\mathbb{R}} G_J(y)\,\mathrm{d}y = 1. \tag{3.147}$$

Equation (3.147) is a partition of unity (in a more abstract sense, since the integration is regarded here as a kind of a continuous summation in contrast to the discrete summation, e.g., in (3.16)). In the proof of Theorem 3.3, we have already observed how helpful such structures can be in Constructive Approximation. The "trick" is similar here.

(3) The main step:
Due to our preliminary work, we have

$$\begin{aligned}\left|\int_{\mathbb{R}} G_J(y)F(x-y)\,\mathrm{d}y - F(x)\right| &= \left|\int_{\mathbb{R}} G_J(y)F(x-y)\,\mathrm{d}y - \int_{\mathbb{R}} G_J(y)\,\mathrm{d}y\,F(x)\right| \\ &= \left|\int_{\mathbb{R}} G_J(y)(F(x-y)-F(x))\,\mathrm{d}y\right| \\ &\le \int_{\mathbb{R}} G_J(y)\,|F(x-y)-F(x)|\,\mathrm{d}y. \end{aligned} \tag{3.148}$$

We haven't used any property of F, yet. So, it's time to do so, since the theorem cannot be valid for every arbitrary function on $\mathbb{R}$.
Let $\varepsilon > 0$ be given. Since F is continuous in x, there exists $\delta > 0$ such that

$$|F(x-y)-F(x)| \le \varepsilon \tag{3.149}$$

for every $y \in \mathbb{R}$ with $|y| \le \delta$. Hence, the last integral in (3.148) can be estimated as follows:

$$\begin{aligned} &\int_{-\delta}^{\delta} G_J(y)\,|F(x-y)-F(x)|\,\mathrm{d}y + \int_{|y|>\delta} G_J(y)\,|F(x-y)-F(x)|\,\mathrm{d}y \\ &\le \varepsilon \int_{-\delta}^{\delta} G_J(y)\,\mathrm{d}y + \int_{|y|>\delta} G_J(y)\,|F(x-y)|\,\mathrm{d}y + \int_{|y|>\delta} G_J(y)\,|F(x)|\,\mathrm{d}y \\ &\le \varepsilon \int_{\mathbb{R}} G_J(y)\,\mathrm{d}y + \int_{|y|>\delta} \underbrace{\left[\sup_{|\eta|>\delta} G_J(\eta)\right]}_{=G_J(\delta)\,(=G_J(-\delta))} |F(x-y)|\,\mathrm{d}y + |F(x)| \int_{|y|>\delta} G_J(y)\,\mathrm{d}y \\ &\le \varepsilon + G_J(\delta)\|F\|_{\mathrm{L}^1(\mathbb{R})} + |F(x)|\left(1 - \int_{-\delta}^{\delta} G_J(y)\,\mathrm{d}y\right). \end{aligned} \tag{3.150}$$

(4) The final steps:
For the very $\varepsilon > 0$ chosen above and its associated $\delta > 0$, we have $\lim_{J\to\infty} G_J(\delta) = 0$, that is, there exists J_0 such that for all $J \geq J_0$, the inequality $(0 <)\, G_J(\delta(\varepsilon)) \leq \frac{\varepsilon}{\|F\|_{\mathrm{L}^1(\mathbb{R})}}$ is valid.[6] Moreover, the substitution $z := \frac{Jy}{2}$ leads us to

$$\begin{aligned}\int_{-\delta}^{\delta} G_J(y)\,\mathrm{d}y &= \frac{J}{2\sqrt{\pi}} \int_{-\delta}^{\delta} \mathrm{e}^{-(Jy)^2/4}\,\mathrm{d}y \\ &= \frac{1}{\sqrt{\pi}} \int_{-J\delta/2}^{J\delta/2} \mathrm{e}^{-z^2}\,\mathrm{d}z \stackrel{J\to\infty}{\longrightarrow} \frac{1}{\sqrt{\pi}} \int_{\mathbb{R}} \mathrm{e}^{-z^2}\,\mathrm{d}z = 1. \qquad (3.151)\end{aligned}$$

Hence, there exists J_1 such that for all $J \geq J_1$, we have

$$(0 <)\, 1 - \int_{-\delta(\varepsilon)}^{\delta(\varepsilon)} G_J(y)\,\mathrm{d}y \leq \frac{\varepsilon}{|F(x)|} \qquad (3.152)$$

(note that x is fixed) in the case of $F(x) \neq 0$. Otherwise, set $J_1 := J_0$.

Finally, since $\varepsilon > 0$ was arbitrary, for every $\varepsilon > 0$, there exists $J_2 := \max(J_0, J_1)$ such that for all $J \geq J_2$, the inequality

$$\left| \int_{\mathbb{R}} G_J(y) F(x-y)\,\mathrm{d}y - F(x) \right| \leq 3\,\varepsilon \qquad (3.153)$$

holds true. □

Note that

$$(G_J * F)(x) := \int_{\mathbb{R}} G_J(y) F(x-y)\,\mathrm{d}y = \int_{\mathbb{R}} G_J(x-y) F(y)\,\mathrm{d}y, \qquad (3.154)$$

is commonly called the **convolution** of G_J and F in the case of functions on the real axis. For further details, please consult, for example, [27]. However, for the inclusion of other domains, the following generalized definition is more useful.

Definition 3.31. Let $D \subset \mathbb{R}^n$ be a measurable set and $\Phi \in \mathrm{L}^2(D \times D)$, $F \in \mathrm{L}^2(D)$. Then

$$(\Phi * F)(x) := \int_D \Phi(x,y) F(y)\,\mathrm{d}y, \quad x \in D, \qquad (3.155)$$

is called the **convolution** of Φ and F.

[6] If $\|F\|_{\mathrm{L}^1(\mathbb{R})} = 0$, that is, $F = 0$ in the sense of $\mathrm{L}^1(\mathbb{R})$, then (3.142) is trivial, and nothing needs to be proved.

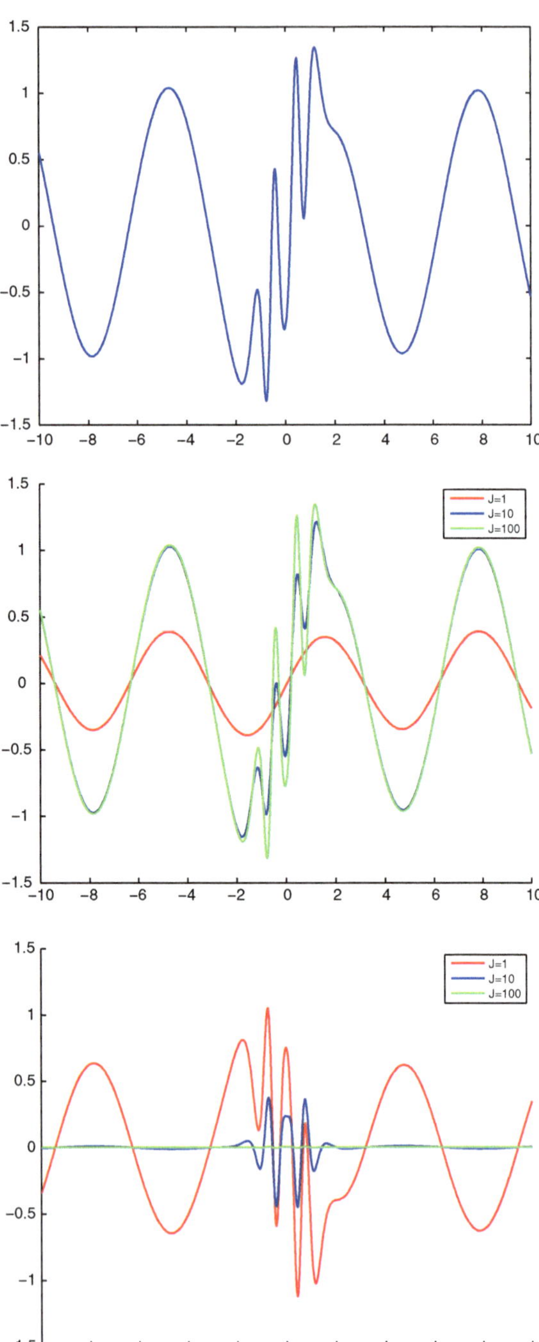

Fig. 3.8 Convolutions $\Phi_J * f$ (*middle*) as approximations to f (*top*) and approximation error $\Phi_J * f - f$ (*bottom*)

Due to the Cauchy–Schwarz inequality, the convolution is always defined in $\mathrm{L}^2(D)$:

$$\begin{aligned}\|\Phi * F\|_{\mathrm{L}^2(D)}^2 &= \int_D \left(\int_D \Phi(x,y)F(y)\,\mathrm{d}y\right)^2 \mathrm{d}x \\ &\le \int_D \int_D \Phi(x,y)^2\,\mathrm{d}y \int_D F(y)^2\,\mathrm{d}y\,\mathrm{d}x \\ &= \|\Phi\|_{\mathrm{L}^2(D\times D)}^2 \|F\|_{\mathrm{L}^2(D)}^2. \end{aligned} \tag{3.156}$$

Theorem 3.32. *If $\Phi \in \mathrm{L}^2(D\times D)$ and $F \in \mathrm{L}^2(D)$, where $D \subset \mathbb{R}^n$ is measurable, then $\Phi * F \in \mathrm{L}^2(D)$.*

For a numerical test, we calculate the convolutions of the Gaussian kernels for $J = 1,10,100$ with $f(x) := \sin x + \frac{1}{1+x^2}\cos\left(\frac{15}{1+x^2}\right)$ on $[-10,10]$. For this purpose, we restrict the integration to $[-20,20]$ and use a composite Simpson's rule with 10,001 grid points. Note that every value $(\Phi_J * f)(x)$, $x \in [-10,10]$, of the plot requires such an integration. The result is shown in Fig. 3.8. For larger J, the approximation is obviously better, whereas for smaller J, the approximation is smoother.

3.4 The Haar Wavelet

The theory and practice of wavelets on the real line has grown immensely since the first work [83]. There are many textbooks which cover some of the main topics such as [13, 20, 27, 28, 40, 112, 114, 143, 189, 201]. Since the focus of this book is on methods on the sphere, we will only study the simplest wavelets: the Haar wavelets, which go back to the paper [86], which was published in 1910, long before anybody thought about wavelets. This study will allow us to get an idea of the principles of the wavelet analysis.

Definition 3.33. The function $H : \mathbb{R} \to \mathbb{R}$ is defined by

$$H(t) = \begin{cases} 1, & 0 \le t < \frac{1}{2} \\ -1, & \frac{1}{2} \le t \le 1 \\ 0 & \text{else} \end{cases}, \quad t \in \mathbb{R}, \tag{3.157}$$

and called the **Haar mother wavelet**. Moreover, the functions $H_{j,k} : \mathbb{R} \to \mathbb{R}$, $j,k \in \mathbb{Z}$, are defined by

$$H_{j,k}(t) := 2^{j/2} H\left(2^j t - k\right), \quad t \in \mathbb{R}, \tag{3.158}$$

and called the **Haar wavelets**.

Here is one principle of the wavelet analysis: We have one function H (which is called the mother wavelet) and construct a whole family of functions by dilation

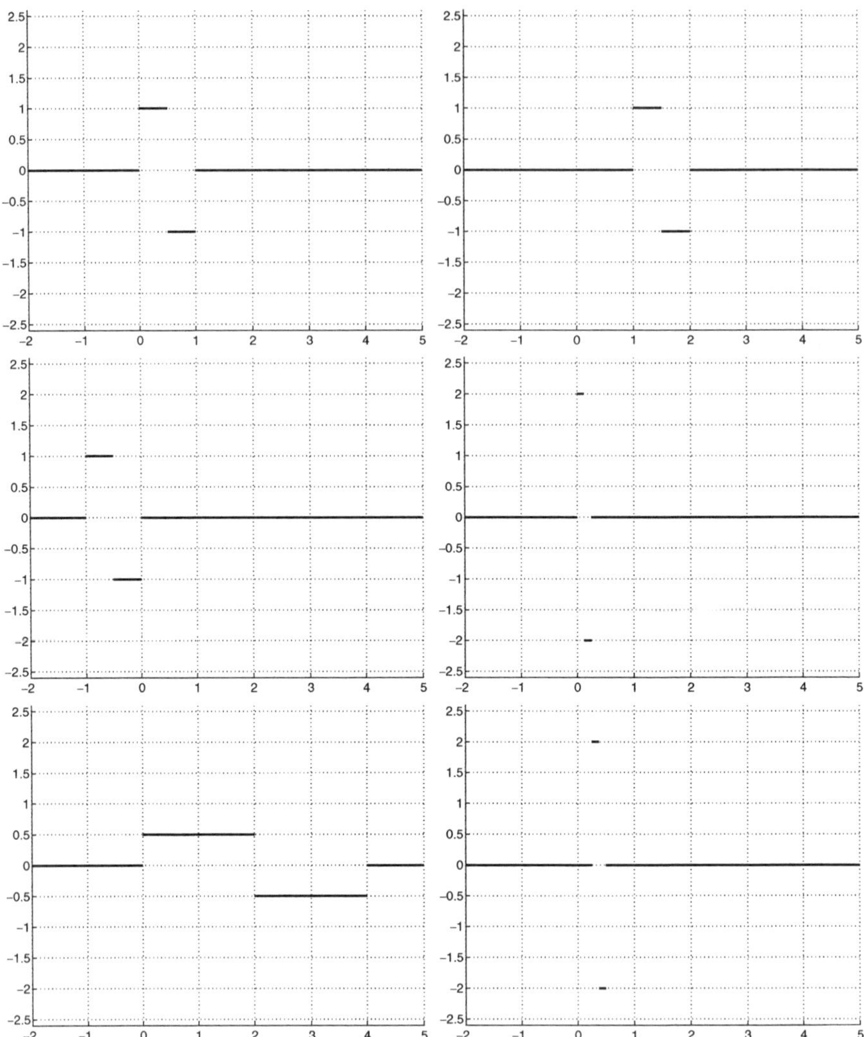

Fig. 3.9 Haar (mother) wavelet $H = H_{0,0}$ (*top left hand*) and translated and dilated versions of it: $H_{0,1}$ (*top right hand*), $H_{0,-1}$ (*middle left hand*), $H_{2,0}$ (*middle right hand*), $H_{-2,0}$ (*bottom left hand*), and $H_{2,1}$ (*bottom right hand*)

(for which j stands) and translation (for which k stands); see Fig. 3.9. The mother wavelet is a "hat function" (OK, it is a weird hat, but let us call it a hat). It concentrates on a certain subset of the real line—in contrast to polynomials. Translation means that this hat moves along the real axis and dilation changes the size of the hat.

Theorem 3.34. $\{H_{j,k}\}_{j\in\mathbb{Z},k\in\mathbb{Z}}$ *is an orthonormal basis of* $\mathrm{L}^2(\mathbb{R})$.

Note that the support of H is obviously $[0,1]$. Hence, the support of $H_{j,k}$ is given by $0 \le 2^j t - k \le 1$ which is equivalent to $2^{-j}k \le t \le 2^{-j}(k+1)$ such that

$$\operatorname{supp} H_{j,k} = \left[2^{-j}k, 2^{-j}(k+1)\right]. \tag{3.159}$$

Consequently, many inner products $\langle H_{j_1,k_1}, H_{j_2,k_2}\rangle_{\mathrm{L}^2(\mathbb{R})}$ simply vanish due to disjoint supports of the involved functions. For the other combinations, the calculation of the inner product is an easy exercise of integration due to finite ranges. The proof of the completeness needs a few more sophisticated arguments; see, for example, [201] for further details.

As a consequence of Theorem 3.34, every $F \in \mathrm{L}^2(\mathbb{R})$ is representable as

$$F = \sum_{j=-\infty}^{+\infty} \sum_{k=-\infty}^{+\infty} \langle F, H_{j,k}\rangle_{\mathrm{L}^2(\mathbb{R})} H_{j,k} \tag{3.160}$$

in the sense of $\mathrm{L}^2(\mathbb{R})$. Moreover, if F has a compact support, for example, $\operatorname{supp} F \subset [-1,1]$, then only such $H_{j,k}$ with $\operatorname{supp} H_{j,k} \cap [-1,1] \neq \emptyset$ are relevant, where even those supports which end at -1 or start at 1 are irrelevant. This implies that only parameter pairs (j,k) with [see (3.159)]

$$2^{-j}(k+1) > -1 \quad \text{and} \quad 2^{-j}k < 1 \tag{3.161}$$

are needed. These requirements are equivalent to

$$k+1 > -2^j \quad \text{and} \quad k < 2^j, \tag{3.162}$$

that is,

$$-2^j - 1 < k < 2^j. \tag{3.163}$$

Hence,

$$F = \sum_{j=-\infty}^{+\infty} \sum_{k=\lceil -2^j-1\rceil}^{\lfloor 2^j \rfloor} \langle F, H_{j,k}\rangle_{\mathrm{L}^2[-1,1]} H_{j,k} \tag{3.164}$$

for all $F \in \mathrm{L}^2[-1,1]$ in the sense of $\mathrm{L}^2[-1,1]$, where

$$\begin{aligned} \lfloor x \rfloor &:= \max\{n \in \mathbb{Z} \,|\, n \le x\}, \\ \lceil x \rceil &:= \min\{n \in \mathbb{Z} \,|\, n \ge x\} \end{aligned} \tag{3.165}$$

represents rounding down and up, respectively. Note that $\lfloor\cdot\rfloor$ is also called the Gaussian bracket.[7] Figure 3.10 shows the subdivision pattern of $[-1,1]$ based on the supports of the wavelets $H_{j,k}$ for different scales j.

---

[7]There exists an alternative notation for the Gaussian bracket, which is $[\cdot]$. However, we will use $\lfloor\cdot\rfloor$ throughout this book.

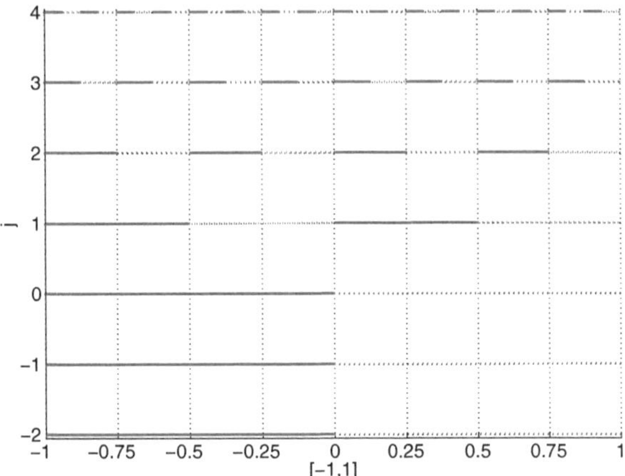

Fig. 3.10 The scales j define how fine the subdivision of $[-1,1]$ based on the supports of the Haar wavelets is. In this figure, every row corresponds to the indicated value of j. Within each row, the changing style and color of the line (between *red-solid* and *blue-dotted*) corresponds to changing basis functions

The representation of $F \in \mathrm{L}^2(\mathbb{R})$ by

$$F = \sum_{j=-\infty}^{+\infty} \sum_{k=-\infty}^{+\infty} \langle F, H_{j,k} \rangle_{\mathrm{L}^2(\mathbb{R})} H_{j,k} \tag{3.166}$$

provides a **multiscale analysis**: The parameter k defines the subinterval of the real axis we are looking at, and the parameter j corresponds to a decomposition of F into parts of different localization. The more local an effect is, the higher is the scale j at which this effect significantly contributes to the expansion coefficients $\langle F, H_{j,k} \rangle_{\mathrm{L}^2(\mathbb{R})}$, which are here called the **wavelet coefficients** of F.

Let us still focus on the interval $[-1,1]$. By definition, $H_{j,k}(t) = 2^{j/2}$ if and only if $0 \leq 2^j t - k < \frac{1}{2}$ and $H_{j,k}(t) = -2^{j/2}$ if and only if $\frac{1}{2} \leq 2^j t - k \leq 1$.

Hence, the Haar wavelet is constant and nonzero on each interval $\left[2^{-j}k, 2^{-j}\left(k+\frac{1}{2}\right)\right[$ and $\left[2^{-j}\left(k+\frac{1}{2}\right), 2^{-j}(k+1)\right]$.	(3.167)

However, since we still look at $[-1,1]$, we can easily derive that

$$\left[2^{-j}k, 2^{-j}\left(k+\frac{1}{2}\right)\right[\cap [-1,1[= [0,1[\text{ for } k=0 \text{ and } j \leq -1 \tag{3.168}$$

and

$$\left[2^{-j}\left(k+\frac{1}{2}\right), 2^{-j}(k+1)\right] \cap [-1,1] = [-1,0] \text{ for } k=-1 \text{ and } j \leq -1, \tag{3.169}$$

where $\operatorname{supp} H_{j,k} \cap]-1,1[= \emptyset$ for $k \notin \{-1,0\}$ in the case of $j \leq 0$ [see inequality (3.163)]. Thus, the part of the expansion (3.164) which belongs to $j \leq -1$ can be reduced as follows:

$$\sum_{j=-\infty}^{-1} \sum_{k=-1}^{0} \langle F, H_{j,k} \rangle_{\mathrm{L}^2[-1,1]} H_{j,k} = a\,\chi_{[-1,0[} + b\,\chi_{[0,1[} \tag{3.170}$$

in the sense of $\mathrm{L}^2[-1,1]$ (note that the particular values at $x = 0$ and $x = 1$ are irrelevant in $\mathrm{L}^2[-1,1]$). We can, for this purpose, introduce the function $\varphi := \chi_{[0,1[}$, which can be dilated and translated as well:

$$\varphi_{j,k}(t) := 2^{j/2} \varphi\left(2^j t - k\right), \quad t \in \mathbb{R}, \tag{3.171}$$

$j,k \in \mathbb{Z}$ (note that $\operatorname{supp} \varphi_{j,k} = [2^{-j}k, 2^{-j}(k+1)]$). For (3.170), we need the two orthogonal functions $\varphi_{0,0}$ and $\varphi_{0,-1}$. Therefore, we can use a modified onb for $\mathrm{L}^2[-1,1]$ such that (3.164) is now, in the sense of $\mathrm{L}^2[-1,1]$,

$$F = \sum_{k=-1}^{0} \langle F, \varphi_{0,k} \rangle_{\mathrm{L}^2[-1,1]} \varphi_{0,k} + \sum_{j=0}^{\infty} \sum_{k=\lceil -2^j - 1 \rceil}^{\lfloor 2^j \rfloor} \langle F, H_{j,k} \rangle_{\mathrm{L}^2[-1,1]} H_{j,k}. \tag{3.172}$$

What have we done here? We observed that it suffices to start at an initial scale j, since all smaller scales are so coarse such that nothing new is gained. This initial step at minimal resolution can be represented by the function φ, which is called the **Haar scaling function**. This Haar scaling function has some other nice properties. It can, for example, generate the Haar mother wavelet:

$$H = \frac{1}{\sqrt{2}} \varphi_{1,0} - \frac{1}{\sqrt{2}} \varphi_{1,1}. \tag{3.173}$$

We change now our point of view and start with the Haar scaling function, but we go back to $\mathrm{L}^2(\mathbb{R})$. Presume that we truncate (3.166) at a (small) scale J_1. The obtained coarse approximation

$$\sum_{j=-\infty}^{J_1} \sum_{k=-\infty}^{+\infty} \langle F, H_{j,k} \rangle_{\mathrm{L}^2(\mathbb{R})} H_{j,k} \tag{3.174}$$

is a simple function, which is constant on all subintervals $]2^{-J_1}k, 2^{-J_1}(k+\frac{1}{2})[$ and $]2^{-J_1}(k+\frac{1}{2}), 2^{-J_1}(k+1)[$, that is, all subintervals $]2^{-(J_1+1)}\tilde{k}, 2^{-(J_1+1)}(\tilde{k}+1)[$, $k \in \mathbb{Z}$ ($\tilde{k} = 2k$ or $\tilde{k} = 2k+1$), since the sections of constant, nonzero values of each $H_{j,k}$ with $j < J_1$ are unions of these sets [see (3.167)].

These constant values can be represented via the scaling function such that

$$\sum_{j=-\infty}^{J_1} \sum_{k=-\infty}^{+\infty} \langle F, H_{j,k} \rangle_{\mathrm{L}^2(\mathbb{R})} H_{j,k} = \sum_{k=-\infty}^{+\infty} \langle F, \varphi_{J_1+1,k} \rangle \varphi_{J_1+1,k}. \tag{3.175}$$

Hence, (3.166) and (3.175) yield for all $F \in \mathrm{L}^2(\mathbb{R})$,

$$F = \sum_{k=-\infty}^{+\infty} \left\langle F, \varphi_{J_1+1,k} \right\rangle \varphi_{J_1+1,k} + \sum_{j=J_1+1}^{+\infty} \sum_{k=-\infty}^{+\infty} \left\langle F, H_{j,k} \right\rangle_{\mathrm{L}^2(\mathbb{R})} H_{j,k} \tag{3.176}$$

in the sense of $\mathrm{L}^2(\mathbb{R})$ for all $J_1 \in \mathbb{Z}$. We will simplify this (including a little index shift) to

$$F = \sum_{k=-\infty}^{+\infty} c_{J_1,k} \varphi_{J_1,k} + \sum_{j=J_1}^{+\infty} \sum_{k=-\infty}^{+\infty} d_{j,k} H_{j,k} \quad \text{for all } J_1 \in \mathbb{Z}. \tag{3.177}$$

Note that the involved functions are orthonormal, since $\varphi_{J_1,k}$ is constant on its whole support and each $H_{j,k}$ has the values $2^{j/2}$ and $-2^{j/2}$ on subintervals of equal lengths on this support of $\varphi_{J_1,k}$ or vanishes on the interior of this support.

Now look at the coefficients of the scaling functions:

$$\begin{aligned} c_{j,k} &= 2^{j/2} \int_{2^{-j}k}^{2^{-j}(k+1)} F(t) \cdot 1 \, \mathrm{d}t = 2^{j/2} \left(\int_{2^{-j}k}^{2^{-j}(k+\frac{1}{2})} F(t) \, \mathrm{d}t + \int_{2^{-j}(k+\frac{1}{2})}^{2^{-j}(k+1)} F(t) \, \mathrm{d}t \right) \\ &= \frac{1}{\sqrt{2}} 2^{(j+1)/2} \left(\int_{2^{-(j+1)}2k}^{2^{-(j+1)}(2k+1)} F(t) \, \mathrm{d}t + \int_{2^{-(j+1)}(2k+1)}^{2^{-(j+1)}(2k+2)} F(t) \, \mathrm{d}t \right) \\ &= \frac{1}{\sqrt{2}} \left(c_{j+1,2k} + c_{j+1,2k+1} \right). \end{aligned} \tag{3.178}$$

Since (3.177) holds for all $J_1 \in \mathbb{Z}$, we can determine the difference between the realizations for $J_1 = J+1$ and $J_1 = J$:

$$\sum_{k=-\infty}^{+\infty} c_{J+1,k} \varphi_{J+1,k} - \sum_{k=-\infty}^{+\infty} c_{J,k} \varphi_{J,k} = \sum_{k=-\infty}^{+\infty} d_{J,k} H_{J,k}. \tag{3.179}$$

We can interpret $\sum_{k=-\infty}^{+\infty} c_{J,k} \varphi_{J,k}$ as a coarse representation of F—an approximation by averages over intervals of the length 2^{-J}. The next approximation $\sum_{k=-\infty}^{+\infty} c_{J+1,k} \varphi_{J+1,k}$ is expected to be finer because the intervals have the half length $2^{-(J+1)}$. The differences between these two approximations—that is, the details that have to be added—are represented by wavelets as we can see in (3.179). This is a characteristic feature of the wavelet analysis: Scaling functions provide us with a coarse approximation (they are sometimes called low-pass filters), and wavelets represent details (wavelets are associated to band-pass filters).

Due to our previous work, we get the following result.

Theorem 3.35. *The scale spaces*

$$V_J := \overline{\operatorname{span} \left\{ \varphi_{J,k} \right\}_{k \in \mathbb{Z}}}^{\|\cdot\|_{\mathrm{L}^2(\mathbb{R})}}, \quad J \in \mathbb{Z}, \tag{3.180}$$

constitute a **multiresolution analysis** *(MRA), that is,*

(i) $V_J \subset V_{J+1} \subset \mathrm{L}^2(\mathbb{R})$ *for all* $J \in \mathbb{Z}$,
(ii) $\overline{\bigcup_{J\in\mathbb{Z}} V_J}^{\|\cdot\|_{\mathrm{L}^2(\mathbb{R})}} = \mathrm{L}^2(\mathbb{R})$.

Moreover, the **detail spaces**

$$W_J := \overline{\operatorname{span}\left\{H_{J,k}\right\}_{k\in\mathbb{Z}}}^{\|\cdot\|_{\mathrm{L}^2(\mathbb{R})}}, \quad J \in \mathbb{Z}, \tag{3.181}$$

have the **scale-step property**

$$\begin{aligned} V_{J_2} &= V_{J_1} \oplus \bigoplus_{j=J_1}^{J_2-1} W_j, \\ \sum_{k=-\infty}^{+\infty} c_{J_2,k}\varphi_{J_2,k} &= \sum_{k=-\infty}^{+\infty} c_{J_1,k}\varphi_{J_1,k} + \sum_{j=J_1}^{J_2-1}\sum_{k=-\infty}^{+\infty} d_{j,k}H_{j,k}, \\ F &= \sum_{k=-\infty}^{+\infty} c_{J_1,k}\varphi_{J_1,k} + \sum_{j=J_1}^{\infty}\sum_{k=-\infty}^{+\infty} d_{j,k}H_{j,k} \end{aligned} \tag{3.182}$$

(in the sense of $\mathrm{L}^2(\mathbb{R})$*) for all* $F \in \mathrm{L}^2(\mathbb{R})$ *and all* $J_1, J_2 \in \mathbb{Z}$ *with* $J_1 < J_2$*, where*

$$c_{j,k} := \left\langle F, \varphi_{j,k}\right\rangle_{\mathrm{L}^2(\mathbb{R})}, \quad d_{j,k} := \left\langle F, H_{j,k}\right\rangle_{\mathrm{L}^2(\mathbb{R})}. \tag{3.183}$$

It would not be reasonable to calculate the coefficients $c_{j,k}$ and $d_{j,k}$ via the inner product in practice, because there is a much faster way: Look at (3.179). The value $c_{J,k}$ occurs on $[2^{-J}k, 2^{-J}(k+1)]$. Only the following functions in the representation (3.179) are relevant for this value [see (3.167) and (3.178)]: $\varphi_{J+1,2k}$ and $\varphi_{J+1,2k+1}$ as well as $H_{J,k}$. In view of (3.179) and the graph of $H_{J,k}$ (see Fig. 3.11), we get

$$c_{J+1,2k}\, 2^{(J+1)/2} - c_{J,k}2^{J/2} = d_{J,k}\cdot 2^{J/2} \tag{3.184}$$

$$c_{J+1,2k+1}\, 2^{(J+1)/2} - c_{J,k}2^{J/2} = d_{J,k}\cdot \left(-2^{J/2}\right) \tag{3.185}$$

and, consequently, by subtracting these two equations:

$$2^{-1/2}\left(c_{J+1,2k} - c_{J+1,2k+1}\right) = d_{J,k}. \tag{3.186}$$

In combination with (3.178),

$$2^{-1/2}\left(c_{J+1,2k} + c_{J+1,2k+1}\right) = c_{J,k} \tag{3.187}$$

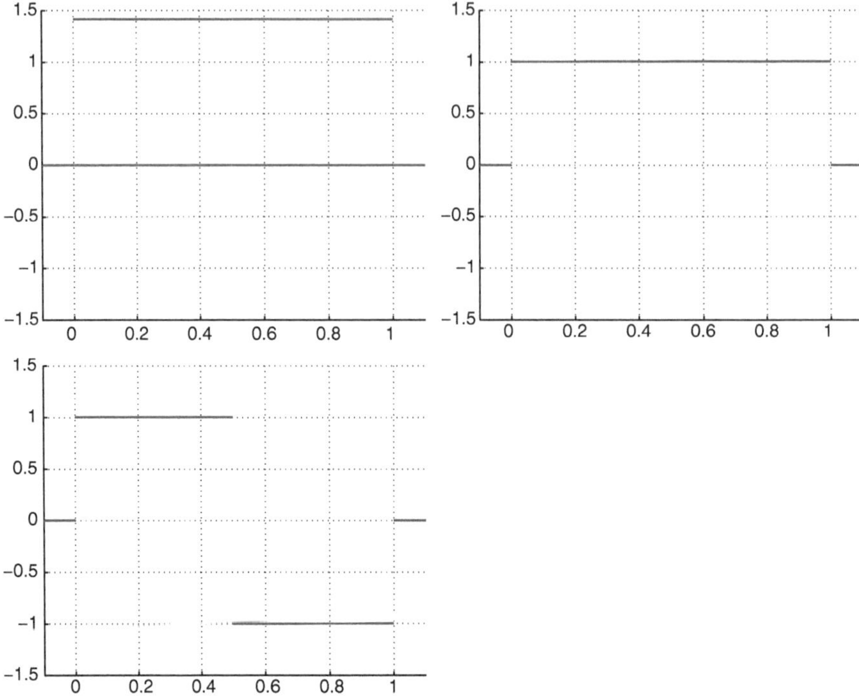

Fig. 3.11 This figure shows the graphs of $\varphi_{J+1,2k}$ *(blue line, top left hand)*, $\varphi_{J+1,2k+1}$ *(red line, top left hand)*, $\varphi_{J,k}$ *(top right hand)*, and $H_{J,k}$ *(bottom left hand)* for $J = k = 0$

$$\begin{array}{ccccc} (c_{J_2,k})_{k\in\mathbb{Z}} & \longrightarrow & (c_{J_2-1,k})_{k\in\mathbb{Z}} & \longrightarrow & (c_{J_2-2,k})_{k\in\mathbb{Z}} \cdots \\ & \searrow & & \searrow & \\ & & (d_{J_2-1,k})_{k\in\mathbb{Z}} & & (d_{J_2-2,k})_{k\in\mathbb{Z}} \cdots \end{array}$$

Fig. 3.12 Decomposition scheme of the fast Haar wavelet transform based on (3.186) and (3.187) with an initial (large) scale J_2

we now have a fast algorithm which allows us to start with $(c_{J_2,k})_{k\in\mathbb{Z}}$ at a sufficiently large (i.e., sufficiently fine) scale $J_2 \in \mathbb{Z}$. Step by step, the coefficients at the lower scales can be computed via (3.186) and (3.187). To obtain $(c_{J_2,k})_{k\in\mathbb{Z}}$, one can, for example, assume that the interval $[2^{-J_2}k, 2^{-J_2}(k+1)]$ is so small such that the mean value of the samples in this interval can be used as an approximation of $c_{J_2,k}\varphi_{J_2,k}$. The scheme of a Haar transform (more precisely, the fast Haar wavelet transform) is shown in Fig. 3.12.

An example is illustrated in Figs. 3.13–3.16. In Figs. 3.13–3.15, the sum of the two functions in one row yields the function on the left-hand side in the row above [see (3.179)]. Note that the details have different amplitudes at different scales. Moreover, the approximations in the left-hand columns yield higher and lower resolutions of the function under investigation.

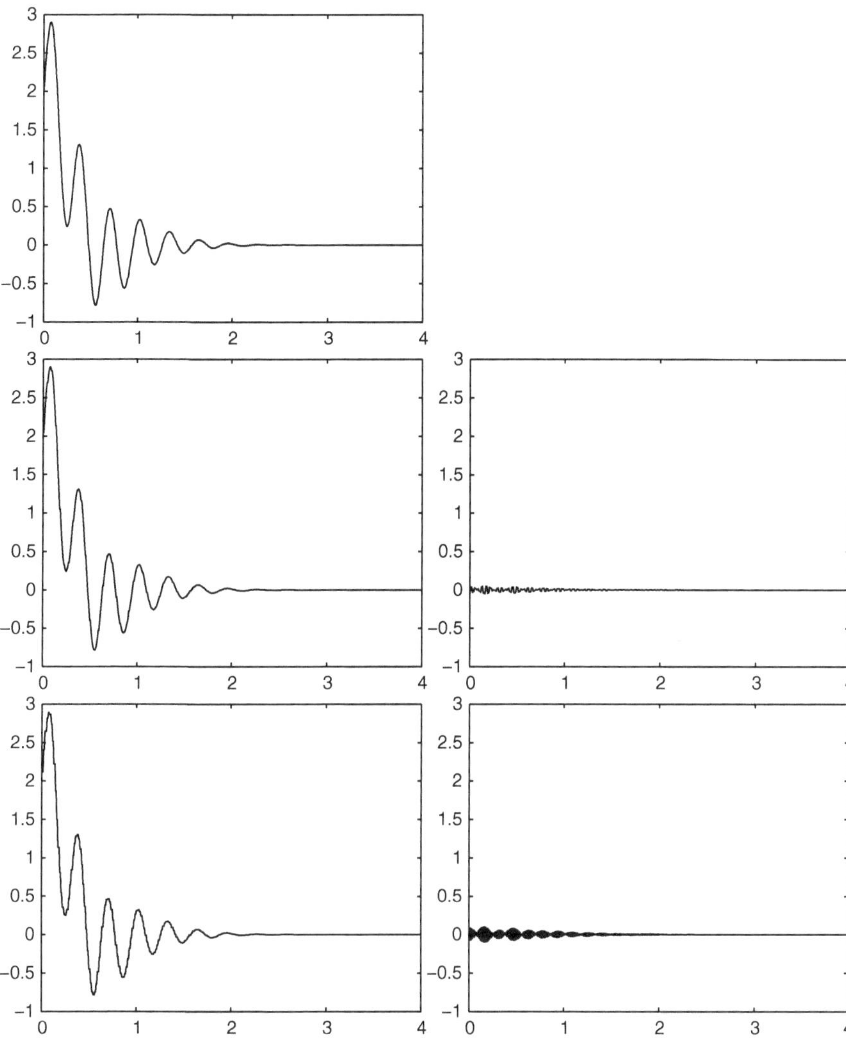

Fig. 3.13 Haar wavelet analysis of the function $f(x) := \exp(-x^2)\sin(20x) + 2\exp(-4x^2)\cos(3x)$ for $x \geq 0$: The *left-hand column* shows the approximations $\sum_k c_{j,k}\varphi_{j,k} \in V_j$, and the *right-hand column* shows the details $\sum_k d_{j,k}H_{j,k} \in W_j$. Each row corresponds to a scale from $j = 8$ (*first row*) to $j = 6$ (*last row*). The samples themselves were used as initial values $c_{8,k}\varphi_{8,k}$, $k = 0, 1, \ldots$

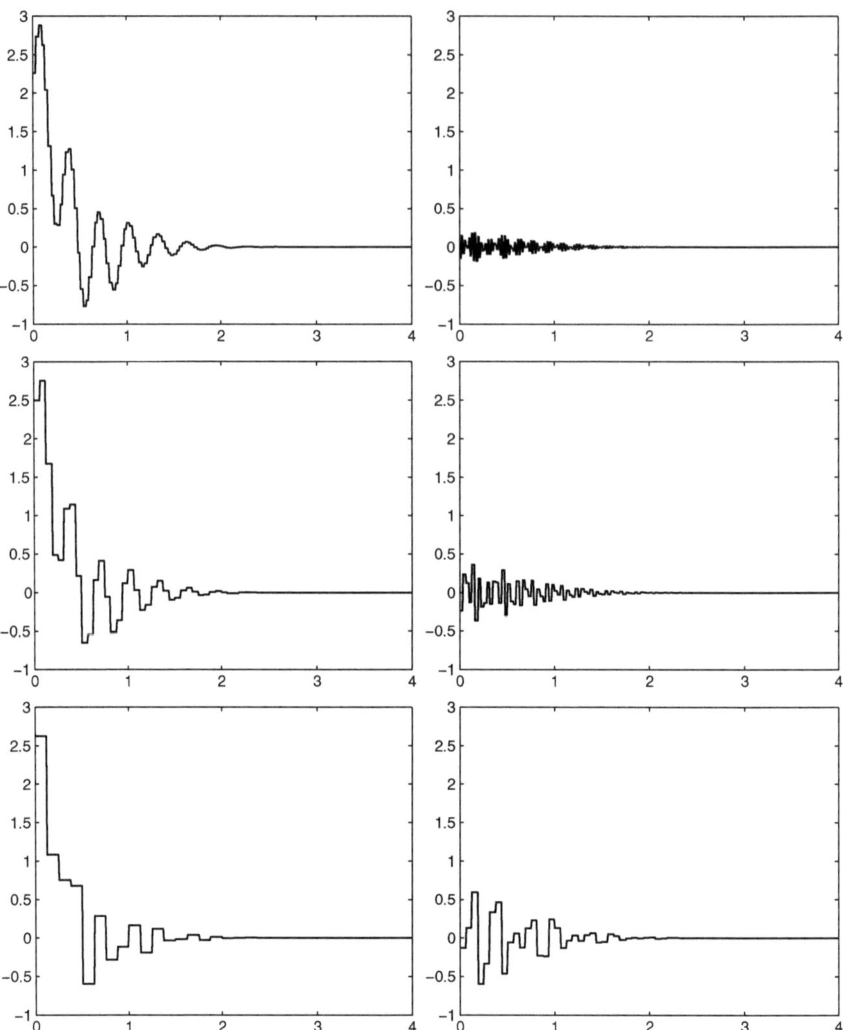

Fig. 3.14 Haar wavelet analysis of the function $f(x) := \exp(-x^2)\sin(20x) + 2\exp(-4x^2)\cos(3x)$ for $x \geq 0$: The *left-hand column* shows the approximations $\sum_k c_{j,k}\varphi_{j,k} \in V_j$, and the *right-hand column* shows the details $\sum_k d_{j,k}H_{j,k} \in W_j$. Each row corresponds to a scale from $j = 5$ (*first row*) to $j = 3$ (*last row*)

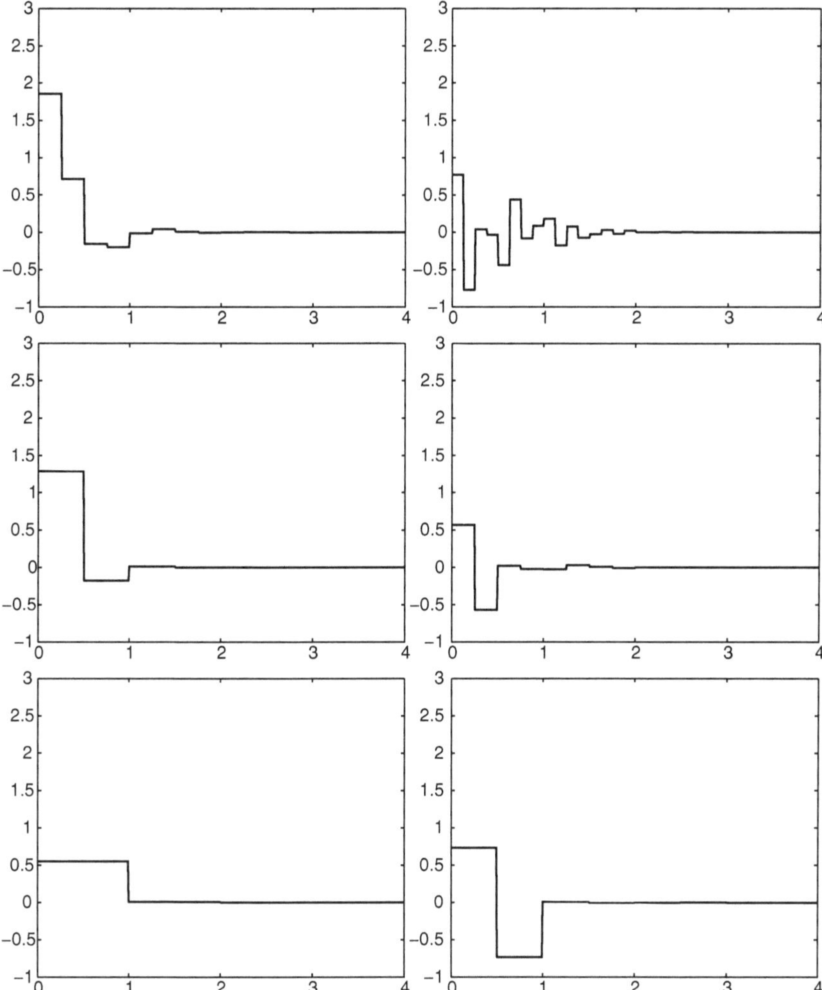

Fig. 3.15 Haar wavelet analysis of the function $f(x) := \exp(-x^2)\sin(20x) + 2\exp(-4x^2)\cos(3x)$ for $x \geq 0$: The *left-hand column* shows the approximations $\sum_k c_{j,k}\varphi_{j,k} \in V_j$, and the *right-hand column* shows the details $\sum_k d_{j,k}H_{j,k} \in W_j$. Each row corresponds to a scale from $j = 2$ (*first row*) to $j = 0$ (*last row*)

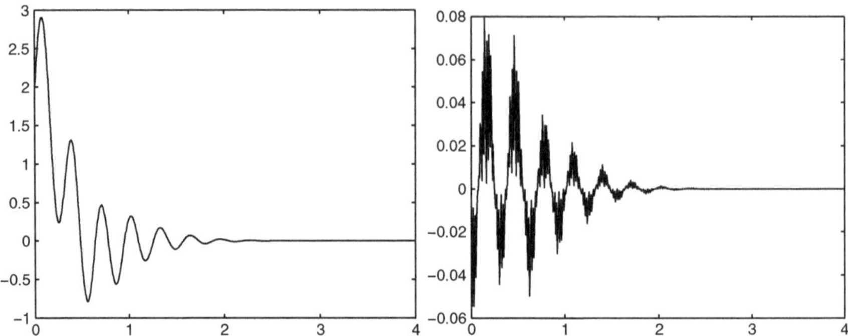

Fig. 3.16 Haar wavelet analysis of the function $f(x) := \exp(-x^2)\sin(20x) + 2\exp(-4x^2)\cos(3x)$ for $x \geq 0$: approximation $\sum_k c_{0,k}\varphi_{0,k} + \sum_{j=0}^{7}\sum_k d_{j,k}H_{j,k}$ (*left-hand side*) and difference to f (*right-hand side*)

3.5 A Remark on Higher Dimensions: The Tensor Product Ansatz

For functions which are given on an n-dimensional cuboid or on $\mathbb{R}^n$, most of the previous results can be transferred by using the so-called **tensor product ansatz**. The idea of this principle gives the following theorem ($\mathrm{L}^2_w(C)$ is defined for $C \subset \mathbb{R}^n$ in analogy to the case $n = 1$).

Theorem 3.36. *Let $C := \prod_{j=1}^n [a_j, b_j]$ be a cuboid in $\mathbb{R}^n$ ($a_j < b_j$ for all $j = 1, \ldots, n$). Then the following holds true: If $\{F_{j,k}\}_{k \in \mathbb{N}_0}$ is an onb of $\mathrm{L}^2_{w_j}[a_j, b_j]$ for each $j = 1, \ldots, n$, then the functions*

$$C \ni x \mapsto G_{k_1,\ldots,k_n}(x) := \prod_{j=1}^{n} F_{j,k_j}(x_j), \tag{3.188}$$

$k_1, \ldots, k_n \in \mathbb{N}_0$, provide an onb of $\mathrm{L}^2_w(C)$, where

$$w(x) := \prod_{j=1}^{n} w_j(x_j), \quad x \in C. \tag{3.189}$$

Proof. Fubini's theorem yields

$$\begin{aligned} \left\langle G_{k_1,\ldots,k_n}, G_{l_1,\ldots,l_n} \right\rangle_{\mathrm{L}^2_w(C)} &= \prod_{j=1}^{n} \int_{a_j}^{b_j} w_j(x_j) F_{j,k_j}(x_j) F_{j,l_j}(x_j)\, \mathrm{d}x_j \\ &= \prod_{j=1}^{n} \delta_{k_j l_j}. \end{aligned} \tag{3.190}$$

Moreover, if $f \in \mathrm{L}^2_w(C)$ with $\langle f, G_{k_1,\ldots,k_n}\rangle_{\mathrm{L}^2_w(C)} = 0$ for all $k_1,\ldots,k_n \in \mathbb{N}_0$, then we get (again by using Fubini's theorem)

$$\int_{a_1}^{b_1} w_1(x_1) \left[\int_{a_2}^{b_2} w_2(x_2) \ldots \right.$$
$$\left. \int_{a_n}^{b_n} w_n(x_n) f(x_1,\ldots,x_n) F_{n,k_n}(x_n)\,\mathrm{d}x_n \ldots F_{2,k_2}(x_2)\,\mathrm{d}x_2 \right] F_{1,k_1}(x_1)\,\mathrm{d}x_1 = 0. \tag{3.191}$$

We conclude, due to the completeness of $\{F_{1,k_1}\}_{k_1\in\mathbb{N}_0}$ in $\mathrm{L}^2_{w_1}[a_1,b_1]$, that

$$\int_{a_2}^{b_2} w_2(x_2) \ldots \int_{a_n}^{b_n} w_n(x_n) f(x_1,\ldots,x_n) F_{n,k_n}(x_n)\,\mathrm{d}x_n \ldots F_{2,k_2}(x_2)\,\mathrm{d}x_2 = 0 \tag{3.192}$$

for almost all $x_1 \in [a_1,b_1]$ and, by induction, that $f = 0$ in $\mathrm{L}^2(C)$. □

Correspondingly, trial functions such as orthogonal polynomials and Haar wavelets can be used to analyze functions on cuboids. However, the tensor product ansatz is not appropriate for domains whose geometry significantly differs from a cuboid such as a sphere or a ball. Therefore, we have to investigate these domains separately.

3.6 Questions for Understanding

- What is $\mathrm{L}^2_w[a,b]$? What kind of a space (in the language of functional analysis) is it?
- What are orthonormal polynomials? Do they always exist and (if yes) are they unique? Why?
- What are orthogonal polynomials? Do they always exist and (if yes) are they unique? Why?
- Which examples of orthogonal polynomials do you know?
- Which ways do you know for the calculation of particular orthogonal polynomials? Which one would you prefer in practice? Why?
- Remember the title of the course: How can you *construct* an *approximation* by means of orthonormal/orthogonal polynomials?
- Let us assume that you calculated the Fourier coefficients of a particular $F \in \mathrm{L}^2_w[a,b]$ (by the way, how would you do this?), what would you have to calculate to plot an approximation of F based on these coefficients? What would be an efficient way of calculating this?
- Which other orthonormal bases for $\mathrm{L}^2[a,b]$ or $\mathrm{L}^2(\mathbb{R})$ do you know?
- Is a polynomial approximation or a trigonometric approximation something you can recommend in general? Which alternatives do you know?
- What is a cubic spline?

- Given arbitrary interpolation conditions, does there always exist an interpolating cubic spline? If yes, is it unique? Why? If it is not unique, how can uniqueness be obtained?
- Why can we expect a natural cubic spline to yield satisfactory results?
- What is, in general, an Approximate Identity?
- What is an example of an Approximate Identity on the real line?
- What can you say about the keyword "partition of unity" in the context of this section?
- In the numerical example for the Gaussian kernel, actually integrals over $\mathbb{R}$ would have to be calculated. However, only $[-20,20]$ was used as the domain of integration. Although $[-20,20]$ is nothing in comparison to $\mathbb{R}$, why can we nevertheless expect to get sufficiently good approximations to the integral?
- What are Haar wavelets?
- What is the relation between the Haar wavelets and the Haar mother wavelet? How can it be interpreted graphically?
- What is a Haar scaling function?
- What is a multiresolution analysis?
- What are scale spaces and detail spaces?
- What is the scale-step property?
- How can you construct an approximation to a given function by means of a Haar wavelet analysis?
- What is the tensor product ansatz?

Part II
Approximation on the Sphere

Chapter 4
Basic Aspects

The focus of Part II is on classical and modern approximation methods on the unit sphere in $\mathbb{R}^3$. We will first learn how orthogonal polynomials on the unit sphere are constructed. These polynomials—the spherical harmonics—have particular properties which are also essential for the construction of localized trial functions, such as splines, wavelets, and Slepian functions, on the sphere. We will also see that the spherical splines and wavelets have analogous properties in comparison to their 1D counterparts.

Before we can discuss the really interesting stuff, we need some definitions, notations, and basic propositions of spherical analysis.

4.1 Some Fundamental Tools

Definition 4.1. The **unit sphere** in $\mathbb{R}^3$ is denoted[1] by

$$\Omega := \left\{ \xi \in \mathbb{R}^3 \middle| \, |\xi| = 1 \right\}. \tag{4.1}$$

Note that every $x \in \mathbb{R}^3$ can be represented as $x = r\xi$, $(|x| =) r \in \mathbb{R}_0^+$, $\xi \in \Omega$, where this representation is unique, if $x \neq 0$.

Definition 4.2. The canonical orthonormal basis in $\mathbb{R}^3$ is abbreviated[2] by

$$\varepsilon^1 := \begin{pmatrix} 1 \\ 0 \\ 0 \end{pmatrix}, \varepsilon^2 := \begin{pmatrix} 0 \\ 1 \\ 0 \end{pmatrix}, \varepsilon^3 := \begin{pmatrix} 0 \\ 0 \\ 1 \end{pmatrix}. \tag{4.2}$$

[1] Note that (in particular, in non-geomathematical literature) the notation $\mathbb{S}^2$ is also common for the unit sphere in $\mathbb{R}^3$.

[2] In the literature, one also finds the notation e_1, e_2, and e_3 for this basis.

V. Michel, *Lectures on Constructive Approximation*, Applied and Numerical Harmonic Analysis, DOI 10.1007/978-0-8176-8403-7_4,

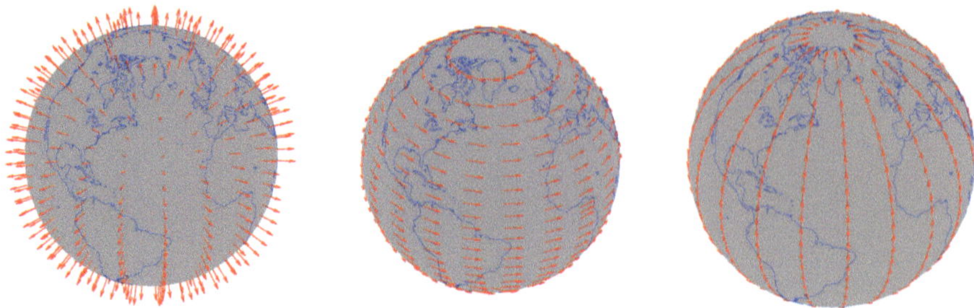

Fig. 4.1 Illustration of the local orthonormal basis system ε^r, ε^φ, ε^t on the sphere: ε^r represents an outer normal vector (*left hand*), whereas ε^φ (*middle*) and ε^t (*right hand*) span the corresponding tangential plane with ε^φ pointing eastwards and ε^t pointing northwards. For a better visibility, the lengths of the arrows are not to scale

Definition 4.3. The polar coordinate representation of $x \in \mathbb{R}^3$ is

$$x(r,\varphi,t) = \begin{pmatrix} r\sqrt{1-t^2}\cos\varphi \\ r\sqrt{1-t^2}\sin\varphi \\ rt \end{pmatrix}, \tag{4.3}$$

where $r \in \mathbb{R}_0^+$ is the distance to the origin, $\varphi \in [0, 2\pi[$ is the **longitude**, and $t \in [-1,1]$ is the **polar distance**.

Furthermore, we define

$$\varepsilon^r(\varphi,t) := \begin{pmatrix} \sqrt{1-t^2}\cos\varphi \\ \sqrt{1-t^2}\sin\varphi \\ t \end{pmatrix},$$

$$\varepsilon^\varphi(\varphi) := \begin{pmatrix} -\sin\varphi \\ \cos\varphi \\ 0 \end{pmatrix},$$

$$\varepsilon^t(\varphi,t) := \varepsilon^r(\varphi,t) \wedge \varepsilon^\varphi(\varphi) = \begin{pmatrix} -t\cos\varphi \\ -t\sin\varphi \\ \sqrt{1-t^2} \end{pmatrix}. \tag{4.4}$$

Note that the vectors ε^r, ε^φ, and ε^t are orthonormal. Moreover, the vectors $\varepsilon^r(\varphi,t)$, $\varepsilon^\varphi(\varphi)$, and $\varepsilon^t(\varphi,t)$ represent a local tripod which is installed at the point of Ω which corresponds to the polar coordinates (φ,t); see Fig. 4.1. ε^r is always the outer unit normal, whereas ε^φ and ε^t span the tangential plane in this point. More precisely, ε^φ is directed eastwards, and ε^t is directed northwards.

Remark 4.4. The use of the polar distance t as a parameter hides the arbitrariness behind the choice of the latitude. Mathematicians commonly prefer $t = \cos\vartheta$ with the **latitude** $\vartheta \in [0,\pi]$, whereas geoscientists usually choose $t = \sin\theta$ with

$\theta \in [-\frac{\pi}{2}, \frac{\pi}{2}]$. This corresponds to the latitude $\theta = 0$ representing the equator—as we all know from our geography lessons at school. The transfer between both systems is represented by $\theta = \frac{\pi}{2} - \vartheta$. Moreover, some geoscientists also prefer the range $\varphi \in [-\pi, \pi]$ for the longitude (which requires no change of the formulae due to the periodicity). Negative longitudes are western longitudes, and positive longitudes represent eastern longitudes—this corresponds again to the maps in our school atlas. Furthermore, $\sin\varphi$ and $\cos\varphi$ are often interchanged in this context. Many mathematicians have faced weird jigsaw puzzles with misplaced continents after using a data set from a geoscientist. If you ever get such figures, too, or if you are, for example, desperately searching South America in a data set but cannot find it, remember the remark you have just read to solve your problem.

Theorem 4.5. *The gradient* ∇ *in* $\mathbb{R}^3$ *can be decomposed into a radial and an angular part:*

$$\nabla = \varepsilon^r \frac{\partial}{\partial r} + \frac{1}{r} \nabla^*, \tag{4.5}$$

where

$$\nabla^* = \varepsilon^\varphi \frac{1}{\sqrt{1-t^2}} \frac{\partial}{\partial \varphi} + \varepsilon^t \sqrt{1-t^2} \frac{\partial}{\partial t} \tag{4.6}$$

is the ***surface gradient****.*

The ***surface curl gradient*** L^* *is defined by*

$$\mathrm{L}^*_\xi F(\xi) := \xi \wedge \nabla^*_\xi F(\xi), \quad \xi \in \Omega, F \in \mathrm{C}^{(1)}(\Omega). \tag{4.7}$$

Its local coordinate expression is

$$\mathrm{L}^* = -\varepsilon^\varphi \sqrt{1-t^2} \frac{\partial}{\partial t} + \varepsilon^t \frac{1}{\sqrt{1-t^2}} \frac{\partial}{\partial \varphi}. \tag{4.8}$$

Moreover, the ***Laplace operator***

$$\Delta_x = \left(\frac{\partial}{\partial x_1}\right)^2 + \left(\frac{\partial}{\partial x_2}\right)^2 + \left(\frac{\partial}{\partial x_3}\right)^2 \tag{4.9}$$

can be decomposed such that

$$\Delta = \left(\frac{\partial}{\partial r}\right)^2 + \frac{2}{r}\frac{\partial}{\partial r} + \frac{1}{r^2}\Delta^*, \tag{4.10}$$

where

$$\Delta^* = \frac{\partial}{\partial t}\left(1-t^2\right)\frac{\partial}{\partial t} + \frac{1}{1-t^2}\left(\frac{\partial}{\partial \varphi}\right)^2 \tag{4.11}$$

is the ***Beltrami operator****, which is also sometimes called the* ***Laplace–Beltrami operator****.*

Proof. We start by using the chain rule. Let $F \in \mathrm{C}^{(1)}(\mathbb{R}^3)$. Then

$$\begin{pmatrix} \frac{\partial F}{\partial r} \\ \frac{\partial F}{\partial \varphi} \\ \frac{\partial F}{\partial t} \end{pmatrix} = \begin{pmatrix} \frac{\partial x_1}{\partial r} & \frac{\partial x_2}{\partial r} & \frac{\partial x_3}{\partial r} \\ \frac{\partial x_1}{\partial \varphi} & \frac{\partial x_2}{\partial \varphi} & \frac{\partial x_3}{\partial \varphi} \\ \frac{\partial x_1}{\partial t} & \frac{\partial x_2}{\partial t} & \frac{\partial x_3}{\partial t} \end{pmatrix} \begin{pmatrix} \frac{\partial F}{\partial x_1} \\ \frac{\partial F}{\partial x_2} \\ \frac{\partial F}{\partial x_3} \end{pmatrix}$$

$$= \underbrace{\begin{pmatrix} (\varepsilon^r)^{\mathrm{T}} \\ r\sqrt{1-t^2}\,(\varepsilon^\varphi)^{\mathrm{T}} \\ \frac{r}{\sqrt{1-t^2}}\,(\varepsilon^t)^{\mathrm{T}} \end{pmatrix}}_{=:M} \underbrace{\begin{pmatrix} \frac{\partial F}{\partial x_1} \\ \frac{\partial F}{\partial x_2} \\ \frac{\partial F}{\partial x_3} \end{pmatrix}}_{=\nabla_x F}. \tag{4.12}$$

Due to the orthonormality of ε^r, ε^φ, and ε^t, the matrix M is easy to invert:

$$M^{-1} = \left(\varepsilon^r, \frac{1}{r\sqrt{1-t^2}}\,\varepsilon^\varphi, \frac{\sqrt{1-t^2}}{r}\,\varepsilon^t \right). \tag{4.13}$$

Hence,

$$\nabla_x F = \left(\varepsilon^r, \frac{1}{r\sqrt{1-t^2}}\,\varepsilon^\varphi, \frac{\sqrt{1-t^2}}{r}\,\varepsilon^t \right) \begin{pmatrix} \frac{\partial F}{\partial r} \\ \frac{\partial F}{\partial \varphi} \\ \frac{\partial F}{\partial t} \end{pmatrix}$$

$$= \varepsilon^r \frac{\partial F}{\partial r} + \varepsilon^\varphi \frac{1}{r\sqrt{1-t^2}} \frac{\partial F}{\partial \varphi} + \varepsilon^t \frac{\sqrt{1-t^2}}{r} \frac{\partial F}{\partial t}. \tag{4.14}$$

This proves (4.5) and (4.6). Keeping in mind that $\xi = \varepsilon^r$, that is, the position vectors of Ω are also the corresponding outer unit normal vectors, we can use (4.5) and (4.4) to prove (4.8) as follows:

$$\mathrm{L}^*_\xi F(\xi) = \varepsilon^r \wedge \left(\varepsilon^\varphi \frac{1}{\sqrt{1-t^2}} \frac{\partial}{\partial \varphi} F + \varepsilon^t \sqrt{1-t^2} \frac{\partial}{\partial t} F \right)$$

$$= \varepsilon^t \frac{1}{\sqrt{1-t^2}} \frac{\partial}{\partial \varphi} F + \underbrace{\varepsilon^r \wedge \varepsilon^t}_{=-\varepsilon^\varphi} \sqrt{1-t^2} \frac{\partial}{\partial t} F. \tag{4.15}$$

Finally, let $F \in \mathrm{C}^{(2)}(\mathbb{R}^3)$. Note that ΔF is the same as the divergence of the gradient of F, that is,

$$
\begin{aligned}
\Delta F &= \nabla\cdot(\nabla F) \\
&= \left(\varepsilon^r \frac{\partial}{\partial r} + \varepsilon^\varphi \frac{1}{r\sqrt{1-t^2}}\frac{\partial}{\partial \varphi} + \varepsilon^t \frac{1}{r}\sqrt{1-t^2}\frac{\partial}{\partial t}\right) \\
&\quad \cdot \left(\varepsilon^r \frac{\partial}{\partial r} + \varepsilon^\varphi \frac{1}{r\sqrt{1-t^2}}\frac{\partial}{\partial \varphi} + \varepsilon^t \frac{1}{r}\sqrt{1-t^2}\frac{\partial}{\partial t}\right) F.
\end{aligned}
\tag{4.16}
$$

For the further proceeding, we have to be aware of the fact that ε^r, ε^φ, and ε^t are not constant but depend on φ and t (more precisely, ε^φ only depends on φ). This essentially complicates the application of the differentiation operators.

$$
\begin{aligned}
\Delta F = {} & \underbrace{\varepsilon^r\cdot\varepsilon^r}_{=1} \frac{\partial^2}{\partial r^2}F + \underbrace{\varepsilon^r\cdot\varepsilon^\varphi}_{=0} \frac{\partial}{\partial r}\left(\frac{1}{r\sqrt{1-t^2}}\frac{\partial}{\partial\varphi}F\right) + \underbrace{\varepsilon^r\cdot\varepsilon^t}_{=0}\frac{\partial}{\partial r}\left(\frac{1}{r}\sqrt{1-t^2}\frac{\partial}{\partial t}F\right) \\
& + \varepsilon^\varphi\cdot\left(\frac{\partial}{\partial\varphi}\varepsilon^r\right)\frac{1}{r\sqrt{1-t^2}}\frac{\partial}{\partial r}F + \underbrace{\varepsilon^\varphi\cdot\varepsilon^r}_{=0}\frac{1}{r\sqrt{1-t^2}}\frac{\partial^2}{\partial\varphi\partial r}F \\
& + \varepsilon^\varphi\cdot\left(\frac{\partial}{\partial\varphi}\varepsilon^\varphi\right)\frac{1}{r^2(1-t^2)}\frac{\partial}{\partial\varphi}F + \underbrace{\varepsilon^\varphi\cdot\varepsilon^\varphi}_{=1}\frac{1}{r^2(1-t^2)}\frac{\partial^2}{\partial\varphi^2}F \\
& + \varepsilon^\varphi\cdot\left(\frac{\partial}{\partial\varphi}\varepsilon^t\right)\frac{1}{r^2}\frac{\partial}{\partial t}F + \underbrace{\varepsilon^\varphi\cdot\varepsilon^t}_{=0}\frac{1}{r^2}\frac{\partial^2}{\partial\varphi\partial t}F \\
& + \varepsilon^t\cdot\left(\frac{\partial}{\partial t}\varepsilon^r\right)\frac{1}{r}\sqrt{1-t^2}\frac{\partial}{\partial r}F + \underbrace{\varepsilon^t\cdot\varepsilon^r}_{=0}\frac{1}{r}\sqrt{1-t^2}\frac{\partial^2}{\partial t\partial r}F \\
& + \underbrace{\varepsilon^t\cdot\varepsilon^\varphi}_{=0}\frac{1}{r^2}\sqrt{1-t^2}\frac{\partial}{\partial t}\left(\frac{1}{\sqrt{1-t^2}}\frac{\partial}{\partial\varphi}F\right) + \varepsilon^t\cdot\left(\frac{\partial}{\partial t}\varepsilon^t\right)\frac{1-t^2}{r^2}\frac{\partial}{\partial t}F \\
& + \underbrace{\varepsilon^t\cdot\varepsilon^t}_{=1}\frac{1}{r^2}\sqrt{1-t^2}\left(-\frac{t}{\sqrt{1-t^2}}\frac{\partial}{\partial t}F + \sqrt{1-t^2}\frac{\partial^2}{\partial t^2}F\right).
\end{aligned}
\tag{4.17}
$$

A small auxiliary calculation yields

$$
\frac{\partial}{\partial\varphi}\varepsilon^r = \begin{pmatrix} -\sqrt{1-t^2}\sin\varphi \\ \sqrt{1-t^2}\cos\varphi \\ 0 \end{pmatrix} = \sqrt{1-t^2}\,\varepsilon^\varphi,
$$

$$
\frac{\partial}{\partial t}\varepsilon^r = \begin{pmatrix} -\frac{t}{\sqrt{1-t^2}}\cos\varphi \\ -\frac{t}{\sqrt{1-t^2}}\sin\varphi \\ 1 \end{pmatrix} = \frac{1}{\sqrt{1-t^2}}\varepsilon^t,
$$

$$\frac{\partial}{\partial\varphi}\varepsilon^{\varphi} = \begin{pmatrix} -\cos\varphi \\ -\sin\varphi \\ 0 \end{pmatrix},$$

$$\frac{\partial}{\partial\varphi}\varepsilon^{t} = \begin{pmatrix} t\sin\varphi \\ -t\cos\varphi \\ 0 \end{pmatrix} = -t\varepsilon^{\varphi},$$

$$\frac{\partial}{\partial t}\varepsilon^{t} = \begin{pmatrix} -\cos\varphi \\ -\sin\varphi \\ \frac{-t}{\sqrt{1-t^2}} \end{pmatrix}. \tag{4.18}$$

Inserting these results in (4.17), we get

$$\begin{aligned}\Delta F &= \frac{\partial^2}{\partial r^2}F + \frac{1}{r}\frac{\partial}{\partial r}F + \frac{1}{r^2(1-t^2)}\frac{\partial^2}{\partial\varphi^2}F - \frac{t}{r^2}\frac{\partial}{\partial t}F \\ &\quad + \frac{1}{r}\frac{\partial}{\partial r}F - \frac{t}{r^2}\frac{\partial}{\partial t}F + \frac{1-t^2}{r^2}\frac{\partial^2}{\partial t^2}F \\ &= \frac{\partial^2}{\partial r^2}F + \frac{2}{r}\frac{\partial}{\partial r}F + \frac{1}{r^2}\left(\frac{1}{1-t^2}\frac{\partial^2}{\partial\varphi^2}F - 2t\frac{\partial}{\partial t}F + \left(1-t^2\right)\frac{\partial^2}{\partial t^2}F\right) \\ &= \frac{\partial^2}{\partial r^2}F + \frac{2}{r}\frac{\partial}{\partial r}F + \frac{1}{r^2}\left[\frac{1}{1-t^2}\frac{\partial^2}{\partial\varphi^2}F + \frac{\partial}{\partial t}\left(\left(1-t^2\right)\frac{\partial}{\partial t}F\right)\right]. \end{aligned} \tag{4.19}$$

This proves (4.10) and (4.11). □

Equation (4.19) shows how to read $\frac{\partial}{\partial t}\left(1-t^2\right)\frac{\partial}{\partial t}$ in (4.11), namely, as a shorthand notation for the differential operator:

$$F \mapsto -2t\frac{\partial F}{\partial t} + \left(1-t^2\right)\frac{\partial^2 F}{\partial t^2}. \tag{4.20}$$

Definition 4.6. The differential operator $\mathscr{L} : \mathrm{C}^{(2)}[-1,1] \to \mathrm{C}[-1,1]$ defined by

$$\mathscr{L}F := -2t\frac{\partial F}{\partial t} + \left(1-t^2\right)\frac{\partial^2 F}{\partial t^2}, \quad F \in \mathrm{C}^{(2)}[-1,1], \tag{4.21}$$

is called the **Legendre operator**.

Note that the operators ∇^*, L^*, and Δ^* become singular at the poles ($t^2 = 1$). A singularity-free coordinate system for the sphere cannot be constructed (see [140, p. 256]). Therefore, coordinate-free representations should always be preferred.

Theorem 4.7. *The differentiation operators satisfy*

$$\nabla^* \cdot \nabla^* = \mathrm{L}^* \cdot \mathrm{L}^* = \Delta^*. \tag{4.22}$$

The proof is left as an exercise. Note that the proof of the identity $\nabla^* \cdot \nabla^* = \Delta^*$ is a part of the proof of the decomposition of Δ in Theorem 4.5.

Theorem 4.8. *For every $F \in \mathrm{C}^{(1)}(\Omega)$, the following identities hold true:*

$$(\xi F(\xi)) \cdot \left(\nabla_\xi^* F(\xi)\right) = (\xi F(\xi)) \cdot \left(\mathrm{L}_\xi^* F(\xi)\right) = \left(\nabla_\xi^* F(\xi)\right) \cdot \left(\mathrm{L}_\xi^* F(\xi)\right) = 0.$$

Proof. Since ∇^* and L^* only contain components with respect to ε^φ and ε^t, the orthogonality to ε^r, that is, the identity

$$(\xi F(\xi)) \cdot \left(\nabla_\xi^* F(\xi)\right) = 0 = (\xi F(\xi)) \cdot \left(\mathrm{L}_\xi^* F(\xi)\right), \tag{4.23}$$

is trivial. Furthermore, (4.6) and (4.8) yield

$$\left(\nabla_\xi^* F(\xi)\right) \cdot \left(\mathrm{L}_\xi^* F(\xi)\right) = -\frac{\sqrt{1-t^2}}{\sqrt{1-t^2}} \frac{\partial F}{\partial \varphi} \frac{\partial F}{\partial t} + \frac{\sqrt{1-t^2}}{\sqrt{1-t^2}} \frac{\partial F}{\partial t} \frac{\partial F}{\partial \varphi} = 0. \tag{4.24}$$

□

The following important properties are well known.

Theorem 4.9. *Let $1 \leq q \leq p < +\infty$. Then*

$$\mathrm{L}^p(\Omega) \subset \mathrm{L}^q(\Omega). \tag{4.25}$$

Theorem 4.9 is a particular case of Theorem 2.6 on p. 17.

Theorem 4.10. *For every $F \in \mathrm{C}(\Omega)$, the inequality*

$$\|F\|_{\mathrm{L}^p(\Omega)} \leq (4\pi)^{1/p} \|F\|_{\mathrm{C}(\Omega)} \tag{4.26}$$

holds for all $p \in [1,\infty[$.

Proof. The theorem is a result of the following considerations:

$$\begin{aligned} \int_\Omega |F(\xi)|^p \, \mathrm{d}\omega(\xi) &\leq \int_\Omega \max_{\eta \in \Omega} |F(\eta)|^p \, \mathrm{d}\omega(\xi) \\ &= \|F\|_{\mathrm{C}(\Omega)}^p \int_\Omega 1 \, \mathrm{d}\omega(\xi) \\ &= \|F\|_{\mathrm{C}(\Omega)}^p \cdot 4\pi. \end{aligned} \tag{4.27}$$

□

Note that Theorem 4.10 can be regarded as a particular case of Theorem 2.14 on p. 20, more precisely (2.28).

Theorem 4.11.

$$\overline{\mathrm{C}(\Omega)}^{\|\cdot\|_{\mathrm{L}^2(\Omega)}} = \mathrm{L}^2(\Omega). \tag{4.28}$$

For a proof, see, for example, [119, p. 229].

The following theorem is also well known. The proof is omitted (see also [7, pp. 459–461], [66, p. 16]).

Theorem 4.12. *Let $F,G \in \mathrm{C}^{(2)}(\overline{\Gamma})$, and $\Gamma \subset \Omega$ be a subset with sufficiently smooth boundary $\partial\Gamma$. Moreover, let the vector ν denote the outward unit normal vector field to $\partial\Gamma$. Then the following identities are valid.*

(a) Green's first surface identity

$$\int_\Gamma \nabla^*_\xi G(\xi)\cdot\nabla^*_\xi F(\xi)\,\mathrm{d}\omega(\xi) + \int_\Gamma F(\xi)\Delta^*_\xi G(\xi)\,\mathrm{d}\omega(\xi) = \int_{\partial\Gamma} F(\xi)\frac{\partial}{\partial\nu(\xi)}G(\xi)\,\mathrm{d}\sigma(\xi). \tag{4.29}$$

(b) Green's second surface identity

$$\int_\Gamma \left(F(\xi)\Delta^*_\xi G(\xi) - G(\xi)\Delta^*_\xi F(\xi)\right)\mathrm{d}\omega(\xi) = \int_{\partial\Gamma}\left(F(\xi)\frac{\partial}{\partial\nu(\xi)}G(\xi) - G(\xi)\frac{\partial}{\partial\nu(\xi)}F(\xi)\right)\mathrm{d}\sigma(\xi). \tag{4.30}$$

A consequence of the substitution rule for volume integrals is the following theorem.

Theorem 4.13. *Let $0\le\alpha<\beta\le+\infty$ and $D := \{x\in\mathbb{R}^3 \,|\, \alpha\le|x|\le\beta\}$. Then*

$$\int_D F(x)\,\mathrm{d}x = \int_\alpha^\beta r^2\int_\Omega F(r\xi)\,\mathrm{d}\omega(\xi)\,\mathrm{d}r \tag{4.31}$$

for all $F\in\mathrm{C}(D)$, provided that the integrals exist (if $\beta=\infty$).

The spline basis functions and the scaling functions as well as the wavelets on the sphere are zonal functions. We will study such functions for this reason.

Definition 4.14. Let $\xi\in\Omega$. A function of the form

$$G_\xi : \Omega\to\mathbb{R},\quad \eta\mapsto G_\xi(\eta) = G(\xi\cdot\eta), \tag{4.32}$$

where $G : [-1,1] \to \mathbb{R}$, is called a ξ-**zonal function** on Ω.

For zonal functions, the following differentiation formulae are often useful.

Theorem 4.15. *Let $F \in \mathrm{C}^{(1)}[-1,1]$ and $\xi, \eta \in \Omega$. Then*

$$\begin{aligned} \nabla_\xi^* F(\xi\cdot\eta) &= F'(\xi\cdot\eta)[\eta - (\xi\cdot\eta)\xi], \\ \mathrm{L}_\xi^* F(\xi\cdot\eta) &= F'(\xi\cdot\eta)\xi\wedge\eta. \end{aligned} \tag{4.33}$$

Proof. Using the chain rule, we obtain

$$\nabla_\xi^* F(\xi\cdot\eta) = F'(\xi\cdot\eta)\,\nabla_\xi^*(\xi\cdot\eta). \tag{4.34}$$

If we write $x = r\xi$, $r > 0$, then we get

$$\nabla_x(x\cdot\eta) = \eta, \tag{4.35}$$

where Theorem 4.5 yields

$$\begin{aligned} \nabla_x(x\cdot\eta) &= \left(\xi\frac{\partial}{\partial r} + \frac{1}{r}\nabla_\xi^*\right)(r\xi\cdot\eta) \\ &= \xi(\xi\cdot\eta) + \nabla_\xi^*(\xi\cdot\eta). \end{aligned} \tag{4.36}$$

Hence,

$$\nabla_\xi^*(\xi\cdot\eta) = \eta - (\xi\cdot\eta)\xi. \tag{4.37}$$

The definition of L^*, finally, yields

$$\begin{aligned} \mathrm{L}_\xi^* F(\xi\cdot\eta) &= \xi\wedge\nabla_\xi^* F(\xi\cdot\eta) \\ &= F'(\xi\cdot\eta)\xi\wedge[\eta - (\xi\cdot\eta)\xi] \\ &= F'(\xi\cdot\eta)\xi\wedge\eta. \end{aligned} \tag{4.38}$$

□

Moreover, there is also a valuable formula for integrals of zonal functions.

Theorem 4.16. *Let $G : [-1,1] \to \mathbb{R}$ be integrable. Then*

$$\int_\Omega G(\xi\cdot\eta)\,\mathrm{d}\omega(\eta) = 2\pi\int_{-1}^{1} G(t)\,\mathrm{d}t \tag{4.39}$$

for all $\xi \in \Omega$.

Proof. Since G is a zonal function, we can write

$$\int_\Omega G_\xi(\eta)\,\mathrm{d}\omega(\eta) = \int_\Omega G(\xi\cdot\eta)\,\mathrm{d}\omega(\eta). \tag{4.40}$$

We first consider the case $\xi = \varepsilon^3$:

$$\int_\Omega G_{\varepsilon^3}(\eta)\,\mathrm{d}\omega(\eta) = \int_\Omega G\left(\varepsilon^3\cdot\eta\right)\,\mathrm{d}\omega(\eta). \tag{4.41}$$

By using the polar coordinate representation

$$\eta = t\varepsilon^3 + \sqrt{1-t^2}\left(\cos\varphi\varepsilon^1 + \sin\varphi\varepsilon^2\right), \tag{4.42}$$

where $-1\le t\le 1,\ 0\le\varphi\le 2\pi$, we obtain

$$\int_\Omega G\left(\varepsilon^3\cdot\eta\right)\,\mathrm{d}\omega(\eta) = \int_{-1}^{1}\int_0^{2\pi} G(t)\left|\frac{\partial\eta}{\partial t}\wedge\frac{\partial\eta}{\partial\varphi}\right|\,\mathrm{d}\varphi\,\mathrm{d}t \tag{4.43}$$

with (see (4.18), note that the formulae for η and ε^r are identical here)

$$\frac{\partial\eta}{\partial\varphi} = \sqrt{1-t^2}\,\varepsilon^\varphi,\quad \frac{\partial\eta}{\partial t} = \frac{1}{\sqrt{1-t^2}}\varepsilon^t, \tag{4.44}$$

$$\frac{\partial\eta}{\partial t}\wedge\frac{\partial\eta}{\partial\varphi} = -\varepsilon^r, \tag{4.45}$$

$$\left|\frac{\partial\eta}{\partial t}\wedge\frac{\partial\eta}{\partial\varphi}\right| = 1. \tag{4.46}$$

Consequently,

$$\int_\Omega G_{\varepsilon^3}(\eta)\,\mathrm{d}\omega(\eta) = 2\pi\int_{-1}^{1} G(t)\,\mathrm{d}t. \tag{4.47}$$

Now let $A\in\mathrm{SO}(3)$, that is, A is a real 3×3 matrix with $A^\mathrm{T}A = I$ and $\det A = 1$, be chosen such that $A^{-1}\varepsilon^3 = \xi$. Then

$$\int_\Omega G(\xi\cdot\eta)\,\mathrm{d}\omega(\eta) = \int_\Omega G\left(A^{-1}\varepsilon^3\cdot\eta\right)\,\mathrm{d}\omega(\eta) = \int_\Omega G\Big(\varepsilon^3\cdot\underbrace{\Big(\left(A^\mathrm{T}\right)^{-1}\eta\Big)}_{=A\eta}\Big)\,\mathrm{d}\omega(\eta). \tag{4.48}$$

We now substitute $\eta = A^{-1}\zeta$ and get[3]

$$\begin{aligned}
\int_{\Omega} G(\xi \cdot \eta)\,\mathrm{d}\omega(\eta) &= \int_{A\Omega} G\left(\varepsilon^3 \cdot \zeta\right)\,\mathrm{d}\omega(\zeta) \\
&= \int_{\Omega} G\left(\varepsilon^3 \cdot \zeta\right)\,\mathrm{d}\omega(\zeta) \\
&= 2\pi \int_{-1}^{1} G(t)\,\mathrm{d}t. \qquad (4.54)
\end{aligned}$$

□

---

[3]If $\eta = \Phi(u,v)$ is a parametrization of Ω, we have $\eta = A^{-1}\Psi(u,v)$, where Ψ is a corresponding transformed parametrization of $A\Omega = \Omega$. Since the set is Ω again, we can use the same parameter range again. In view of (2.66), we calculate

$$\begin{aligned}
\left|\frac{\partial \Phi}{\partial u} \wedge \frac{\partial \Phi}{\partial v}\right| &= \left|\frac{\partial\left(A^{-1}\Psi\right)}{\partial u} \wedge \frac{\partial\left(A^{-1}\Psi\right)}{\partial v}\right| \\
&= \left|\frac{\partial\left(A^{-1}\Psi\right)}{\partial u}\right| \left|\frac{\partial\left(A^{-1}\Psi\right)}{\partial v}\right| \sin\sphericalangle\left(\frac{\partial\left(A^{-1}\Psi\right)}{\partial u}, \frac{\partial\left(A^{-1}\Psi\right)}{\partial v}\right) \\
&= \left|A^{-1}\frac{\partial \Psi}{\partial u}\right| \left|A^{-1}\frac{\partial \Psi}{\partial v}\right| \sin\sphericalangle\left(A^{-1}\frac{\partial \Psi}{\partial u}, A^{-1}\frac{\partial \Psi}{\partial v}\right). \qquad (4.49)
\end{aligned}$$

Since A is orthogonal, we get

$$\left|A^{-1}\frac{\partial \Psi}{\partial u}\right|^2 = \left\langle A^{\mathrm{T}}\frac{\partial \Psi}{\partial u}, A^{\mathrm{T}}\frac{\partial \Psi}{\partial u}\right\rangle = \left\langle \frac{\partial \Psi}{\partial u}, AA^{\mathrm{T}}\frac{\partial \Psi}{\partial u}\right\rangle = \left|\frac{\partial \Psi}{\partial u}\right|^2 \qquad (4.50)$$

and, analogously,

$$\left|A^{-1}\frac{\partial \Psi}{\partial v}\right| = \left|\frac{\partial \Psi}{\partial v}\right| \qquad (4.51)$$

as well as

$$\begin{aligned}
\sphericalangle\left(A^{-1}\frac{\partial \Psi}{\partial u}, A^{-1}\frac{\partial \Psi}{\partial v}\right) &= \arccos \frac{\left\langle A^{\mathrm{T}}\frac{\partial \Psi}{\partial u}, A^{\mathrm{T}}\frac{\partial \Psi}{\partial v}\right\rangle}{\left|A^{\mathrm{T}}\frac{\partial \Psi}{\partial u}\right| \left|A^{\mathrm{T}}\frac{\partial \Psi}{\partial v}\right|} \\
&= \arccos \frac{\left\langle \frac{\partial \Psi}{\partial u}, \frac{\partial \Psi}{\partial v}\right\rangle}{\left|\frac{\partial \Psi}{\partial u}\right| \left|\frac{\partial \Psi}{\partial v}\right|} \\
&= \sphericalangle\left(\frac{\partial \Psi}{\partial u}, \frac{\partial \Psi}{\partial v}\right). \qquad (4.52)
\end{aligned}$$

Hence,

$$\left|\frac{\partial \Phi}{\partial u} \wedge \frac{\partial \Phi}{\partial v}\right| = \left|\frac{\partial \Psi}{\partial u} \wedge \frac{\partial \Psi}{\partial v}\right|. \qquad (4.53)$$

Zonal functions play an important role throughout the constructive approximation on the sphere as you will see. Note that a ξ-zonal function actually only depends on the distance to ξ, since

$$|\xi - \eta|^2 = |\xi|^2 + |\eta|^2 - 2\xi \cdot \eta = 2(1 - \xi \cdot \eta). \tag{4.55}$$

4.2 Questions for Understanding

- What is the use of the vectors ε^r, ε^φ, and ε^t?
- What are ∇^*, L^*, and Δ^*?
- What is the Legendre operator?
- What do you know about $\mathrm{L}^2(\Omega)$?
- What are the Green's surface identities?
- What are zonal functions?
- Which nice properties of zonal functions do you know? How are they proved?

Chapter 5
Fourier Analysis

5.1 Spherical Harmonics

In this section, we will study the spherical harmonics as an example for a complete orthonormal system in $\mathrm{L}^2(\Omega)$. This system and its theory are well established; see, for example, [66, 75, 91, 94, 139].

Definition 5.1. Let $D \subset \mathbb{R}^3$ be open and connected. A function $F \in \mathrm{C}^{(2)}(D)$ is called **harmonic**, if $\Delta_x F(x) = 0$ for all $x \in D$, when Δ is the Laplace operator [see (2.77)]. The set of all harmonic functions in $\mathrm{C}^{(2)}(D)$ is denoted by $\mathrm{Harm}(D)$.

Definition 5.2. A polynomial P on $\mathbb{R}^n$, $n \in \mathbb{N}$, is called **homogeneous** of degree $m \in \mathbb{N}_0$, if there exist real numbers C_α, which do not all vanish, such that

$$P(x) = \sum_{|\alpha|=m} C_\alpha x^\alpha \quad \forall x \in \mathbb{R}^n, \tag{5.1}$$

where $\alpha = (\alpha_1, \ldots, \alpha_n) \in \mathbb{N}_0^n$ is a multi-index with $|\alpha| := \sum_{i=1}^n \alpha_i$, and $x^\alpha := \prod_{i=1}^n x_i^{\alpha_i}$, $x \in \mathbb{R}^n$. The set of all homogeneous polynomials of degree m on $\mathbb{R}^n$ united with the zero polynomial is denoted by $\mathrm{Hom}_m(\mathbb{R}^n)$. The set of their restrictions to a domain $D \subset \mathbb{R}^n$ is defined as follows:

$$\mathrm{Hom}_m(D) := \left\{P|_D : P \in \mathrm{Hom}_m(\mathbb{R}^n)\right\}. \tag{5.2}$$

Note that $P(x) = x_1^2 + x_2^2 + x_3^2 \in \mathrm{Hom}_2(\mathbb{R}^3)$, and therefore, $P|_\Omega \in \mathrm{Hom}_2(\Omega)$ although $P|_\Omega \equiv 1$, that is, $\deg P|_\Omega = 0$. In other words, the index m of $\mathrm{Hom}_m(\Omega)$ refers to the degree of the (original) polynomial on $\mathbb{R}^3$ and not to the degree of the restricted polynomial.

V. Michel, *Lectures on Constructive Approximation*, Applied and Numerical Harmonic Analysis, DOI 10.1007/978-0-8176-8403-7_5,

Theorem 5.3. *Let $n \in \mathbb{N}_0$ be given. Then*

$$\dim\left(\mathrm{Hom}_n\left(\mathbb{R}^3\right)\right) = \frac{(n+1)(n+2)}{2} \tag{5.3}$$

and the functions

$$\{x \mapsto x^{\alpha}\}_{\alpha \in \mathbb{N}_0^3,\, \alpha_1+\alpha_2+\alpha_3=n} \tag{5.4}$$

form a basis of $\mathrm{Hom}_n(\mathbb{R}^3)$.

Proof. A function $P \in \mathrm{Hom}_n(\mathbb{R}^3)$ can be represented by

$$P(x) = \sum_{|\alpha|=n} C_{\alpha} x^{\alpha} \tag{5.5}$$

[see (5.1)]. At first, we will prove that the functions in (5.4) are linearly independent and, consequently, form a basis for $\mathrm{Hom}_n(\mathbb{R}^3)$. Let

$$\sum_{|\alpha|=n} C_{\alpha} x^{\alpha} = 0 \tag{5.6}$$

for all $x \in \mathbb{R}^3$. We get using the multi-indices

$$\sum_{\alpha_1=0}^{n} \left(\sum_{\alpha_2=0}^{n-\alpha_1} C_{\alpha_1,\alpha_2,n-\alpha_1-\alpha_2} x_2^{\alpha_2} x_3^{n-\alpha_1-\alpha_2} \right) x_1^{\alpha_1} = 0. \tag{5.7}$$

If we keep x_2 and x_3 arbitrary but fixed for the moment, then we get in (5.7) a polynomial in x_1. As this polynomial has to vanish for all x_1 and nontrivial polynomials of degree $\leq n$ can only have at most n roots, we conclude

$$\sum_{\alpha_2=0}^{n-\alpha_1} C_{\alpha_1,\alpha_2,n-\alpha_1-\alpha_2} x_2^{\alpha_2} x_3^{n-\alpha_1-\alpha_2} = 0 \tag{5.8}$$

for all $\alpha_1 \in \{0,\ldots,n\}$ and all $x_2, x_3 \in \mathbb{R}$. If we now keep x_3 fixed, then we get in (5.8) a polynomial in x_2 which vanishes for all x_2. This means that

$$C_{\alpha_1,\alpha_2,n-\alpha_1-\alpha_2} x_3^{n-\alpha_1-\alpha_2} = 0 \tag{5.9}$$

for all $\alpha_1 \in \{0,\ldots,n\}$, $\alpha_2 \in \{0,\ldots,n-\alpha_1\}$ and all $x_3 \in \mathbb{R}$. However, this implies immediately that

$$C_{\alpha_1,\alpha_2,n-\alpha_1-\alpha_2} = 0 \tag{5.10}$$

for all $\alpha_1 \in \{0,\ldots,n\}$ and all $\alpha_2 \in \{0,\ldots,n-\alpha_1\}$. Consequently, $C_{\alpha} = 0$ for all $\alpha \in \mathbb{N}_0^3$ with $|\alpha| = n$ in (5.6). Hence, the functions form a basis.

Let us calculate the dimension. For α_1, we have $n+1$ possible values, namely, $0,\ldots,n$. If α_1 is given, then we have $n+1-\alpha_1$ possible values for α_2, namely, $0,\ldots,n-\alpha_1$. Finally, if α_1 and α_2 are given, no choice is left for $\alpha_3 : \alpha_3 = n-\alpha_1-\alpha_2$. Hence, there are

$$\sum_{\alpha_1=0}^{n}(n+1-\alpha_1)=\sum_{\nu=1}^{n+1}\nu=\frac{(n+1)(n+2)}{2} \tag{5.11}$$

linearly independent functions $\{x\mapsto x^\alpha\}_{\alpha\in\mathbb{N}_0^3,|\alpha|=n}$. □

In analogy to the 3D case, every $P\in\mathrm{Hom}_n(\mathbb{R}^2)$ can be represented as

$$P(x)=\sum_{\alpha_1+\alpha_2=n}C_\alpha x^\alpha=\sum_{\alpha_1=0}^{n}C_{\alpha_1,n-\alpha_1}x_1^{\alpha_1}x_2^{n-\alpha_1}. \tag{5.12}$$

This yields the following result.

Theorem 5.4. *For every $n\in\mathbb{N}_0$,*

$$\dim\left(\mathrm{Hom}_n\left(\mathbb{R}^2\right)\right)=n+1. \tag{5.13}$$

Definition 5.5. The set of all homogeneous harmonic polynomials on $\mathbb{R}^3$ with degree $m\in\mathbb{N}_0$ is denoted by $\mathrm{Harm}_m(\mathbb{R}^3)$, that is,

$$\mathrm{Harm}_m\left(\mathbb{R}^3\right):=\left\{P\in\mathrm{Hom}_m\left(\mathbb{R}^3\right)\middle|\,\Delta P=0\right\}. \tag{5.14}$$

Furthermore, we define

$$\mathrm{Harm}_{0\ldots m}\left(\mathbb{R}^3\right):=\bigoplus_{i=0}^{m}\mathrm{Harm}_i\left(\mathbb{R}^3\right),\quad m\in\mathbb{N}_0,$$

$$\mathrm{Harm}_{0\ldots\infty}\left(\mathbb{R}^3\right):=\bigcup_{i=0}^{\infty}\mathrm{Harm}_{0\ldots i}\left(\mathbb{R}^3\right), \tag{5.15}$$

and for $D\subset\mathbb{R}^3$,

$$\begin{aligned}\mathrm{Harm}_n(D)&:=\left\{P|_D:P\in\mathrm{Harm}_n\left(\mathbb{R}^3\right)\right\},\quad n\in\mathbb{N}_0,\\ \mathrm{Harm}_{0\ldots n}(D)&:=\left\{P|_D:P\in\mathrm{Harm}_{0\ldots n}\left(\mathbb{R}^3\right)\right\},\quad n\in\mathbb{N}_0,\\ \mathrm{Harm}_{0\ldots\infty}(D)&:=\left\{P|_D:P\in\mathrm{Harm}_{0\ldots\infty}\left(\mathbb{R}^3\right)\right\}.\end{aligned} \tag{5.16}$$

The elements of the spaces $\mathrm{Harm}_n(\Omega)$, $n\in\mathbb{N}_0$, are called the (scalar) **spherical harmonics**.

Note the remark after Definition 5.2. The homogeneity and the corresponding degree are verified before the function is restricted. Moreover, the harmonicity also has to be checked before the restriction, since the differential quotient requires an open domain.

Theorem 5.6. *For every $n \in \mathbb{N}_0$,*

$$\dim \mathrm{Harm}_n\left(\mathbb{R}^3\right) = \dim \mathrm{Harm}_n\left(\Omega\right) = 2n+1. \tag{5.17}$$

Proof. Any homogeneous polynomial H_n of degree n on $\mathbb{R}^3$ can be represented by

$$H_n(x) = \sum_{j=0}^{n} x_3^j A_{n-j}\left(x_1,x_2\right), \tag{5.18}$$

where A_{n-j} is a homogeneous polynomial of degree $n-j$ on $\mathbb{R}^2$. If H_n is additionally harmonic, then

$$\begin{aligned}
0 = \Delta_x H_n(x) &= \sum_{j=0}^{n} \left(\frac{\partial^2}{\partial x_1^2} + \frac{\partial^2}{\partial x_2^2} + \frac{\partial^2}{\partial x_3^2}\right) \left(x_3^j A_{n-j}\left(x_1,x_2\right)\right) \\
&= \sum_{j=2}^{n} j(j-1) x_3^{j-2} A_{n-j}\left(x_1,x_2\right) + \sum_{j=0}^{n} x_3^j \underbrace{\left(\frac{\partial^2}{\partial x_1^2} + \frac{\partial^2}{\partial x_2^2}\right) A_{n-j}\left(x_1,x_2\right)}_{=0 \text{ for } j \in \{n-1,n\}} \\
&= \sum_{j=0}^{n-2} (j+2)(j+1) x_3^j A_{n-j-2}\left(x_1,x_2\right) + \sum_{j=0}^{n-2} x_3^j \left(\frac{\partial^2}{\partial x_1^2} + \frac{\partial^2}{\partial x_2^2}\right) A_{n-j}\left(x_1,x_2\right).
\end{aligned} \tag{5.19}$$

Since the functions $\{x_3 \mapsto x_3^j\}_{j=0,\ldots,n-2}$ are linearly independent, we obtain by comparison of the coefficients:

$$\begin{aligned}
\Delta H_n = 0 \Leftrightarrow 0 &= (j+2)(j+1) A_{n-j-2}\left(x_1,x_2\right) \\
&\quad + \left(\frac{\partial^2}{\partial x_1^2} + \frac{\partial^2}{\partial x_2^2}\right) A_{n-j}\left(x_1,x_2\right); \quad j = 0,\ldots,n-2.
\end{aligned} \tag{5.20}$$

Thus, if we know A_n and A_{n-1}, then $A_0,\ldots,A_{n-2}$ are uniquely determined and, consequently, also H_n. The dimension of $\mathrm{Harm}_n(\mathbb{R}^3)$ is, therefore, equal to the total number of coefficients of A_n and A_{n-1}. Since $\dim \mathrm{Hom}_n(\mathbb{R}^2) = n+1$ (see Theorem 5.4), we get

$$\dim \mathrm{Harm}_n\left(\mathbb{R}^3\right) = \underbrace{n+1}_{\#\text{ coeff. of } A_n} + \underbrace{n}_{\#\text{ coeff. of } A_{n-1}}. \tag{5.21}$$

Obviously, $\dim \mathrm{Harm}_n(\Omega) \leq \dim \mathrm{Harm}_n(\mathbb{R}^3)$. Let us assume that there exist linearly independent functions $\{Y_{n,j}\}_{j=1,\ldots,2n+1} \in \mathrm{Harm}_n(\mathbb{R}^3)$ such that their restrictions $\{Y_{n,j}|_\Omega\}_{j=1,\ldots,2n+1}$ are linearly **dependent**. Then there exists one function, without loss of generality (wlog) $Y_{n,1}|_\Omega$, such that $Y_{n,1}|_\Omega = \sum_{j=2}^{2n+1} \alpha_j Y_{n,j}|_\Omega$. From the theory of elliptic partial differential equations, it is known that there exists one and only one $\mathrm{C}^{(2)}$-function F_j on $\overline{B_1(0)}$, which is harmonic in $\Omega_{\mathrm{int}} := B_1(0)$, with $F_j|_\Omega = Y_{n,j}|_\Omega$. Hence, $F_j|_{\overline{\Omega_{\mathrm{int}}}} = Y_{n,j}|_{\overline{\Omega_{\mathrm{int}}}}$. If we additionally require that F_j is a polynomial on $\mathbb{R}^3$, we get $F_j = Y_{n,j}$. Moreover, this elliptic boundary-value problem is linear in the boundary values. Consequently, $F_1 = \sum_{j=2}^{2n+1} \alpha_j F_j$ and $Y_{n,1} = \sum_{j=2}^{2n+1} \alpha_j Y_{n,j}$ in $\mathbb{R}^3$, which is a contradiction to the linear independence. □

Corollary 5.7. *For every $n \in \mathbb{N}_0$,*

$$\dim \mathrm{Harm}_{0\ldots n}\left(\mathbb{R}^3\right) = \dim \mathrm{Harm}_{0\ldots n}(\Omega) = (n+1)^2. \tag{5.22}$$

Proof. From Theorem 5.6, we know that

$$\dim \mathrm{Harm}_j(\Omega) = \dim \mathrm{Harm}_j\left(\mathbb{R}^3\right) = 2j+1. \tag{5.23}$$

Hence, the required dimension can be obtained as follows:

$$\sum_{j=0}^{n}(2j+1) = 2\frac{n(n+1)}{2} + (n+1) = n(n+1) + (n+1) = (n+1)(n+1). \tag{5.24}$$

□

Lemma 5.8. *Any spherical harmonic $Y_n \in \mathrm{Harm}_n(\Omega)$, $n \in \mathbb{N}_0$, is an infinitely often differentiable eigenfunction of the Beltrami operator Δ^* corresponding to the eigenvalue $-n(n+1) =: (\Delta^*)^\wedge(n)$:*

$$\Delta^*_\xi Y_n(\xi) = (\Delta^*)^\wedge(n) Y_n(\xi) \quad \forall \xi \in \Omega. \tag{5.25}$$

The sequence $((\Delta^*)^\wedge(n))_{n\in\mathbb{N}_0}$ is called the ***spherical symbol*** *of the Beltrami operator.*

Proof. Every $Y_n \in \mathrm{Harm}_n(\Omega)$ is a restriction of a polynomial $H_n \in \mathrm{Harm}_n(\mathbb{R}^3)$ to Ω, that is, (see Theorem 4.5)

$$\Delta_x H_n(x) = 0 \forall x \in \mathbb{R}^3$$

$$\Rightarrow \left(\left(\frac{\partial}{\partial r}\right)^2 + \frac{2}{r}\frac{\partial}{\partial r} + \frac{1}{r^2}\Delta^*_\xi\right) H_n(r\xi) = 0 \quad \forall r \in \mathbb{R}^+ \forall \xi \in \Omega. \tag{5.26}$$

We know that the homogeneity of H_n means ($x = r\xi$, $r = |x|$)

$$
\begin{aligned}
H_n(x) &= \sum_{\substack{|\alpha|=n\\ \alpha\in\mathbb{N}_0^3}} C_\alpha x^\alpha = \sum_{\substack{|\alpha|=n\\ \alpha\in\mathbb{N}_0^3}} C_\alpha (r\xi)^\alpha = \sum_{\substack{|\alpha|=n\\ \alpha\in\mathbb{N}_0^3}} C_\alpha (r\xi_1)^{\alpha_1} (r\xi_2)^{\alpha_2} (r\xi_3)^{\alpha_3} \\
&= \sum_{\substack{|\alpha|=n\\ \alpha\in\mathbb{N}_0^3}} C_\alpha r^{|\alpha|} \xi^\alpha = r^n Y_n(\xi). \qquad (5.27)
\end{aligned}
$$

This implies ($r > 0$):

$$
\begin{aligned}
0 = \Delta_x H_n(x) &= n(n-1) r^{n-2} Y_n(\xi) + \frac{2}{r} n r^{n-1} Y_n(\xi) + r^{n-2} \Delta_\xi^* Y_n(\xi) \\
&= r^{n-2} \left(n^2 - n + 2n + \Delta_\xi^*\right) Y_n(\xi). \qquad (5.28)
\end{aligned}
$$

Consequently,

$$
0 = \Delta_\xi^* Y_n(\xi) + \left(n^2 + n\right) Y_n(\xi) \quad \forall \xi \in \Omega. \qquad (5.29)
$$

Hence, we obtain the desired result:

$$
\Delta_\xi^* Y_n(\xi) = -n(n+1) Y_n(\xi) \quad \forall \xi \in \Omega. \qquad (5.30)
$$

□

Theorem 5.9. *If* $Y_n \in \mathrm{Harm}_n(\Omega)$ *and* $Y_m \in \mathrm{Harm}_m(\Omega)$, $n, m \in \mathbb{N}_0$, $n \neq m$, *then*

$$
\langle Y_n, Y_m \rangle_{\mathrm{L}^2(\Omega)} = 0. \qquad (5.31)
$$

Proof. Y_n and Y_m are restrictions of homogeneous harmonic polynomials $H_n \in \mathrm{Harm}_n(\mathbb{R}^3)$ and $H_m \in \mathrm{Harm}_m(\mathbb{R}^3)$ to Ω; see Definition 5.5. If we apply the second Green's identity (the volume version, Theorem 2.33 on p. 29) to H_n and H_m on the unit ball $U := \overline{B_1(0)}$, we get

$$
\begin{aligned}
&\int_U H_n(x) \underbrace{\Delta H_m(x)}_{=0} - H_m(x) \underbrace{\Delta H_n(x)}_{=0} \, \mathrm{d}x \\
&= \int_\Omega Y_n(\xi) \frac{\partial H_m}{\partial \nu}(\xi) - Y_m(\xi) \frac{\partial H_n}{\partial \nu}(\xi) \, \mathrm{d}\omega(\xi), \qquad (5.32)
\end{aligned}
$$

where ν is the outer unit normal to Ω. Using Theorems 4.5 and 4.8 as well as the fact that $\nu(\xi) = \xi$ on Ω, we obtain

$$
\begin{aligned}
\frac{\partial H_p}{\partial \nu}(\xi) &= \left(\xi \cdot \left(\xi \frac{\partial}{\partial r} + \frac{1}{r} \nabla_\xi^*\right) H_p(r\xi)\right)\Big|_{r=1} \\
&= \left(\frac{\partial}{\partial r} H_p(r\xi)\right)\Big|_{r=1} = \left(\frac{\partial}{\partial r} r^p Y_p(\xi)\right)\Big|_{r=1} \\
&= p Y_p(\xi) \qquad \forall \xi \in \Omega \qquad (5.33)
\end{aligned}
$$

for all $H_p \in \mathrm{Harm}_p(\Omega)$ (for $H_p(r\xi) = r^p Y_p(\xi)$, $Y_p \in \mathrm{Harm}_p(\Omega)$, see the proof of Lemma 5.8) and all $p \in \mathbb{N}_0$. Hence,

$$0 = \int_\Omega Y_n(\xi) Y_m(\xi) \, \mathrm{d}\omega(\xi) \, (m-n). \tag{5.34}$$

□

Definition 5.10. For every fixed $n \in \mathbb{N}_0$, $\{Y_{n,j}\}_{j=1,\ldots,2n+1}$ shall always denote an orthonormal system in $(\mathrm{Harm}_n(\Omega), \langle\cdot,\cdot\rangle_{\mathrm{L}^2(\Omega)})$.

Note that Theorem 5.6 implies that this system is also complete in $\mathrm{Harm}_n(\Omega)$. Hence, this definition yields a system with the following properties:

1. $\langle Y_{n,j}, Y_{n,k}\rangle_{\mathrm{L}^2(\Omega)} = \delta_{jk} \; \forall\, j,k \in \{1,\ldots,2n+1\}$.
2. If $\langle F, Y_{n,j}\rangle_{\mathrm{L}^2(\Omega)} = 0 \; \forall\, j = 1,\ldots,2n+1$ is satisfied by $F \in \mathrm{Harm}_n(\Omega)$, then $F = 0$.

The function system obtained by taking such an ons for each n, that is, the set $\{Y_{n,j}\}_{n\in\mathbb{N}_0, j=1,\ldots,2n+1}$, consequently forms an $\mathrm{L}^2(\Omega)$-orthonormal system in the space $\mathrm{Harm}_{0\ldots\infty}(\Omega)$ due to Theorem 5.9:

$$\left\langle Y_{n,j}, Y_{m,k}\right\rangle_{\mathrm{L}^2(\Omega)} = \delta_{nm}\delta_{jk}. \tag{5.35}$$

We call n the **degree** of $Y_{n,j}$ and j the **order** of $Y_{n,j}$. An alternative enumeration is $\{Y_{n,j}\}_{j=-n,\ldots,n}$. Note that the convention introduced in Definition 5.10 is only valid for Y with a double index.

Another important property of spherical harmonics is the following identity.

Theorem 5.11 (Addition Theorem for Spherical Harmonics). *If* $\{Y_{n,j}\}_{j=1,\ldots,2n+1}$ *is an orthonormal system in* $(\mathrm{Harm}_n(\Omega), \langle\cdot,\cdot\rangle_{\mathrm{L}^2(\Omega)})$, $n \in \mathbb{N}_0$, *then*

$$\sum_{j=1}^{2n+1} Y_{n,j}(\xi) \, Y_{n,j}(\eta) = \frac{2n+1}{4\pi} P_n(\xi\cdot\eta) \tag{5.36}$$

for all $(\xi,\eta) \in \Omega^2$, *where* P_n *is the Legendre polynomial of degree* n.

Proof.

(1) Invariance with respect to orthogonal transformations:
Let

$$F(\xi,\eta) := \sum_{j=1}^{2n+1} Y_{n,j}(\xi) Y_{n,j}(\eta); \quad \xi,\eta \in \Omega; \tag{5.37}$$

be the expression of interest and $A \in \mathbb{R}^{3\times 3}$ be an arbitrary orthogonal matrix, that is, $A^{\mathrm{T}} = A^{-1}$. Moreover, let $H_{n,j} \in \mathrm{Harm}_n(\mathbb{R}^3)$, $j = 1,\ldots,2n+1$, with $H_{n,j}\big|_\Omega = Y_{n,j}$ for all $j = 1,\ldots,2n+1$. Then the functions $\mathbb{R}^3 \ni x \mapsto H_{n,j}(Ax)$ are also homogeneous harmonic polynomials for the following reasons: Note that a polynomial P is homogeneous of degree n if and only if $P(\lambda x) = \lambda^n P(x)$ for all $\lambda \in \mathbb{R}$ and all $x \in \mathbb{R}^3$.

Since $H_{n,j}(A(\lambda x)) = H_{n,j}(\lambda Ax) = \lambda^n H_{n,j}(Ax)$ for all $\lambda \in \mathbb{R}$ and all $x \in \mathbb{R}^3$, the homogeneity is not influenced by the transformation. For the harmonicity, consider

$$\begin{aligned}\frac{\partial}{\partial x_k} H_{n,j}(Ax) &= \left(\nabla H_{n,j}(Ax)\right) \cdot \frac{\partial}{\partial x_k}(Ax) \\ &= \sum_{i=1}^{3} \left(\frac{\partial}{\partial y_i} H_{n,j}(y)\Big|_{y=Ax} \frac{\partial}{\partial x_k} \sum_{l=1}^{3} a_{il} x_l \right) \\ &= \sum_{i=1}^{3} \frac{\partial}{\partial y_i} H_{n,j}(y)\Big|_{y=Ax} a_{ik} \end{aligned} \tag{5.38}$$

and

$$\frac{\partial^2}{\partial x_k^2} H_{n,j}(Ax) = \sum_{l=1}^{3}\sum_{i=1}^{3} \frac{\partial^2}{\partial y_l \partial y_i} H_{n,j}(y)\Big|_{y=Ax} a_{lk} a_{ik}. \tag{5.39}$$

Hence,

$$\begin{aligned}\Delta_x H_{n,j}(Ax) &= \sum_{l-1}^{3}\sum_{i=1}^{3} \frac{\partial^2}{\partial y_l \partial y_i} H_{n,j}(y)\Big|_{y=Ax} \underbrace{\sum_{k=1}^{3} a_{lk} a_{ik}}_{=\left(AA^{\mathrm{T}}\right)_{l,i}=\delta_{li}} \\ &= \Delta_y H_{n,j}(y)\Big|_{y=Ax} \\ &= 0. \end{aligned} \tag{5.40}$$

Consequently, each $\Omega \ni \xi \mapsto Y_{n,j}(A\xi)$, $j = 1,\ldots,2n+1$, is a spherical harmonic of degree n. Thus, these functions can be represented in the orthonormal basis of $\mathrm{Harm}_n(\Omega)$: There exist constants $\{b_k^{(j)}\}_{k=1,\ldots,2n+1}$ for each $j = 1,\ldots,2n+1$ such that

$$Y_{n,j}(A\xi) = \sum_{k=1}^{2n+1} b_k^{(j)} Y_{n,k}(\xi) \quad \forall \xi \in \Omega. \tag{5.41}$$

We now consider the integral

$$\int_\Omega Y_{n,j}(A\xi) Y_{n,l}(A\xi)\, \mathrm{d}\omega(\xi) = \sum_{k,m=1}^{2n+1} b_k^{(j)} b_m^{(l)} \underbrace{\int_\Omega Y_{n,k}(\xi) Y_{n,m}(\xi)\, \mathrm{d}\omega(\xi)}_{=\delta_{km}}. \tag{5.42}$$

With the substitution $A\xi =: \eta$ on the left-hand side (see also the considerations in the proof of Theorem 4.16), we get

$$\underbrace{\int_\Omega Y_{n,j}(\eta) Y_{n,l}(\eta)\, \mathrm{d}\omega(\eta)}_{=\delta_{jl}} = \sum_{k=1}^{2n+1} b_k^{(j)} b_k^{(l)} \tag{5.43}$$

for all $j,l \in \{1,\ldots,2n+1\}$. As a consequence, the matrix $B := (b_k^{(j)})_{j,k=1,\ldots,2n+1}$ is orthogonal. For this reason, we also get

$$\sum_{j=1}^{2n+1} b_k^{(j)} b_l^{(j)} = \delta_{kl} \quad \forall k,l = 1,\ldots,2n+1. \tag{5.44}$$

For our function F, this means that

$$\begin{aligned} F(A\xi, A\eta) &= \sum_{j=1}^{2n+1} Y_{n,j}(A\xi) Y_{n,j}(A\eta) \\ &= \sum_{j=1}^{2n+1} \sum_{k,l=1}^{2n+1} b_k^{(j)} b_l^{(j)} Y_{n,k}(\xi) Y_{n,l}(\eta) \\ &= \sum_{k,l=1}^{2n+1} \delta_{kl} Y_{n,k}(\xi) Y_{n,l}(\eta) \\ &= \sum_{k=1}^{2n+1} Y_{n,k}(\xi) Y_{n,k}(\eta) \\ &= F(\xi,\eta) \end{aligned} \tag{5.45}$$

for all $\xi,\eta \in \Omega$. Hence, F is invariant with respect to orthogonal transformations.

(2) *F* is a 1D-function:
Let $\xi,\eta \in \Omega$ be arbitrary but fixed. We consider the set

$$C_{\xi,\eta} := \{\zeta \in \Omega \,|\, \xi \cdot \zeta = \xi \cdot \eta\}. \tag{5.46}$$

Since $|\xi - \zeta|^2 = 2 - 2\xi \cdot \zeta$, the set $C_{\xi,\eta}$ is the circle which is obtained by intersecting Ω and the sphere with center ξ and radius $\sqrt{2 - 2\xi \cdot \eta}$ (see Fig. 5.1). Hence, every element $\zeta \in C_{\xi,\eta}$ is representable as $\zeta = A\eta$ where A is an orthogonal matrix with $A\xi = \xi$ (i.e., we rotate Ω around the axis ξ). Due to part (1) of this proof, we, consequently, have

$$F(\xi,\eta) = F(\xi,\zeta) \quad \forall \zeta \in C_{\xi,\eta}. \tag{5.47}$$

Hence, F depends on the inner product of its two arguments only.

(3) A particular spherical harmonic:
We set

$$Y(\eta) := F\left(\varepsilon^3, \eta\right), \quad \eta \in \Omega. \tag{5.48}$$

Obviously, the definition of F yields that $Y \in \mathrm{Harm}_n(\Omega)$. Let $H \in \mathrm{Harm}_n(\mathbb{R}^3)$ with $H|_\Omega = Y$. In analogy to the proof of Theorem 5.6, we can represent H as

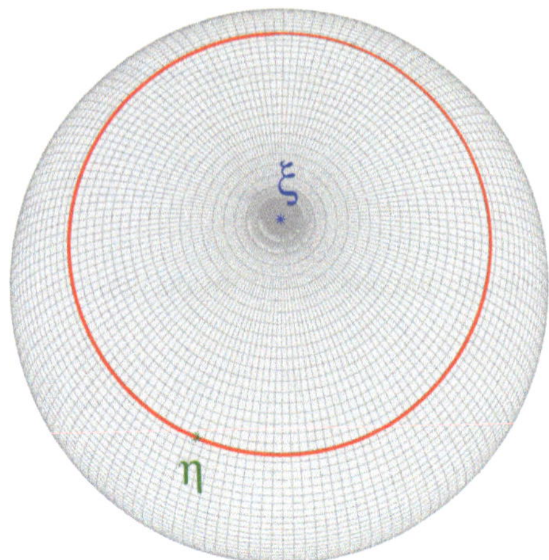

Fig. 5.1 Illustration of the circle $C_{\xi,\eta}$ (*red*) on the unit sphere Ω

$$H(x) = \sum_{j=0}^{n} x_3^j A_{n-j}(x_1,x_2) \quad \forall x = (x_1,x_2,x_3)^{\mathrm{T}} \in \mathbb{R}^3, \tag{5.49}$$

where $A_0,\ldots,A_{n-2}$ are uniquely determined by A_{n-1} and A_n. Since our particular Y actually depends on $\varepsilon^3 \cdot \eta = t_\eta$ only, where t_η is the polar distance as one of the spherical coordinates of η, we conclude that

$$Y(\eta) = \sum_{j=0}^{n} x_3^j A_{n-j}(x_1,x_2)\Big|_{x_1^2+x_2^2+x_3^2=1,\, x_3=t_\eta}. \tag{5.50}$$

Hence, each A_{n-j} is constant for all $(x_1,x_2)^{\mathrm{T}}$ with $x_1^2+x_2^2 = 1-t_\eta^2$. Thus, every A_{n-j} depends on $x_1^2+x_2^2$ only. In particular, this means that $A_{n-j} \equiv 0$, if $n-j$ is odd. We get

$$A_n(x_1,x_2) = \begin{cases} c\left(x_1^2+x_2^2\right)^k = c\left(1-t_\eta^2\right)^k, & \text{if } n=2k,\, k\in\mathbb{N}_0 \\ 0 & \text{else} \end{cases},$$

$$A_{n-1}(x_1,x_2) = \begin{cases} c\left(x_1^2+x_2^2\right)^k = c\left(1-t_\eta^2\right)^k, & \text{if } n=2k+1,\, k\in\mathbb{N}_0 \\ 0 & \text{else} \end{cases}, \tag{5.51}$$

where $c \in \mathbb{R}$ is a constant. Hence, Y is already uniquely determined up to the constant c due to these considerations. Moreover, Y is representable as a 1D-polynomial of t with degree n. Thus, $F(\xi,\eta)$ is a 1D-polynomial of $\xi\cdot\eta$ with degree n. Note that we can rotate the sphere by an orthogonal matrix A such that $A\xi = \varepsilon^3$.

(4) We recognize the Legendre polynomial:
For every $n \in \mathbb{N}_0$, we get such a function F. Let us now introduce the notation

$$G_p(\xi \cdot \eta) := \sum_{j=1}^{2p+1} Y_{p,j}(\xi) Y_{p,j}(\eta); \quad \xi, \eta \in \Omega \quad (\Rightarrow \xi \cdot \eta \in [-1,1]), \quad p \in \mathbb{N}_0. \tag{5.52}$$

We get

$$2\pi \int_{-1}^{1} G_p(t) G_q(t) \, \mathrm{d}t = \int_\Omega G_p\left(\varepsilon^3 \cdot \eta\right) G_q\left(\varepsilon^3 \cdot \eta\right) \mathrm{d}\omega(\eta) = 0 \tag{5.53}$$

for $p \neq q$ due to Theorems 4.16 and 5.9. Moreover,

$$G_n(1) = G_n(\eta \cdot \eta) = \sum_{j=1}^{2n+1} Y_{n,j}(\eta) Y_{n,j}(\eta) \tag{5.54}$$

for all $\eta \in \Omega$ and all $n \in \mathbb{N}_0$ such that an integration over Ω yields

$$\begin{aligned} 4\pi G_n(1) &= \sum_{j=1}^{2n+1} \int_\Omega Y_{n,j}(\eta) Y_{n,j}(\eta) \, \mathrm{d}\omega(\eta) \\ &= 2n+1. \end{aligned} \tag{5.55}$$

We summarize our results: The sequence $\{G_n\}_{n \in \mathbb{N}_0}$ is a sequence of polynomials which is orthogonal in $\mathrm{L}^2[-1,1]$ and satisfies $\deg G_n = n$ and $G_n(1) = \frac{2n+1}{4\pi}$ for all $n \in \mathbb{N}_0$. Due to Theorem 3.9 and Definition 3.10 (see p. 40), we conclude

$$G_n(t) = \frac{2n+1}{4\pi} P_n(t) \quad \forall t \in [-1,1] \quad \forall n \in \mathbb{N}_0. \tag{5.56}$$

□

Remark 5.12. The name "addition theorem" probably reminds you of a property of trigonometric functions. This is not a coincidence. The reason is the fact that the theory of spherical harmonics can be established for a general $(n-1)$-sphere $\mathbb{S}^{n-1} := \{x \in \mathbb{R}^n \mid |x| = 1\}$ in $\mathbb{R}^n$, $n \geq 2$ (see, e.g., [139]). In the simplest case, the 1-sphere, we have a unit circle with one polar coordinate: $\varphi \in [0, 2\pi]$. A corresponding orthonormal basis of spherical harmonics is

$$\begin{aligned} Y_{0,1} &:= \frac{1}{\sqrt{2\pi}}, \\ Y_{n,1}\left(\zeta(\varphi)\right) &:= \frac{1}{\sqrt{\pi}} \cos(n\varphi), \\ Y_{n,2}\left(\zeta(\varphi)\right) &:= \frac{1}{\sqrt{\pi}} \sin(n\varphi), \end{aligned} \tag{5.57}$$

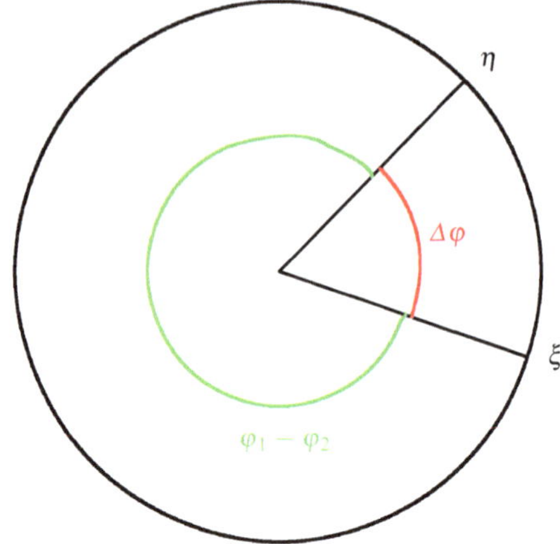

Fig. 5.2 Illustration corresponding to Remark 5.12: instead of $\varphi_1 - \varphi_2$, the angle $\Delta\varphi = 2\pi - (\varphi_1 - \varphi_2)$ can be chosen, since $\cos[(n\Delta\varphi)] = \cos[2n\pi - n(\varphi_1 - \varphi_2)] = \cos[-n(\varphi_1 - \varphi_2)] = \cos[n(\varphi_1 - \varphi_2)]$ for all $n \in \mathbb{N}_0$

$n \in \mathbb{N}$, where $\zeta(\varphi)$ refers to the representation of $\zeta \in \mathbb{S}^1$ in terms of the polar coordinate φ. This is the orthonormal basis for $\mathrm{L}^2[0,2\pi]$ which we already know from Theorem 3.23 on p. 53. For a corresponding addition theorem (which is one of the common addition theorems for trigonometric functions), we have to consider the term (for $n \geq 1$)

$$\frac{1}{\sqrt{\pi}}\cos(n\varphi_1)\frac{1}{\sqrt{\pi}}\cos(n\varphi_2) + \frac{1}{\sqrt{\pi}}\sin(n\varphi_1)\frac{1}{\sqrt{\pi}}\sin(n\varphi_2) = \frac{1}{\pi}\cos[n(\varphi_1 - \varphi_2)] \tag{5.58}$$

with $\varphi_1, \varphi_2 \in [0,2\pi]$. If $\xi = \zeta(\varphi_1)$ and $\eta = \zeta(\varphi_2)$, the inner product is, consequently,

$$\xi \cdot \eta = \begin{pmatrix} \cos\varphi_1 \\ \sin\varphi_1 \end{pmatrix} \cdot \begin{pmatrix} \cos\varphi_2 \\ \sin\varphi_2 \end{pmatrix} = \cos(\varphi_1 - \varphi_2). \tag{5.59}$$

Note that the right-hand side depends on $|\varphi_1 - \varphi_2|$ only, where $|\varphi_1 - \varphi_2|$ can be assumed to be in $[0,\pi]$ (see Fig. 5.2). Hence, there is a one-to-one relation between $\xi \cdot \eta$ and $|\varphi_1 - \varphi_2|$ such that the right-hand side of (5.58), namely, $\frac{1}{\pi}\cos[n(\varphi_1 - \varphi_2)] = \frac{1}{\pi}\cos(n|\varphi_1 - \varphi_2|)$, is a function of $\xi \cdot \eta$. Therefore, the Chebyshev polynomials (see Theorem 3.11) on p. 43 are the corresponding analogues of the Legendre polynomials:

$$\sum_{j=1}^{2} Y_{n,j}(\zeta(\varphi_1))Y_{n,j}(\zeta(\varphi_2)) = \frac{1}{\pi}T_n(\zeta(\varphi_1)\cdot\zeta(\varphi_2)) \quad \forall n \in \mathbb{N}. \tag{5.60}$$

For the trivial case $n = 0$, one gets

$$Y_{0,1}(\zeta(\varphi_1))Y_{0,1}(\zeta(\varphi_2)) = \frac{1}{2\pi} = \frac{1}{2\pi}T_0(\cos(\varphi_1 - \varphi_2)). \tag{5.61}$$

Note that the coefficient in front of T_n is the quotient of the number of basis functions for the degree n (i.e., $2-\delta_{n0}$) and the length of $\mathbb{S}^1$ (i.e., $\int_{\mathbb{S}^1} 1\,\mathrm{dl} = 2\pi$) in analogy to the addition theorem on $\mathbb{S}^2 = \Omega$.

The addition theorem for spherical harmonics has many advantages. One of them is connected to the calculation of splines and wavelets on the sphere. In both cases, we will have to calculate kernels on $\Omega \times \Omega$ which have the form

$$K(\xi,\eta) = \sum_{n=0}^{\infty} \sum_{j=1}^{2n+1} k_n Y_{n,j}(\xi) Y_{n,j}(\eta), \quad \xi,\eta \in \Omega. \tag{5.62}$$

Due to the addition theorem, this 4D-function on $\Omega \times \Omega$ can be simplified to a 1D-function on $[-1,1]$:

$$K(\xi,\eta) = \sum_{n=0}^{\infty} k_n \frac{2n+1}{4\pi} P_n(\xi \cdot \eta); \quad \xi,\eta \in \Omega. \tag{5.63}$$

Moreover, the addition theorem is also a helpful tool in many proofs as we will see on the next pages. One of these consequences of the addition theorem is connected to the theory of reproducing kernels (see also [43]).

Definition 5.13. Let $(H, \langle\cdot,\cdot\rangle)$ be a (real) Hilbert space of functions which are defined on the domain $D \subset \mathbb{R}^n$, $n \in \mathbb{N}$. A function $K_H : D \times D \to \mathbb{R}$ is called a **reproducing kernel** of H, if it satisfies the following properties:

(i) $K_H(x,\cdot) \in H$ for all $x \in D$.
(ii) $\langle K_H(x,\cdot), F\rangle = F(x)$ for all $F \in H$ and all $x \in D$.

In this case, H is called a **reproducing kernel Hilbert space**.

The second property, which is called the "reproducing property," is the reason for the name "reproducing kernel" because the value of F is reproduced at an arbitrary point x by means of the kernel. The first property is only technical. It merely guarantees that the inner product in the reproducing property can be calculated.

An equivalent definition of a reproducing kernel would require:

1. $K_H(\cdot,x) \in H$ for all $x \in D$,
2. $\langle K_H(\cdot,x), F\rangle = F(x)$ for all $F \in H$ and all $x \in D$,

as the following theorem shows.

Theorem 5.14 (Symmetry of Reproducing Kernels). *Let H be a Hilbert space of functions on $D \subset \mathbb{R}^n$ with the reproducing kernel K_H. Then*

$$K_H(x,y) = K_H(y,x) \tag{5.64}$$

for all $x,y \in D$.

Proof. We know that $K_H(y,\cdot)$, $K_H(x,\cdot) \in H$ for all $x,y \in D$. In the reproducing property

$$\begin{aligned} \langle K_H(y,\cdot), F\rangle &= F(y) \qquad \forall F \in H \quad \forall y \in D \\ \langle K_H(x,\cdot), G\rangle &= G(x) \qquad \forall G \in H \quad \forall x \in D, \end{aligned} \tag{5.65}$$

we can insert, consequently, $F = K_H(x,\cdot)$ and $G = K_H(y,\cdot)$. This yields, in combination with the symmetry of the inner product,

$$\begin{aligned} F(y) &= K_H(x,y) \\ &= \langle K_H(y,\cdot), K_H(x,\cdot)\rangle \\ &= \langle K_H(x,\cdot), K_H(y,\cdot)\rangle \\ &= K_H(y,x) \\ &= G(x), \end{aligned} \tag{5.66}$$

where $x, y \in D$ are arbitrary. □

Theorem 5.15. *The reproducing kernel of a reproducing kernel Hilbert space is always unique.*

The proof is very easy and is, therefore, a good exercise.

Let us go back to the addition theorem for spherical harmonics:

$$\frac{2n+1}{4\pi} P_n(\xi\cdot\eta) = \sum_{j=1}^{2n+1} Y_{n,j}(\xi) Y_{n,j}(\eta) \qquad \forall \xi, \eta \in \Omega. \tag{5.67}$$

We multiply both sides by an arbitrary spherical harmonic[1] $Y_n \in \mathrm{Harm}_n(\Omega)$ and get

$$\frac{2n+1}{4\pi} P_n(\xi\cdot\eta) Y_n(\eta) = \sum_{j=1}^{2n+1} Y_{n,j}(\xi) Y_n(\eta) Y_{n,j}(\eta) \qquad \forall \xi, \eta \in \Omega. \tag{5.68}$$

Now, we integrate with respect to $\eta \in \Omega$:

$$\int_\Omega \frac{2n+1}{4\pi} P_n(\xi\cdot\eta) Y_n(\eta)\,\mathrm{d}\omega(\eta) = \sum_{j=1}^{2n+1} Y_{n,j}(\xi) \int_\Omega Y_n(\eta) Y_{n,j}(\eta)\,\mathrm{d}\omega(\eta) \quad \forall \xi \in \Omega. \tag{5.69}$$

Note that the right-hand side is the Fourier series for $Y_n \in \mathrm{Harm}_n(\Omega)$, since the system $\{Y_{n,j}\}_{j=1,\ldots,2n+1}$ is an orthonormal basis for $\mathrm{Harm}_n(\Omega)$. Hence,

$$\int_\Omega \frac{2n+1}{4\pi} P_n(\xi\cdot\eta) Y_n(\eta)\,\mathrm{d}\omega(\eta) = Y_n(\xi) \qquad \forall \xi \in \Omega. \tag{5.70}$$

This proves the following theorem.

[1]Note that we do not require that Y_n is an element of the onb because we only have one index.

Theorem 5.16. *The kernel*

$$\Omega \times \Omega \ni (\xi, \eta) \mapsto \frac{2n+1}{4\pi} P_n(\xi \cdot \eta) \tag{5.71}$$

is the reproducing kernel of $(\mathrm{Harm}_n(\Omega), \langle\cdot,\cdot\rangle_{\mathrm{L}^2(\Omega)})$.

Reproducing kernels play a more important role in the development of the spherical spline method, as we will see later.

Two further theorems which are based on the addition theorem yield estimates for the maxima of spherical harmonics and of the Legendre polynomials.

Theorem 5.17. *Every* $Y_n \in \mathrm{Harm}_n(\Omega)$, $n \in \mathbb{N}_0$, *satisfies*

$$\|Y_n\|_{\mathrm{C}(\Omega)} \leq \sqrt{\frac{2n+1}{4\pi}}\, \|Y_n\|_{\mathrm{L}^2(\Omega)} . \tag{5.72}$$

In particular,

$$\left\|Y_{n,j}\right\|_{\mathrm{C}(\Omega)} \leq \sqrt{\frac{2n+1}{4\pi}}. \tag{5.73}$$

Proof. We use again the completeness of the orthonormal system $\{Y_{n,j}\}_{j=1,\ldots,2n+1}$ in $\mathrm{Harm}_n(\Omega)$. For every $\xi \in \Omega$, we get using the Cauchy–Schwarz inequality in $\mathbb{R}^{2n+1}$ the result:

$$\begin{aligned} |Y_n(\xi)| &= \left| \sum_{j=1}^{2n+1} \langle Y_n, Y_{n,j} \rangle_{\mathrm{L}^2(\Omega)} Y_{n,j}(\xi) \right| \\ &\leq \left(\sum_{j=1}^{2n+1} \langle Y_n, Y_{n,j} \rangle^2_{\mathrm{L}^2(\Omega)} \right)^{1/2} \left(\sum_{j=1}^{2n+1} (Y_{n,j}(\xi))^2 \right)^{1/2} . \end{aligned} \tag{5.74}$$

The use of the addition theorem was already announced above. Can you see where we can do this? The last factor is one-hand side of the addition theorem. We get

$$\sum_{j=1}^{2n+1} (Y_{n,j}(\xi))^2 = \frac{2n+1}{4\pi} P_n(\xi \cdot \xi) = \frac{2n+1}{4\pi} P_n(1) = \frac{2n+1}{4\pi} \tag{5.75}$$

due to the definition of the Legendre polynomials (see Definition 3.10 on p. 41).

Moreover, the penultimate factor in (5.74) is simply the Parseval identity for $\|Y_n\|_{\mathrm{L}^2(\Omega)}$. Hence, we get

$$|Y_n(\xi)| \leq \|Y_n\|_{\mathrm{L}^2(\Omega)} \sqrt{\frac{2n+1}{4\pi}} \qquad \forall \xi \in \Omega. \tag{5.76}$$

□

We now prove Theorem 3.18 on p. 49 for the particular case $\alpha = \beta = 0$.

Theorem 5.18. *For every $n \in \mathbb{N}_0$, the Legendre polynomial P_n satisfies*

$$\|P_n\|_{\mathrm{C}[-1,1]} = 1 = P_n(1). \tag{5.77}$$

Proof. We start with the addition theorem (with absolute values on both sides):

$$\frac{2n+1}{4\pi}\,|P_n(\xi\cdot\eta)| = \left|\sum_{j=1}^{2n+1} Y_{n,j}(\xi)Y_{n,j}(\eta)\right| \qquad \forall \xi,\eta \in \Omega. \tag{5.78}$$

We apply again the Cauchy–Schwarz inequality in $\mathbb{R}^{2n+1}$:

$$\begin{aligned}\frac{2n+1}{4\pi}\,|P_n(\xi\cdot\eta)| &\le \left(\sum_{j=1}^{2n+1}(Y_{n,j}(\xi))^2\right)^{1/2}\left(\sum_{j=1}^{2n+1}(Y_{n,j}(\eta))^2\right)^{1/2}\\ &= \left(\frac{2n+1}{4\pi}\right)^{1/2}\cdot\left(\frac{2n+1}{4\pi}\right)^{1/2}\\ &= \frac{2n+1}{4\pi} \qquad \forall \xi,\eta\in\Omega.\end{aligned} \tag{5.79}$$

Since every $t \in [-1,1]$ can be represented as $t = \xi\cdot\eta$ for an appropriate choice of $\xi,\eta \in \Omega$, our work is done. □

Let us not forget what the purpose of this section is. We are looking for a Fourier analysis tool on Ω. For this purpose, we need a complete orthonormal system. We already have a system which is $\langle\cdot,\cdot\rangle_{\mathrm{L}^2(\Omega)}$-orthonormal. Moreover, we know that it is complete in $\mathrm{Harm}_{0\ldots\infty}(\Omega)$, which is, however, not a practicable space. The good news is we will see that the system $\{Y_{n,j}\}_{n\in\mathbb{N}_0;j=1,\ldots,2n+1}$ is also complete in $\mathrm{L}^2(\Omega)$—this will be the main result of this section. The bad news is the way to get there is not easy. The next essential milestone on our way is the Poisson integral formula.

Theorem 5.19 (Poisson Integral Formula). *Let $F \in \mathrm{C}(\Omega)$, then*

$$\lim_{h\to 1-}\sup_{\xi\in\Omega}\left|\frac{1}{4\pi}\int_\Omega \frac{(1-h^2)\,F(\eta)}{(1+h^2-2h(\xi\cdot\eta))^{3/2}}\,\mathrm{d}\omega(\eta) - F(\xi)\right| = 0. \tag{5.80}$$

Proof.

(1) The case $F \equiv 1$:
In this case, we get the integral of a zonal function (see Theorem 4.16):

$$\frac{1}{4\pi}\int_\Omega \frac{1-h^2}{(1+h^2-2h(\xi\cdot\eta))^{3/2}}\,\mathrm{d}\omega(\eta) = \frac{1}{2}\int_{-1}^{1}\frac{1-h^2}{(1+h^2-2ht)^{3/2}}\,\mathrm{d}t. \tag{5.81}$$

For the integrand of the right-hand side, we already know an expansion in Legendre polynomials from Theorem 3.16. Moreover, we find that

$$\frac{1-h^2}{(1+h^2-2ht)^{3/2}} = \sum_{n=0}^{\infty}(2n+1)h^n P_n(t) \tag{5.82}$$

is uniformly convergent with respect to $t \in [-1,1]$ for every fixed $h \in [0,1[$ since

$$|(2n+1)h^n P_n(t)| \le (2n+1)h^n \qquad \forall n \in \mathbb{N}_0 \tag{5.83}$$

and

$$\sum_{n=0}^{\infty}(2n+1)h^n < +\infty. \tag{5.84}$$

Hence, we are allowed to interchange the integration with the series, and we get from (5.81) and Theorem 3.14

$$\begin{aligned}\frac{1}{4\pi}\int_{\Omega}\frac{1-h^2}{(1+h^2-2h(\xi\cdot\eta))^{3/2}}\,\mathrm{d}\omega(\eta) &= \sum_{n=0}^{\infty}\frac{2n+1}{2}h^n\int_{-1}^{1}P_n(t)\mathrm{d}t \\ &= \sum_{n=0}^{\infty}\frac{2n+1}{2}h^n\underbrace{\int_{-1}^{1}P_n(t)\overbrace{P_0(t)}^{\equiv 1}\,\mathrm{d}t}_{=\delta_{n0}\frac{2}{2n+1}} \\ &= 1.\end{aligned} \tag{5.85}$$

The partition of unity (5.85) is helpful to prove the formula for other F.

(2) Subdivision of Ω:
We keep $\xi \in \Omega$ and $h \in [\frac{1}{2},1[$ fixed (smaller values of h are irrelevant for the limit). We divide the sphere now into two parts:

$$\begin{aligned}D_1 &:= \left\{\eta\in\Omega \,\middle|\, (-1\le)\xi\cdot\eta\le 1-\sqrt[3]{1-h}\right\}, \\ D_2 &:= \left\{\eta\in\Omega \,\middle|\, 1-\sqrt[3]{1-h}<\xi\cdot\eta(\le 1)\right\} = \Omega\setminus D_1\end{aligned} \tag{5.86}$$

(see Fig. 5.3). We will now show that the following expression tends to zero as $h \to 1-$ (i.e., as h approaches 1 with values $h < 1$):

$$\begin{aligned}&\frac{1}{4\pi}\int_{\Omega}\frac{(1-h^2)\,F(\eta)}{(1+h^2-2h(\xi\cdot\eta))^{3/2}}\,\mathrm{d}\omega(\eta) - F(\xi) \\ &\overset{(5.85)}{=} \frac{1}{4\pi}\int_{\Omega}\frac{(1-h^2)\,(F(\eta)-F(\xi))}{(1+h^2-2h(\xi\cdot\eta))^{3/2}}\,\mathrm{d}\omega(\eta)\end{aligned}$$

Fig. 5.3 Illustration of the sets D_1 (*green*) and D_2 (*yellow*) with respect to a fixed point ξ (*black dot*, here: Vienna) for $h = 0.9$ (*left hand*), $h = 0.999$ (*middle*), and $h = 0.99999$ (*right hand*)

$$= \frac{1}{4\pi}\int_{D_1} \frac{(1-h^2)\,(F(\eta)-F(\xi))}{(1+h^2-2h(\xi\cdot\eta))^{3/2}}\,\mathrm{d}\omega(\eta)$$

$$+\frac{1}{4\pi}\int_{D_2} \frac{(1-h^2)\,(F(\eta)-F(\xi))}{(1+h^2-2h(\xi\cdot\eta))^{3/2}}\,\mathrm{d}\omega(\eta). \tag{5.87}$$

(3) The integral over D_1:
For $t \in \left[-1, 1-\sqrt[3]{1-h}\right]$, we find

$$1+h^2-2ht = \underbrace{(1-h)^2}_{\geq 0} + 2h\underbrace{(1-t)}_{\geq 1-\left(1-\sqrt[3]{1-h}\right)} \geq 2h\sqrt[3]{1-h}. \tag{5.88}$$

Consequently, the integrand (without F) can be estimated by

$$\frac{1-h^2}{(1+h^2-2ht)^{3/2}} \leq \frac{1-h^2}{\left(2h\sqrt[3]{1-h}\right)^{3/2}} = \frac{\overbrace{1+h}^{\leq 2}}{\underbrace{(2h)^{3/2}}_{\geq 1}}\cdot\frac{1-h}{\sqrt{1-h}} \leq 2\sqrt{1-h}. \tag{5.89}$$

This estimate is now inserted in the integral over D_1 in (5.87) after the application of the triangle inequality and Theorem 4.16:

$$\left|\int_{D_1} \frac{(1-h^2)\,(F(\eta)-F(\xi))}{(1+h^2-2h(\xi\cdot\eta))^{3/2}}\,\mathrm{d}\omega(\eta)\right|$$

$$\leq \int_{D_1} \frac{(1-h^2)\left(\|F\|_{\mathrm{C}(\Omega)}+\|F\|_{\mathrm{C}(\Omega)}\right)}{(1+h^2-2h(\xi\cdot\eta))^{3/2}}\,\mathrm{d}\omega(\eta)$$

$$= 2\|F\|_{\mathrm{C}(\Omega)}2\pi\int_{-1}^{1-\sqrt[3]{1-h}} \frac{1-h^2}{(1+h^2-2ht)^{3/2}}\,\mathrm{d}t$$

$$
\begin{aligned}
&\le 4\pi\|F\|_{\mathrm{C}(\Omega)} \int_{-1}^{1-\sqrt[3]{1-h}} 2\sqrt{1-h}\,\mathrm{d}t \\
&= 8\pi\|F\|_{\mathrm{C}(\Omega)}\sqrt{1-h}\left(2-\sqrt[3]{1-h}\right) \\
&\le 16\pi\|F\|_{\mathrm{C}(\Omega)}\sqrt{1-h} \underset{h\to 1-}{\longrightarrow} 0. \qquad (5.90)
\end{aligned}
$$

(4) The integral over D_2:
F is a continuous function on the compact domain Ω. Therefore, it is uniformly continuous. This means, first of all, the following:

$$
\forall \varepsilon > 0 \quad \exists \delta = \delta(\varepsilon) > 0 : (|\xi - \eta| < \delta;\, \xi, \eta \in \Omega \Rightarrow |F(\xi) - F(\eta)| \le \varepsilon). \qquad (5.91)
$$

This implies the existence of a function $h \mapsto \mu(h)$ with $\lim_{h\to 1-} \mu(h) = 0$ and

$$
|F(\xi) - F(\eta)| \le \mu(h) \quad \text{for all } \eta \in D_2 \quad (\text{more precisely: } D_2(h)). \qquad (5.92)
$$

Since

$$
\begin{aligned}
\eta \in D_2 &\Leftrightarrow 1 - \sqrt[3]{1-h} < \xi \cdot \eta \\
&\Leftrightarrow \sqrt[3]{1-h} > 1 - \xi \cdot \eta \\
&\Leftrightarrow \sqrt[3]{1-h} > \frac{1}{2}|\xi - \eta|^2 \\
&\Leftrightarrow \sqrt[6]{8(1-h)} > |\xi - \eta|, \qquad (5.93)
\end{aligned}
$$

such a function can be obtained by choosing $\varepsilon \mapsto \delta(\varepsilon)$ in (5.91) as a monotonically increasing function (which is obviously possible) and setting

$$
\mu(h) := \begin{cases} \inf\left(\left\{\varepsilon > 0 \,\middle|\, \delta(\varepsilon) \ge \sqrt[6]{8(1-h)}\right\}\right), & \text{if } \sup_{\varepsilon>0}\delta(\varepsilon) > \sqrt[6]{8(1-h)} \\ 2\|F\|_{\mathrm{C}(\Omega)} & \text{else} \end{cases}. \qquad (5.94)
$$

Since $\lim_{h\to 1-} \sqrt[6]{8(1-h)} = 0$, only the first case in (5.94) occurs, if h has exceeded a certain threshold. Property (5.91) implies that $\lim_{h\to 1-}\mu(h) = 0$. Finally, for an arbitrary h (without loss of generality $\sqrt[6]{8(1-h)} < \sup_{\varepsilon>0}\delta(\varepsilon)$), we have, due to (5.93),

$$
\begin{aligned}
\eta \in D_2 &\Leftrightarrow |\xi - \eta| < \sqrt[6]{8(1-h)} \\
&\Rightarrow \exists \varepsilon > 0 : |\xi - \eta| < \delta(\varepsilon). \qquad (5.95)
\end{aligned}
$$

By taking the infimum over all such $\varepsilon > 0$, we get

$$|F(\xi) - F(\eta)| \leq \mu(h) \qquad \forall \eta \in D_2. \tag{5.96}$$

Thus, the function $\mu(h)$ really exists.[2] We use it to estimate the integral over D_2:

$$\begin{aligned}
\left| \int_{D_2} \frac{(1-h^2)\,(F(\eta)-F(\xi))}{(1+h^2-2h(\xi\cdot\eta))^{3/2}}\,\mathrm{d}\omega(\eta) \right| &\leq \int_{D_2} \frac{(1-h^2)\,|F(\eta)-F(\xi)|}{(1+h^2-2h(\xi\cdot\eta))^{3/2}}\,\mathrm{d}\omega(\eta) \\
&\leq \int_{D_2} \frac{(1-h^2)\,\mu(h)}{(1+h^2-2h(\xi\cdot\eta))^{3/2}}\,\mathrm{d}\omega(\eta) \\
&\leq \int_{\Omega} \frac{(1-h^2)\,\mu(h)}{(1+h^2-2h(\xi\cdot\eta))^{3/2}}\,\mathrm{d}\omega(\eta) \\
&\overset{(5.85)}{=} \mu(h)\cdot 4\pi.
\end{aligned} \tag{5.97}$$

(5) Recombining $\boldsymbol{\Omega}$:
Finally, we get, due to (5.87), (5.90), and (5.97),

$$\left| \frac{1}{4\pi} \int_{\Omega} \frac{(1-h^2)\,(F(\eta)-F(\xi))}{(1+h^2-2h(\xi\cdot\eta))^{3/2}}\,\mathrm{d}\omega(\eta) \right| \leq 4\|F\|_{\mathrm{C}(\Omega)}\sqrt{1-h} + \mu(h). \tag{5.98}$$

The right-hand side of (5.98) converges uniformly to 0 as $h \to 1-$ since it is independent of $\xi \in \Omega$. □

Remark 5.20. Two comments might be interesting here:

(a) The kernel

$$\Omega \times \Omega \ni (\xi,\eta) \mapsto \frac{1}{4\pi} \frac{1-h^2}{(1+h^2-2h(\xi\cdot\eta))^{3/2}}, \tag{5.99}$$

which was used in the Poisson integral formula, is called the **Abel–Poisson kernel**. If one keeps $\xi \in \Omega$ fixed, one will observe that the graph of the resulting function of $\eta \in \Omega$ is a kind of a "hat" around ξ which gets thinner and thinner as h tends to 1 (see Fig. 5.4). We say the kernel is a **localized function**. Moreover, the Poisson integral formula is our first encounter with a spherical Approximate Identity. The integral represents a spherical convolution of the Abel–Poisson kernel and F, and we saw that this convolution uniformly converges to F, if F is continuous.

[2] Note that μ is independent of ξ. This is the characteristic property of the uniform continuity.

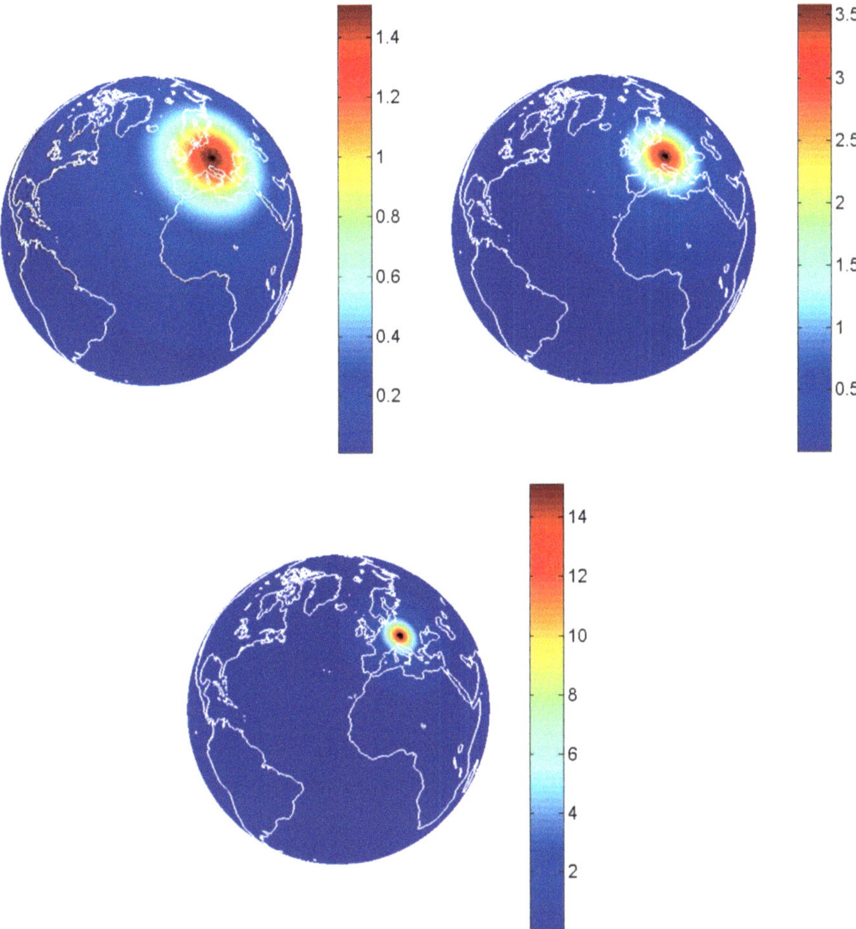

Fig. 5.4 The pictures show the graphs of the Abel–Poisson kernel as a function of $\eta \in \Omega$ with a fixed unit vector ξ (here again Vienna) for $h = 0.7$ (*top left hand*), $h = 0.8$ (*top right hand*), and $h = 0.9$ (*bottom*). With increasing h, the kernel gets more and more localized

(b) You might know the name "Poisson integral formula" from the potential theory or the theory of elliptic partial differential equations (see, e.g., [101, p. 240], [136, p. 263], [155, p. 112], [190, p. 43]). In this context, the formula for solving the Dirichlet boundary-value problem for the Laplace equation in the case of a spherical boundary[3] is called the Poisson integral formula or Poisson's formula. Indeed, both formulae are quasi the same. Equation (5.80) is obtained from the formula in the potential theory by calculating the solution of the (inner)

[3]The problem is as follows: Given a continuous function F on a sphere, find a harmonic function U inside the sphere which equals F on the spherical boundary. The Poisson integral formula is then a formula which represents U by a spherical integral, which involves F in the integrand.

boundary-value problem (with the boundary Ω) at a sphere with the radius $h < 1$. In the limit $h \to 1-$, one gets the boundary value, that is, the function F, again. That's what the Poisson integral formula (5.80) claims.

We will meet the Abel–Poisson kernel again—in the spline theory as well as in the wavelet theory.

Remember again what we are looking for: an orthonormal basis of $\mathrm{L}^2(\Omega)$. The Poisson integral formula really helps us to reach this goal. The next intermediate step is to identify the system $\{Y_{n,j}\}_{n\in\mathbb{N}_0,\, j=1,\ldots,2n+1}$ as a closed system in the Banach space $(\mathrm{C}(\Omega), \|\cdot\|_{\mathrm{C}(\Omega)})$. Remember Definition 3.4 and Theorem 3.5 in this context.

Lemma 5.21. *Let $F \in \mathrm{C}(\Omega)$. Then the series*

$$\sum_{n=0}^{\infty} h^n \sum_{j=1}^{2n+1} \langle F, Y_{n,j} \rangle_{\mathrm{L}^2(\Omega)} Y_{n,j}(\xi) \tag{5.100}$$

converges uniformly with respect to all $\xi \in \Omega$ for each fixed $h \in]0,1[$. Moreover,

$$\lim_{h\to 1-} \sum_{n=0}^{\infty} h^n \sum_{j=1}^{2n+1} \langle F, Y_{n,j} \rangle_{\mathrm{L}^2(\Omega)} Y_{n,j}(\xi) = F(\xi), \tag{5.101}$$

where this convergence is also uniform with respect to all $\xi \in \Omega$.

Proof.

(1) The uniform convergence of the series:
Since $\{Y_{n,j}\}_{n\in\mathbb{N}_0,\, j=1,\ldots,2n+1}$ is an orthonormal system with respect to $\langle \cdot, \cdot \rangle_{\mathrm{L}^2(\Omega)}$ and $F \in \mathrm{C}(\Omega) \subset \mathrm{L}^2(\Omega)$, the Bessel inequality (see Lemma 2.17 on p. 21) yields

$$\sum_{n=0}^{\infty} \sum_{j=1}^{2n+1} \langle F, Y_{n,j} \rangle_{\mathrm{L}^2(\Omega)}^2 \le \|F\|_{\mathrm{L}^2(\Omega)}^2 < +\infty. \tag{5.102}$$

This implies, in particular, the existence of the maximum

$$\max_{\substack{m\in\mathbb{N}_0,\\ k\in\{1,\ldots,2m+1\}}} \left| \langle F, Y_{m,k} \rangle_{\mathrm{L}^2(\Omega)} \right| < +\infty. \tag{5.103}$$

This result in combination with the triangle inequality and Theorem 5.17 yields

$$\begin{aligned}
&\left| \sum_{n=0}^{\infty} h^n \sum_{j=1}^{2n+1} \langle F, Y_{n,j} \rangle_{\mathrm{L}^2(\Omega)} Y_{n,j}(\xi) - \sum_{n=0}^{N} h^n \sum_{j=1}^{2n+1} \langle F, Y_{n,j} \rangle_{\mathrm{L}^2(\Omega)} Y_{n,j}(\xi) \right| \\
&= \left| \sum_{n=N+1}^{\infty} h^n \sum_{j=1}^{2n+1} \langle F, Y_{n,j} \rangle_{\mathrm{L}^2(\Omega)} Y_{n,j}(\xi) \right| \\
&\le \sum_{n=N+1}^{\infty} h^n \sum_{j=1}^{2n+1} \left| \langle F, Y_{n,j} \rangle_{\mathrm{L}^2(\Omega)} \right| \left| Y_{n,j}(\xi) \right|
\end{aligned}$$

$$\leq \sum_{n=N+1}^{\infty} h^n \sum_{j=1}^{2n+1} \max_{\substack{m\in\mathbb{N}_0,\\ k\in\{1,\ldots,2m+1\}}} \left|\left\langle F, Y_{m,k}\right\rangle_{\mathrm{L}^2(\Omega)}\right| \sqrt{\frac{2n+1}{4\pi}}$$

$$= \max_{\substack{m\in\mathbb{N}_0,\\ k\in\{1,\ldots,2m+1\}}} \left|\left\langle F, Y_{m,k}\right\rangle_{\mathrm{L}^2(\Omega)}\right| \sum_{n=N+1}^{\infty} h^n \underbrace{\sum_{j=1}^{2n+1} \sqrt{\frac{2n+1}{4\pi}}}_{=\frac{(2n+1)^{3/2}}{\sqrt{4\pi}}}$$

$$\longrightarrow 0 \quad \text{as } N\to\infty, \tag{5.104}$$

since $|h| < 1$. Since the last term in (5.104) is independent of ξ, the convergence is uniform.

(2) The convergence for $h \to 1-$:
Let us have a closer look at the series. The definition of the inner product and the addition theorem (Theorem 5.11) yield

$$\begin{aligned}
&\sum_{n=0}^{\infty} h^n \sum_{j=1}^{2n+1} \left\langle F, Y_{n,j}\right\rangle_{\mathrm{L}^2(\Omega)} Y_{n,j}(\xi)\\
&= \sum_{n=0}^{\infty} h^n \sum_{j=1}^{2n+1} \int_{\Omega} F(\eta) Y_{n,j}(\eta)\, \mathrm{d}\omega(\eta) Y_{n,j}(\xi)\\
&= \sum_{n=0}^{\infty} h^n \int_{\Omega} F(\eta) \sum_{j=1}^{2n+1} Y_{n,j}(\eta) Y_{n,j}(\xi)\, \mathrm{d}\omega(\eta)\\
&= \sum_{n=0}^{\infty} h^n \int_{\Omega} F(\eta) \frac{2n+1}{4\pi} P_n(\eta\cdot\xi)\, \mathrm{d}\omega(\eta).
\end{aligned} \tag{5.105}$$

From Theorem 3.16 (see p. 46) and the proof of the Poisson integral formula (Theorem 5.19), we know that the series

$$\sum_{n=0}^{\infty} h^n \frac{2n+1}{4\pi} P_n(t) \tag{5.106}$$

converges uniformly with respect to all $t \in [-1,1]$, if $h \in]-1,1[$ is kept fixed (which is still the case here). Therefore, we may interchange the series and the integration in (5.105). We obtain

$$\begin{aligned}
\sum_{n=0}^{\infty} h^n \sum_{j=1}^{2n+1} \left\langle F, Y_{n,j}\right\rangle_{\mathrm{L}^2(\Omega)} Y_{n,j}(\xi) &= \int_{\Omega} F(\eta) \sum_{n=0}^{\infty} h^n \frac{2n+1}{4\pi} P_n(\xi\cdot\eta)\, \mathrm{d}\omega(\eta) \quad (5.107)\\
&= \frac{1}{4\pi} \int_{\Omega} F(\eta) \frac{1-h^2}{\left(1+h^2-2h(\xi\cdot\eta)\right)^{3/2}}\, \mathrm{d}\omega(\eta).
\end{aligned}$$

The Poisson integral formula (Theorem 5.19) says that the latter expression uniformly converges to $F(\xi)$, $\xi \in \Omega$, as $h \to 1-$, and we are done. □

Theorem 5.22. *The system* $\{Y_{n,j}\}_{n\in\mathbb{N}_0,\, j=1,\ldots,2n+1}$ *is closed in* $(\mathrm{C}(\Omega), \|\cdot\|_{\mathrm{C}(\Omega)})$ *(in the sense of the approximation theory).*

Proof. We have to show the following property (see also Definition 3.4 on p. 38): For every $F \in \mathrm{C}(\Omega)$ and every $\varepsilon > 0$, there exists a finite linear combination:

$$\sum_{n=0}^{N} \sum_{j=1}^{2n+1} d_{n,j} Y_{n,j} \tag{5.108}$$

such that

$$\left\| F - \sum_{n=0}^{N} \sum_{j=1}^{2n+1} d_{n,j} Y_{n,j} \right\|_{\mathrm{C}(\Omega)} \leq \varepsilon. \tag{5.109}$$

Now, let $F \in \mathrm{C}(\Omega)$ and $\varepsilon > 0$ be arbitrary. From Lemma 5.21, we know that there exists a real number $h = h(\varepsilon) < 1$ such that

$$\left\| F - \sum_{n=0}^{\infty} h^n \sum_{j=1}^{2n+1} \langle F, Y_{n,j} \rangle_{\mathrm{L}^2(\Omega)} Y_{n,j} \right\|_{\mathrm{C}(\Omega)} \leq \frac{\varepsilon}{2}. \tag{5.110}$$

Consulting again Lemma 5.21, we observe that there exists an upper limit $N = N(\varepsilon) \in \mathbb{N}_0$ for the summation such that

$$\left\| \sum_{n=0}^{\infty} h^n \sum_{j=1}^{2n+1} \langle F, Y_{n,j} \rangle_{\mathrm{L}^2(\Omega)} Y_{n,j} - \sum_{n=0}^{N} h^n \sum_{j=1}^{2n+1} \langle F, Y_{n,j} \rangle_{\mathrm{L}^2(\Omega)} Y_{n,j} \right\|_{\mathrm{C}(\Omega)} \leq \frac{\varepsilon}{2}. \tag{5.111}$$

The rest is a simple application of the triangle inequality:

$$\begin{aligned}
&\left\| F - \sum_{n=0}^{N} \sum_{j=1}^{2n+1} \overbrace{h^n \langle F, Y_{n,j} \rangle_{\mathrm{L}^2(\Omega)}}^{=d_{n,j}} Y_{n,j} \right\|_{\mathrm{C}(\Omega)} \\
&\leq \left\| F - \sum_{n=0}^{\infty} \sum_{j=1}^{2n+1} h^n \langle F, Y_{n,j} \rangle_{\mathrm{L}^2(\Omega)} Y_{n,j} \right\|_{\mathrm{C}(\Omega)} \\
&\quad + \left\| \sum_{n=0}^{\infty} \sum_{j=1}^{2n+1} h^n \langle F, Y_{n,j} \rangle_{\mathrm{L}^2(\Omega)} Y_{n,j} - \sum_{n=0}^{N} \sum_{j=1}^{2n+1} h^n \langle F, Y_{n,j} \rangle_{\mathrm{L}^2(\Omega)} Y_{n,j} \right\|_{\mathrm{C}(\Omega)} \\
&\leq \frac{\varepsilon}{2} + \frac{\varepsilon}{2} = \varepsilon.
\end{aligned} \tag{5.112}$$

□

We will now switch over to $\mathrm{L}^2(\Omega)$ in two steps.

Corollary 5.23. *The system* $\{Y_{n,j}\}_{n\in\mathbb{N}_0;\, j=1,\ldots,2n+1}$ *is closed (in the sense of the approximation theory) in the normed space* $(\mathrm{C}(\Omega), \|\cdot\|_{\mathrm{L}^2(\Omega)})$.

Proof. Let $F \in \mathrm{C}(\Omega)$ and $\varepsilon > 0$ be arbitrary but fixed. Note that we now have to show that there exists a finite linear combination $\sum_{n=0}^{N}\sum_{j=1}^{2n+1} b_{n,j}Y_{n,j}$ such that

$$\left\| F - \sum_{n=0}^{N}\sum_{j=1}^{2n+1} b_{n,j}Y_{n,j} \right\|_{\mathrm{L}^2(\Omega)} \leq \varepsilon, \tag{5.113}$$

that is, the chosen norm is the only difference. From Theorem 5.22, we already know that there exists a finite linear combination $\sum_{n=0}^{N}\sum_{j=1}^{2n+1} b_{n,j}Y_{n,j}$ such that

$$\left\| F - \sum_{n=0}^{N}\sum_{j=1}^{2n+1} b_{n,j}Y_{n,j} \right\|_{\mathrm{C}(\Omega)} \leq \frac{\varepsilon}{\sqrt{4\pi}}. \tag{5.114}$$

Using Theorem 4.10 on p. 91, we can immediately conclude that

$$\left\| F - \sum_{n=0}^{N}\sum_{j=1}^{2n+1} b_{n,j}Y_{n,j} \right\|_{\mathrm{L}^2(\Omega)} \leq \sqrt{4\pi}\left\| F - \sum_{n=0}^{N}\sum_{j=1}^{2n+1} b_{n,j}Y_{n,j} \right\|_{\mathrm{C}(\Omega)} \leq \varepsilon. \tag{5.115}$$

□

Corollary 5.24. *The system* $\{Y_{n,j}\}_{n\in\mathbb{N}_0;\, j=1,\ldots,2n+1}$ *is closed in* $(\mathrm{L}^2(\Omega), \|\cdot\|_{\mathrm{L}^2(\Omega)})$ *(in the sense of the approximation theory).*

Proof. Let $F \in \mathrm{L}^2(\Omega)$ and $\varepsilon > 0$ be arbitrary but fixed. Due to Theorem 4.11 (see p. 92), there exists $G \in \mathrm{C}(\Omega)$ such that

$$\|F - G\|_{\mathrm{L}^2(\Omega)} \leq \frac{\varepsilon}{2}. \tag{5.116}$$

Moreover, Corollary 5.23 guarantees the existence of $\sum_{n=0}^{N}\sum_{j=1}^{2n+1} c_{n,j}Y_{n,j}$, corresponding to G, with

$$\left\| G - \sum_{n=0}^{N}\sum_{j=1}^{2n+1} c_{n,j}Y_{n,j} \right\|_{\mathrm{L}^2(\Omega)} \leq \frac{\varepsilon}{2}. \tag{5.117}$$

The rest is again a simple application of the triangle inequality:

$$\begin{aligned} \left\| F - \sum_{n=0}^{N}\sum_{j=1}^{2n+1} c_{n,j}Y_{n,j} \right\|_{\mathrm{L}^2(\Omega)} &\leq \|F - G\|_{\mathrm{L}^2(\Omega)} + \left\| G - \sum_{n=0}^{N}\sum_{j=1}^{2n+1} c_{n,j}Y_{n,j} \right\|_{\mathrm{L}^2(\Omega)} \\ &\leq \frac{\varepsilon}{2} + \frac{\varepsilon}{2} = \varepsilon. \end{aligned} \tag{5.118}$$

□

That's it. Due to Theorem 3.5 (see p. 38), $\{Y_{n,j}\}_{n\in\mathbb{N}_0;\,j=1,\ldots,2n+1}$ is complete in the Hilbert space $(\mathrm{L}^2(\Omega), \langle\cdot,\cdot\rangle_{\mathrm{L}^2(\Omega)})$! So, according to Theorem 2.18 on p. 21, we can expand every $F\in\mathrm{L}^2(\Omega)$ in a Fourier series, and we have the Parseval identities—the basic tools of Fourier analysis. Let us summarize this.

Theorem 5.25. *The functions $\{Y_{n,j}\}_{n\in\mathbb{N}_0;\,j=1,\ldots,2n+1}$ represent a complete orthonormal system in the Hilbert space $(\mathrm{L}^2(\Omega), \langle\cdot,\cdot\rangle_{\mathrm{L}^2(\Omega)})$, that is,*

*(1) Every $F\in\mathrm{L}^2(\Omega)$ can be expanded in a **Fourier series**:*

$$\lim_{N\to\infty}\left\| F-\sum_{n=0}^{N}\sum_{j=1}^{2n+1}\langle F,Y_{n,j}\rangle_{\mathrm{L}^2(\Omega)}Y_{n,j}\right\|_{\mathrm{L}^2(\Omega)}=0; \tag{5.119}$$

often, we will briefly write

$$\text{“}F=\sum_{n=0}^{\infty}\sum_{j=1}^{2n+1}\langle F,Y_{n,j}\rangle_{\mathrm{L}^2(\Omega)}Y_{n,j}\ \textit{in the sense of}\ \mathrm{L}^2(\Omega)\text{”}. \tag{5.120}$$

*(2) For every $F, G\in\mathrm{L}^2(\Omega)$, the **Parseval identities** are valid:*

$$\begin{aligned}\|F\|^2_{\mathrm{L}^2(\Omega)} &= \sum_{n=0}^{\infty}\sum_{j=1}^{2n+1}\langle F,Y_{n,j}\rangle^2_{\mathrm{L}^2(\Omega)}, \\ \langle F,G\rangle_{\mathrm{L}^2(\Omega)} &= \sum_{n=0}^{\infty}\sum_{j=1}^{2n+1}\langle F,Y_{n,j}\rangle_{\mathrm{L}^2(\Omega)}\langle G,Y_{n,j}\rangle_{\mathrm{L}^2(\Omega)}.\end{aligned} \tag{5.121}$$

(3) The following density property holds true:

$$\overline{\mathrm{span}\left\{Y_{n,j}\,\middle|\,n\in\mathbb{N}_0,\ j=1,\ldots,2n+1\right\}}^{\|\cdot\|_{\mathrm{L}^2(\Omega)}}=\mathrm{L}^2(\Omega). \tag{5.122}$$

Note that the convergence of the Fourier series (5.119) is valid in the $\mathrm{L}^2(\Omega)$-sense. This is less than a pointwise convergence. Therefore, *never ever* write a spherical Fourier series by inserting points in F and $Y_{n,j}$. Fortunately, Gronwall [85] proved that a uniform convergence is given at least for Lipschitz continuous functions.

Theorem 5.26. *Let the function $F:\Omega\to\mathbb{R}$ be Lipschitz continuous. Then its Fourier series is uniformly convergent:*

$$\lim_{N\to\infty}\left\| F-\sum_{n=0}^{N}\sum_{j=1}^{2n+1}\langle F,Y_{n,j}\rangle_{\mathrm{L}^2(\Omega)}Y_{n,j}\right\|_{\mathrm{C}(\Omega)}=0. \tag{5.123}$$

Definition 5.27. The **spherical Fourier transform** is defined by the mapping

$$\mathrm{SFT} : \mathrm{L}^2(\Omega) \ni F \mapsto \left(\langle F, Y_{n,j} \rangle_{\mathrm{L}^2(\Omega)} \right)_{n \in \mathbb{N}_0,\, j=1,\ldots,2n+1}. \tag{5.124}$$

Moreover, the following abbreviation for the (spherical) Fourier transform is common:

$$F^{\wedge}(n,j) := \langle F, Y_{n,j} \rangle_{\mathrm{L}^2(\Omega)}, \quad n \in \mathbb{N}_0,\ j \in \{1,\ldots,2n+1\}. \tag{5.125}$$

The corresponding **inverse spherical Fourier transform** is defined by

$$(\mathrm{SFT})^{-1} : (a_{n,j})_{n \in \mathbb{N}_0,\, j=1,\ldots,2n+1} \mapsto \sum_{n=0}^{\infty} \sum_{j=1}^{2n+1} a_{n,j} Y_{n,j} \tag{5.126}$$

for all sequences $(a_{n,j})$ with

$$\sum_{n=0}^{\infty} \sum_{j=1}^{2n+1} a_{n,j}^2 < +\infty. \tag{5.127}$$

Obviously,

$$(\mathrm{SFT})^{-1}(\mathrm{SFT})(F) = F \tag{5.128}$$

for all $F \in \mathrm{L}^2(\Omega)$ and

$$(\mathrm{SFT})(\mathrm{SFT})^{-1}(a_{n,j}) = (a_{n,j}) \tag{5.129}$$

for all sequences $(a_{n,j})$ satisfying (5.127).

In Lemma 5.8, we saw that every spherical harmonic $Y_n \in \mathrm{Harm}_n(\Omega)$ is an eigenfunction of the Beltrami operator Δ^* corresponding to the eigenvalue $-n(n+1)$, that is,

$$\Delta_\xi^* Y_n(\xi) = -n(n+1) Y_n(\xi) \qquad \forall \xi \in \Omega. \tag{5.130}$$

Now, as we have an orthonormal $\mathrm{L}^2(\Omega)$-basis of spherical harmonics, we are able to prove that there are no other eigenvalues and no other eigenfunctions.

Theorem 5.28 (Eigenfunctions of the Beltrami Operator). *The eigenvalues of the Beltrami operator $\Delta^* : \mathrm{C}^{(2)}(\Omega) \to \mathrm{C}(\Omega)$ are given by $\{-n(n+1) \,|\, n \in \mathbb{N}_0\}$. Moreover, the eigenspace of Δ^* corresponding to the eigenvalue $-n(n+1)$, $n \in \mathbb{N}_0$, is* $\mathrm{Harm}_n(\Omega)$.

Proof.

(1) Determination of the eigenvalues:
Let us assume that there exists a function $F \in \mathrm{C}^{(2)}(\Omega)$ and a value $\lambda \in \mathbb{R}$ such that $\Delta^* F = \lambda F$ and $\lambda \neq -n(n+1)$ for all $n \in \mathbb{N}_0$. We have to show that $F = 0$. For this purpose, we apply Green's second surface identity (see Theorem 4.12 on p. 92) to

the functions $Y_{n,j}$ (for arbitrary indices $n \in \mathbb{N}_0$ and $j \in \{1,\ldots,2n+1\}$) and F. This yields

$$\int_\Omega Y_{n,j}(\xi)\Delta^*_\xi F(\xi) - F(\xi)\Delta^*_\xi Y_{n,j}(\xi)\,\mathrm{d}\omega(\xi) = 0. \tag{5.131}$$

The right-hand side vanishes for the following reason: In Theorem 4.12, the more general case of an integral over a subsurface $\Gamma \subset \Omega$ is considered, where the right-hand side consists of an integral over $\partial\Gamma$. Imagine a sequence of subsurfaces $\Gamma_k \subset \Omega$ which in some sense tends to Ω as $k \to \infty$. Then $\partial\Gamma_k$ tends to the empty set $\emptyset$. Therefore, the right-hand side of (5.131) vanishes.

We proceed by using Lemma 5.8, which gives $\Delta^* Y_{n,j} = -n(n+1)Y_{n,j}$. Hence,

$$\int_\Omega Y_{n,j}(\xi)\lambda F(\xi) - F(\xi)[-n(n+1)]Y_{n,j}(\xi)\,\mathrm{d}\omega(\xi) = 0 \tag{5.132}$$

and, consequently,

$$[\lambda + n(n+1)]\underbrace{\int_\Omega Y_{n,j}(\xi)F(\xi)\,\mathrm{d}\omega(\xi)}_{=F^\wedge(n,j)} = 0 \tag{5.133}$$

for all $n \in \mathbb{N}_0$ and all $j \in \{1,\ldots,2n+1\}$. Since the left-hand factor never (i.e., not for any $n \in \mathbb{N}_0$) vanishes due to our assumption at the beginning of the proof, the right-hand factor must vanish for all $n \in \mathbb{N}_0$ and all $j \in \{1,\ldots,2n+1\}$. However, since the ons $\{Y_{n,j}\}_{n\in\mathbb{N}_0;\,j=1,\ldots,2n+1}$ is complete (here is the argument that we could not use before), this implies that $F = 0$ (in the sense of $\mathrm{L}^2(\Omega)$).

(2) Determination of the eigenspace:
Let λ be an eigenvalue of Δ^*, that is, there exists an integer $k \in \mathbb{N}_0$ such that $\lambda = -k(k+1)$. Starting with $F \in \mathrm{C}^{(2)}(\Omega)$ satisfying $\Delta^* F = \lambda F$ in the same way as we started the first part of this proof, we again end up with (5.133) in the sense that

$$[-k(k+1) + n(n+1)]F^\wedge(n,j) = 0 \qquad \forall n \in \mathbb{N}_0 \quad \forall j \in \{1,\ldots,2n+1\}. \tag{5.134}$$

Hence, all Fourier coefficients of F corresponding to degrees $n \neq k$ vanish. This essentially shortens the Fourier series of F (note that we use here again the completeness):

$$F = \sum_{n=0}^{\infty}\sum_{j=1}^{2n+1} F^\wedge(n,j)Y_{n,j} = \sum_{j=1}^{2k+1} F^\wedge(k,j)Y_{k,j} \in \mathrm{Harm}_k(\Omega). \tag{5.135}$$

□

A similar result can be proved for the Legendre operator (see Definition 4.6).

Theorem 5.29 (Eigenfunctions of the Legendre Operator). *The eigenvalues of the Legendre operator* $\mathscr{L} : \mathrm{C}^{(2)}[-1,1] \to \mathrm{C}[-1,1]$ *are given by* $\{-n(n+1) \mid n \in \mathbb{N}_0\}$. *Moreover, the eigenspace of* $\mathscr{L}$ *corresponding to the eigenvalue* $-n(n+1)$, $n \in \mathbb{N}_0$, *is given by* $\{\lambda P_n \mid \lambda \in \mathbb{R}\}$, *where* P_n *is the Legendre polynomial of degree* n.

Proof.

(1) $\mathscr{L}$ is self-adjoint:
Let $F,G \in \mathrm{C}^{(2)}[-1,1]$. Then, integrating by parts twice, we obtain

$$\begin{aligned}
\langle \mathscr{L}F,G\rangle_{\mathrm{L}^2[-1,1]} &= \int_{-1}^{1} \frac{\mathrm{d}}{\mathrm{d}t}\left[\left(1-t^2\right)\frac{\mathrm{d}}{\mathrm{d}t}F(t)\right]G(t)\,\mathrm{d}t \\
&= \underbrace{\left(1-t^2\right)\left(\frac{\mathrm{d}}{\mathrm{d}t}F(t)\right)G(t)\Big|_{-1}^{1}}_{=0} - \int_{-1}^{1}\left(1-t^2\right)\left(\frac{\mathrm{d}}{\mathrm{d}t}F(t)\right)\frac{\mathrm{d}}{\mathrm{d}t}G(t)\,\mathrm{d}t \\
&= \underbrace{-\left(1-t^2\right)\left(\frac{\mathrm{d}}{\mathrm{d}t}G(t)\right)F(t)\Big|_{-1}^{1}}_{=0} + \int_{-1}^{1}\frac{\mathrm{d}}{\mathrm{d}t}\left[\left(1-t^2\right)\frac{\mathrm{d}}{\mathrm{d}t}G(t)\right]F(t)\,\mathrm{d}t \\
&= \langle \mathscr{L}G,F\rangle_{\mathrm{L}^2[-1,1]}\,. \qquad (5.136)
\end{aligned}$$

(2) The Legendre polynomials are eigenfunctions of $\mathscr{L}$:
We apply the Beltrami operator to the addition theorem (Theorem 5.11) with $\eta = \varepsilon^3$:

$$\Delta_\xi^*\left(\sum_{j=1}^{2n+1} Y_{n,j}(\xi)Y_{n,j}\left(\varepsilon^3\right)\right) = \frac{2n+1}{4\pi}\Delta_\xi^* P_n\left(\xi\cdot\varepsilon^3\right) \quad \forall \xi\in\Omega. \qquad (5.137)$$

On the left-hand side, we will now use the linearity of Δ^* and Theorem 5.28. On the right-hand side, we will use Theorem 4.5 on p. 87 and the fact that simply $\xi\cdot\varepsilon^3 = t$ in polar coordinates. We get

$$\sum_{j=1}^{2n+1} \underbrace{\Delta_\xi^* Y_{n,j}(\xi)}_{=-n(n+1)Y_{n,j}(\xi)}\, Y_{n,j}\left(\varepsilon^3\right) = \frac{2n+1}{4\pi}\left(\underbrace{\frac{\partial}{\partial t}\left(1-t^2\right)\frac{\partial}{\partial t}}_{=\mathscr{L}_t} + \frac{1}{1-t^2}\frac{\partial^2}{\partial\varphi^2}\right)P_n(t). \qquad (5.138)$$

Finally, the addition theorem yields

$$-n(n+1)\frac{2n+1}{4\pi}P_n\Big(\underbrace{\xi\cdot\varepsilon^3}_{=t}\Big) = \frac{2n+1}{4\pi}\mathscr{L}_t P_n(t) \quad \forall t\in[-1,1], \qquad (5.139)$$

which can be simplified to

$$-n(n+1)P_n = \mathscr{L}P_n \qquad \text{on } [-1,1]. \qquad (5.140)$$

Due to the linearity of Δ^*, every λP_n, $\lambda\in\mathbb{R}\setminus\{0\}$, also is an eigenfunction corresponding to the eigenvalue $-n(n+1)$.

(3) The spectrum of $\mathscr{L}$:
Let $F \in \mathrm{C}^{(2)}[-1,1]$ be an eigenfunction of $\mathscr{L}$, that is, there exists $\lambda \in \mathbb{R}$ with $\mathscr{L}F = \lambda F$ and $F \neq 0$. Then we get, due to part 1 of this proof, for each $n \in \mathbb{N}_0$,

$$\langle \mathscr{L}F, P_n \rangle_{\mathrm{L}^2[-1,1]} = \langle F, \mathscr{L}P_n \rangle_{\mathrm{L}^2[-1,1]} . \tag{5.141}$$

Consequently, part 2 yields

$$\langle \lambda F, P_n \rangle_{\mathrm{L}^2[-1,1]} = \langle F, -n(n+1)P_n \rangle_{\mathrm{L}^2[-1,1]} \quad \forall n \in \mathbb{N}_0 \tag{5.142}$$

such that (see Definition 3.15 on p. 45)

$$\frac{\lambda}{2\pi} F^\wedge(n) = \frac{-n(n+1)}{2\pi} F^\wedge(n) \quad \forall n \in \mathbb{N}_0. \tag{5.143}$$

Due to Theorem 3.9 (i.e., the property of the Legendre polynomials as an orthogonal basis in $\mathrm{L}^2[-1,1]$), this implies that there exists $k \in \mathbb{N}_0$ with $\lambda = -k(k+1)$ or $F = 0$ in the sense of $\mathrm{L}^2[-1,1]$.

(4) There are no other eigenfunctions corresponding to these eigenvalues:
We follow part 3 again until (5.143). If $\lambda = -k(k+1)$ for an arbitrary but fixed $k \in \mathbb{N}_0$, then

$$F^\wedge(n) = 0 \quad \forall n \in \mathbb{N}_0 \setminus \{k\}. \tag{5.144}$$

Inserting this conclusion in (3.60), we, eventually, obtain

$$F = \sum_{n=0}^{\infty} F^\wedge(n) \frac{2n+1}{4\pi} P_n = F^\wedge(k) \frac{2k+1}{4\pi} P_k \in \{ \lambda P_k | \, \lambda \in \mathbb{R} \} . \tag{5.145}$$

□

Theorems 5.28 and 5.29 provide us with alternative ways of defining the spherical harmonics and the Legendre polynomials. For instance, P_n can be defined as the $\mathrm{C}^{(2)}[-1,1]$-eigenfunction of $\mathscr{L}$ corresponding to the eigenvalue $-n(n+1)$, where $P_n(1) = 1$.

5.2 Fully Normalized Spherical Harmonics

In Definition 5.10, we required that, for each $n \in \mathbb{N}_0$, $Y_{n,j}$ is an element of an orthonormal basis of $\mathrm{Harm}_n(\Omega)$. This does not provide us with unique basis functions. In practice, the common system of the fully normalized spherical harmonics is used (see, e.g., [35, Appendix B], [91, Sect. 1–14]). For this system, the index range $j = -n, \ldots, n$ instead of $j = 1, \ldots, 2n+1$ is common. We will see here how they can be constructed.

We use a separation ansatz

$$Y_{n,j}\left(\xi\left(\varphi,t\right)\right)=F_{n,j}(\varphi)G_{n,j}(t), \tag{5.146}$$

where $\xi\left(\varphi,t\right)$ stands for the representation of ξ in terms of the polar coordinates φ and t (see Definition 4.3 on p. 86). As we have seen, an alternative way of defining spherical harmonics is the use of their property as eigenfunctions of the Beltrami operator Δ^* (see Theorem 5.28). For applying Δ^* to (5.146), we need the representation of the differential operator in polar coordinates (see Theorem 4.5):

$$\left(\frac{\partial}{\partial t}\left(1-t^2\right)\frac{\partial}{\partial t}+\frac{1}{1-t^2}\frac{\partial^2}{\partial\varphi^2}\right)\left(F_{n,j}(\varphi)G_{n,j}(t)\right)=-n(n+1)F_{n,j}(\varphi)G_{n,j}(t). \tag{5.147}$$

This yields

$$\left[\left(\frac{\mathrm{d}}{\mathrm{d}t}\left(1-t^2\right)\frac{\mathrm{d}}{\mathrm{d}t}\right)G_{n,j}(t)\right]F_{n,j}(\varphi)+\frac{1}{1-t^2}\left(\frac{\mathrm{d}^2}{\mathrm{d}\varphi^2}F_{n,j}(\varphi)\right)G_{n,j}(t)$$
$$=-n(n+1)F_{n,j}(\varphi)G_{n,j}(t). \tag{5.148}$$

Dividing the last equation by $F_{n,j}G_{n,j}$ (outside the roots of this product) and multiplying by $1-t^2$, we obtain

$$\left(1-t^2\right)\frac{\left(\frac{\mathrm{d}}{\mathrm{d}t}\left(1-t^2\right)\frac{\mathrm{d}}{\mathrm{d}t}\right)G_{n,j}(t)}{G_{n,j}(t)}+\frac{\frac{\mathrm{d}^2}{\mathrm{d}\varphi^2}F_{n,j}(\varphi)}{F_{n,j}(\varphi)}=-n(n+1)\left(1-t^2\right), \tag{5.149}$$

which is equivalent to

$$\left(1-t^2\right)\left[\frac{\left(\frac{\mathrm{d}}{\mathrm{d}t}\left(1-t^2\right)\frac{\mathrm{d}}{\mathrm{d}t}\right)G_{n,j}(t)}{G_{n,j}(t)}+n(n+1)\right]=-\frac{\frac{\mathrm{d}^2}{\mathrm{d}\varphi^2}F_{n,j}(\varphi)}{F_{n,j}(\varphi)}. \tag{5.150}$$

Since the left-hand side only depends on t and the right-hand side only depends on φ, both sides must be constant. Let $\lambda\in\mathbb{R}$ be this constant. For the right-hand side, we obtain

$$F''_{n,j}(\varphi)=-\lambda F_{n,j}(\varphi)\quad\forall\varphi\in[0,2\pi] \tag{5.151}$$

(for roots φ of $F_{n,j}$, which had to be omitted above, we can use here a continuity argument to show the validity of (5.151) for all $\varphi\in[0,2\pi]$). Since φ is the longitude, the obtained result has to be periodic in the sense that $F_{n,j}^{(k)}(0)=F_{n,j}^{(k)}(2\pi)$ for all $k\in\mathbb{N}_0$. Hence, only constants $\lambda\geq 0$ with $\sqrt{\lambda}\in\mathbb{N}_0$ yield solutions of the differential equation with this boundary-value condition. Since we only need $2n+1$ functions per degree $n\in\mathbb{N}_0$, we choose

$$F_{n,j}(\varphi):=\begin{cases}\sin(j\varphi), & j=1,\ldots,n\\ \cos(j\varphi), & j=-n,\ldots,0\end{cases}. \tag{5.152}$$

For $G_{n,j}(t)$, we, consequently, obtain the ordinary differential equation

$$\left(1-t^2\right)\left[\frac{\mathrm{d}}{\mathrm{d}t}\left(1-t^2\right)\frac{\mathrm{d}}{\mathrm{d}t}G_{n,j}(t)+n(n+1)G_{n,j}(t)\right]=j^2G_{n,j}(t). \tag{5.153}$$

For $j=0$, we already know the solutions (see Theorem 5.29):

$$G_{n,0}=\mu P_n \tag{5.154}$$

for arbitrary constants $\mu\in\mathbb{R}$. The more general equation in (5.153) is solved by the associated Legendre functions $P_{n,|j|}$ (and their multiples); see, for example, [1, p. 332], [35, p. 847], [91, p. 21], and [113, p. 198]. We set $G_{n,j}=\gamma_{n,j}P_{n,|j|}$, where $\gamma_{n,j}$ is a constant. Note that there exists an alternative system with the same name but the (commonly different[4]) notation $P_n^{|j|}$. The relation is

$$P_n^{|j|}=(-1)^{|j|}P_{n,|j|}. \tag{5.155}$$

Up to now, we only used that $Y_{n,j}\in\mathrm{Harm}_n(\Omega)$. However, a further requirement is the orthonormality. Whereas Theorem 5.9 guarantees that $\langle Y_{n,j},Y_{m,k}\rangle_{\mathrm{L}^2(\Omega)}=0$ for $n\neq m$, we have to make sure that

$$\left\langle Y_{n,j},Y_{n,k}\right\rangle_{\mathrm{L}^2(\Omega)}=0 \quad \text{for } j\neq k \tag{5.156}$$

and

$$\left\|Y_{n,j}\right\|_{\mathrm{L}^2(\Omega)}=1. \tag{5.157}$$

Equation (5.156) is already fulfilled, since

$$\begin{aligned}\int_\Omega Y_{n,j}(\xi)Y_{n,k}(\xi)\,\mathrm{d}\omega(\xi)&=\int_{-1}^1\int_0^{2\pi}G_{n,j}(t)G_{n,k}(t)F_{n,j}(\varphi)F_{n,k}(\varphi)\,\mathrm{d}\varphi\,\mathrm{d}t\\&=\int_{-1}^1 G_{n,j}(t)G_{n,k}(t)\,\mathrm{d}t\underbrace{\int_0^{2\pi}F_{n,j}(\varphi)F_{n,k}(\varphi)\,\mathrm{d}\varphi}_{=0,\text{ if } j\neq k}\end{aligned} \tag{5.158}$$

due to Theorem 3.23 (see p. 53). Hence, only (5.157) requires further care. We get (see again Theorem 3.23)

$$\int_\Omega Y_{n,j}(\xi)^2\,\mathrm{d}\omega(\xi)=\gamma_{n,j}^2\int_{-1}^1 P_{n,|j|}(t)^2\,\mathrm{d}t\underbrace{\int_0^{2\pi}F_{n,j}(\varphi)^2\,\mathrm{d}\varphi}_{=\frac{2\pi}{2-\delta_{j0}}}. \tag{5.159}$$

---

[4]However, do not rely on the location of the index. Better check in each case how an author defined his associated Legendre functions.

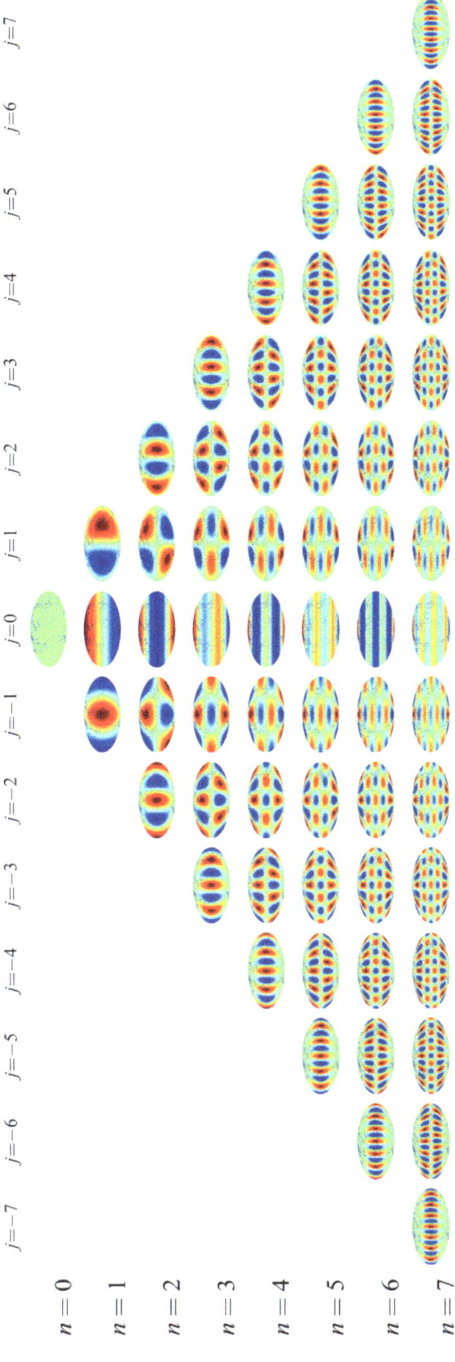

Fig. 5.5 Fully normalized spherical harmonics $Y_{n,j}$ of degrees 0–7

From [113, p. 198], we get

$$\int_{-1}^{1} P_{n,|j|}(t)^2 \, \mathrm{d}t = \frac{(n+|j|)!}{\left(n+\frac{1}{2}\right)(n-|j|)!} = \frac{2(n+|j|)!}{(2n+1)(n-|j|)!}. \tag{5.160}$$

This allows us now to give a formula for the fully normalized spherical harmonics[5]:

$$\boxed{Y_{n,j}(\xi(\varphi,t)) = \sqrt{\frac{(2n+1)(n-|j|)!(2-\delta_{j0})}{4\pi(n+|j|)!}} \, P_{n,|j|}(t) \begin{cases} \sin(j\varphi), & j=1,\ldots,n \\ \cos(j\varphi), & j=-n,\ldots,0 \end{cases}.} \tag{5.161}$$

Note that in some geoscientific works, the functions are normalized such that $\|Y_{n,j}\|_{\mathrm{L}^2(\Omega)} = \sqrt{4\pi}$. This corresponds to the omission of 4π in the denominator of the square root in (5.161).

Figure 5.5 shows some fully normalized spherical harmonics.

Several ways of calculating the associated Legendre functions, which can be defined by

$$P_{n,m}(t) := \left(1-t^2\right)^{m/2} \frac{\mathrm{d}^m}{\mathrm{d}t^m} P_n(t), \quad t \in [-1,1], \tag{5.162}$$

for $n \in \mathbb{N}_0$, $m \in \{0,\ldots,n\}$, are known. In particular, there also exist recurrence relations. For further details, see the references listed above. There are also efficient numerical algorithms for computing spherical harmonics; see, for example, [116] and the recurrence relations in [144, p. 81] and [193, p. 245].

Moreover, there exist fast algorithms for computing the spherical Fourier transform and its inverse; see, for example, [104, 137, 187].

5.3 Point Grids

We now have a complete orthonormal system for the Hilbert space $\mathrm{L}^2(\Omega)$. Following Principle 1 described in Chap. 1, we can now determine an approximate expansion $\sum_{n=0}^{N}\sum_{j=1}^{2n+1} a_{n,j}Y_{n,j}$ for a function $F \in \mathrm{L}^2(\Omega)$ based on given values $\{F(\eta_k)\}_{k=1,\ldots,M}$. We "only" have to solve the system of linear equations:

$$\sum_{n=0}^{N}\sum_{j=1}^{2n+1} a_{n,j}Y_{n,j}(\eta_k) = F(\eta_k) \quad \forall k = 1,\ldots,M \tag{5.163}$$

for the $(N+1)^2$ unknowns $a_{0,1},\ldots,a_{N,2N+1}$. Behind this "only," several theoretical and practical problems are hidden.

---

[5] Though one encounters the name "fully normalized" in the literature, it is not really a good choice, since all $Y_{n,j}$ have an $\mathrm{L}^2(\Omega)$-norm which is 1, no matter if this particular system is used or not.

Problem 1: Is (5.163) (uniquely) solvable?
For the sake of simplicity, we assume that the matrix $A := (Y_{n,j}(\eta_k))_{k,(n,j)}$ is quadratic, that is, $(N+1)^2 = M$. In this case, the problem above can be stated as follows: Is A a regular matrix? This question is connected to the concept of unisolvence (see also [43, p. 31]).

Definition 5.30. Let $S \subset \mathbb{R}^n$, $n \in \mathbb{N}$, be a given domain, and let $F_1, \ldots, F_k$ with $k \in \mathbb{N}$ be a given finite sequence of functions on S. Then the system $\{F_1, \ldots, F_k\}$ is called **unisolvent**, if, for every point system $X_k := \{x_1, \ldots, x_k\} \subset S$ of pairwise distinct points, the matrix

$$\operatorname{matr}_{X_k}(F_1, \ldots, F_k) := (F_i(x_j))_{i,j=1,\ldots,k} \tag{5.164}$$

is regular.

The matrix in (5.163) is the transpose of such a matrix. Thus, unisolvence means that interpolation problems on the space $\operatorname{span}\{F_j\}_{j=1,\ldots,k}$ can be uniquely solved, no matter which k pairwise distinct points one chooses on S.

In the 1D case, the monomials have such a nice property.

Example 5.31. Let $k \in \mathbb{N}_0$. Then the monomials $x^0, \ldots, x^k$ are unisolvent on every interval $[a,b]$, $a < b$, because the matrix

$$\left(x_j^i\right)_{i,j=0,\ldots,k} \tag{5.165}$$

is the so-called **Vandermonde matrix**, whose regularity is well known.

Nevertheless, there are also 1D systems which are not unisolvent. Furthermore, the same system can be unisolvent for one domain and non-unisolvent for another one.

Example 5.32. Let $F_0 :\equiv 1$ and $F_1(x) := x^2$. On the domain $[0,1]$, this system is unisolvent, since

$$\det\begin{pmatrix} 1 & 1 \\ x_1^2 & x_2^2 \end{pmatrix} = x_2^2 - x_1^2 = 0 \Leftrightarrow x_1 = x_2. \tag{5.166}$$

However, on $[-1,1]$, the system is non-unisolvent since

$$\det\begin{pmatrix} 1 & 1 \\ x_1^2 & x_2^2 \end{pmatrix} = x_2^2 - x_1^2 \tag{5.167}$$

can also vanish, if $x_2 = -x_1$.

There is some bad news in the case of higher dimensions.

Theorem 5.33 (Haar's Theorem). *Let $S \subset \mathbb{R}^n$ with $\mathbf{n \geq 2}$ be a given domain which contains at least one inner point p. Furthermore, let $F_1, \ldots, F_k$ with $k \in \mathbb{N} \setminus \{1\}$ be given functions on S which are continuous at least in a neighborhood of p. Then this system of functions on S is **not unisolvent**.*

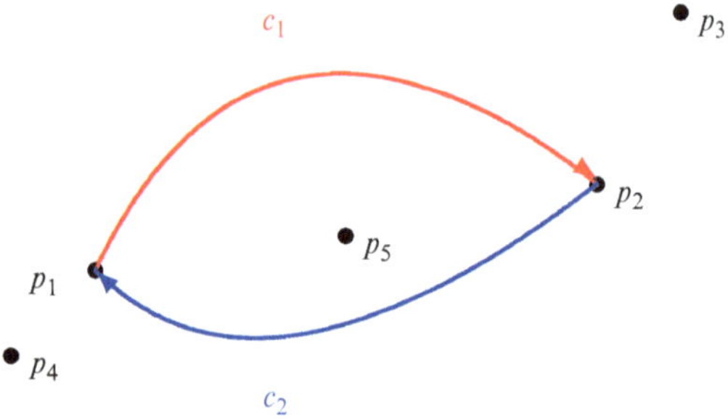

Fig. 5.6 Illustration of the curves c_1 and c_2 in the proof of Haar's theorem

Proof. Due to the assumptions of the theorem, there exists a neighborhood U of p such that every F_j, $j = 1, \ldots, k$, is continuous at U and $U \subset S$. Now, let us choose k arbitrary pairwise distinct points $p_1, \ldots, p_k \in U$. If it turns out that

$$\det\left(\mathrm{matr}_{\{p_1,\ldots,p_k\}}(F_1,\ldots,F_k)\right) = 0, \tag{5.168}$$

we are done. Otherwise, we proceed as follows: We construct two continuous curves $c_1, c_2 : [0,1] \to U$ with

$$c_1(0) = p_1, \tag{5.169a}$$
$$c_1(1) = p_2, \tag{5.169b}$$
$$c_2(0) = p_2, \tag{5.169c}$$
$$c_2(1) = p_1, \tag{5.169d}$$
$$c_1(t) \neq c_2(t) \quad \forall t \in [0,1] \tag{5.169e}$$
$$c_j(t) \notin \{p_3,\ldots,p_k\} \quad \forall t \in [0,1] \quad \forall j \in \{1,2\}. \tag{5.169f}$$

In other words, the points p_1 and p_2 change their places, where they meet neither each other nor any of the other points $p_3, \ldots, p_k$. Moreover, the curves are continuous, that is, Scotty may not beam any of the points. Figure 5.6 shows an example of such curves. The other points $p_3, \ldots, p_k$ are kept fixed.

Now we define the function $G : [0,1] \to \mathbb{R}$ by

$$G(t) = \det\begin{pmatrix} F_1(c_1(t)) & F_1(c_2(t)) & F_1(p_3) & \ldots & F_1(p_k) \\ \vdots & \vdots & \vdots & & \vdots \\ F_k(c_1(t)) & F_k(c_2(t)) & F_k(p_3) & \ldots & F_k(p_k) \end{pmatrix}, \quad t \in [0,1]. \tag{5.170}$$

Since the construction of c_1 and c_2 corresponds to an interchanging of the first and the second column, if we compare $t = 0$ with $t = 1$, we have

$$G(0) = -G(1) \neq 0. \tag{5.171}$$

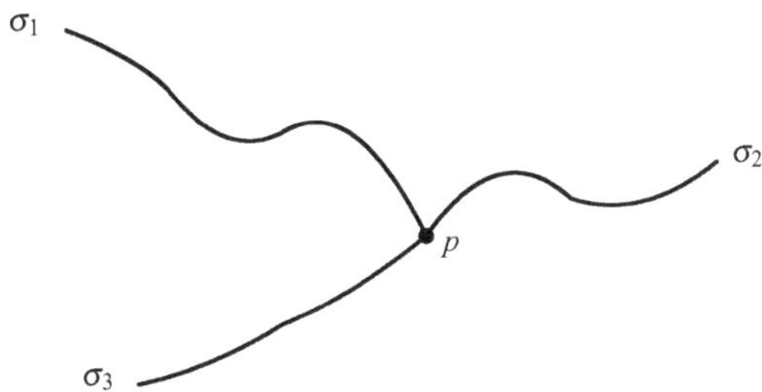

Fig. 5.7 Illustration of the ramification point in the extended Haar's theorem

Moreover, since $F_1, \ldots, F_k, c_1, c_2$ are continuous, G is continuous. Hence, we have a continuous function which changes its sign on the interval $[0,1]$. Due to the intermediate value theorem, there must, consequently, be a zero somewhere on $]0,1[$. In other words, there exists $\tau \in]0,1[$ such that

$$\mathrm{matr}_{\{c_1(\tau), c_2(\tau), p_3, \ldots, p_k\}} (F_1, \ldots, F_k) \tag{5.172}$$

is singular. Due to the construction of c_1 and c_2, the points $c_1(\tau)$, $c_2(\tau)$, $p_3, \ldots, p_k$ are pairwise distinct. Therefore, $\{F_1, \ldots, F_k\}$ is not unisolvent. □

Note that the construction of the intersection-free curves c_1 and c_2 requires that the domain is at least two dimensional.

At the moment, we could still believe that we are safe because the sphere Ω does not have any inner point. However, here is the really bad news.

Theorem 5.34 (Extended Haar's Theorem). *Let $S \subset \mathbb{R}^n$ with $\boldsymbol{n \geq 2}$ be a given domain such that there exists a point $p \in S$ and three arcs σ_1, σ_2, and σ_3 in S which only intersect at p (see Fig. 5.7).[6] Moreover, let $F_1, \ldots, F_k$ with $k \in \mathbb{N} \setminus \{1\}$ be given functions on S which are continuous on $\sigma_1 \cup \sigma_2 \cup \sigma_3$. Then this system of functions on S is **not unisolvent**.*

Proof. The idea of the proof of the extended theorem is the same as for the original theorem. We have to find again two curves c_1 and c_2 which satisfy (5.169). Now they have to remain on $\sigma_1 \cup \sigma_2 \cup \sigma_3$. Without loss of generality, we put p_1 and p_2 both on the same curve, for example, σ_1, and $p_3, \ldots, p_k$ on the other curves (see Fig. 5.8). The question is now as follows: How can p_1 and p_2 change places without meeting each other (and without meeting $p_3, \ldots, p_k$)? The answer is train switching! Figure 5.9 shows how it works. The rest is analogous to the proof of Haar's theorem. □

Every point on Ω can be considered as a ramification point. Hence, any (finite) subsystem of $\{Y_{n,j}\}_{n \in \mathbb{N}_0, j=1,\ldots,2n+1}$ (with more than one element) is non-unisolvent.

[6] Such a point is called a ramification point.

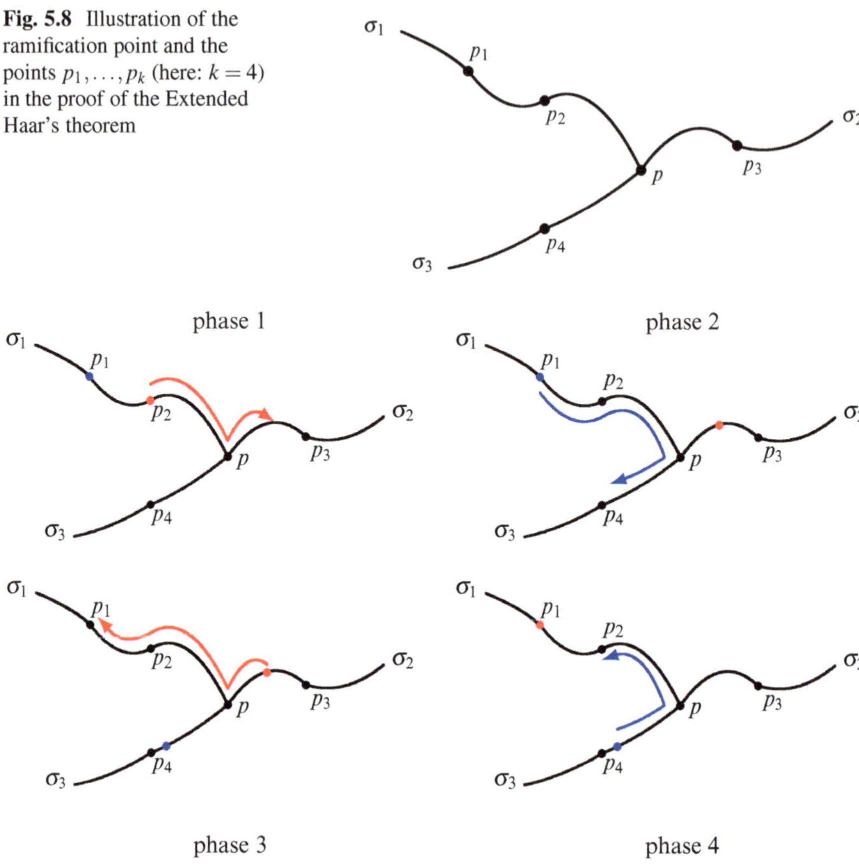

Fig. 5.8 Illustration of the ramification point and the points $p_1, \ldots, p_k$ (here: $k = 4$) in the proof of the Extended Haar's theorem

Fig. 5.9 Illustration of the "train switching" in the proof of the Extended Haar's theorem

Though the sphere has no inner point, the way the points are interchanged can nevertheless also be realized in analogy to the proof of Haar's theorem (Theorem 5.33). This is demonstrated in Fig. 5.10, where we take four points $p_1, \ldots, p_4 \in \Omega$ and two corresponding curves $c_1, c_2 : [0,1] \to \Omega$. The associated function G representing the determinant (where the fully normalized spherical harmonics of degree 0 and 1 were chosen as functions) is shown in Fig. 5.11. The four points corresponding to the zero of G are plotted in Fig. 5.12. Figure 5.11 also shows that it is rather a case of bad luck to catch a singular matrix since G has one zero only in this example.

Nevertheless, there is no reason for being satisfied at this point since a matrix can be theoretically regular but numerically singular.

Problem 2: Can (5.163) be solved in a numerically stable way?
The best way to analyze the stability of the inversion is to have a look at the condition number $\operatorname{cond}(A) := \|A\| \, \|A^{-1}\|$ of the regular matrix A, where $\|\cdot\|$ is a

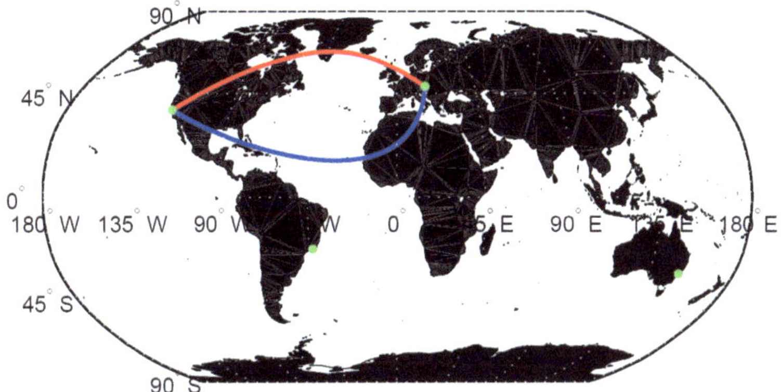

Fig. 5.10 Four points (Vienna, San Francisco, Rio de Janeiro, and Sydney) and two corresponding curves for the verification of the non-unisolvence of the fully normalized spherical harmonics of degree ≤ 1

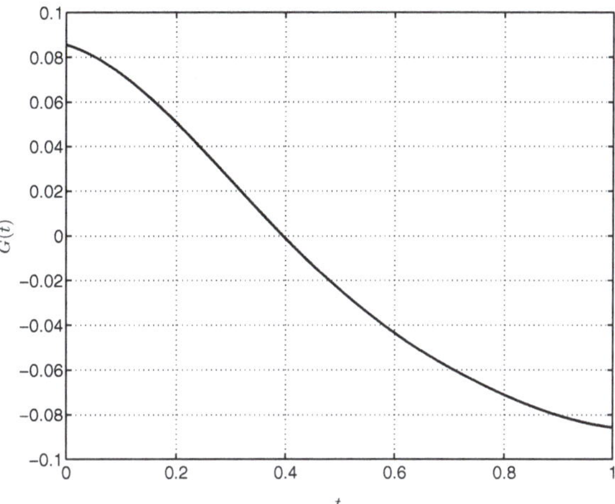

Fig. 5.11 The function G obviously has a zero on $]0,1[$ as a consequence of the intermediate value theorem

chosen matrix norm (see, e.g., [102, p. 80], [147, p. 85], [153, p. 36] for further details). It is an indicator for the numerical instability of the inversion of A. Before we continue investigating this matter, we need examples of point grids (see also [66, pp. 171–177]).

Example 5.35 (Driscoll–Healy Grid). A simple grid is an equiangular grid which is given by the polar coordinates

$$\begin{aligned} \varphi_i &:= \frac{2\pi}{N} i, \qquad i = 0,\ldots,N, \\ \vartheta_j &:= \frac{\pi}{N} j, \qquad j = 0,\ldots,N. \end{aligned} \tag{5.173}$$

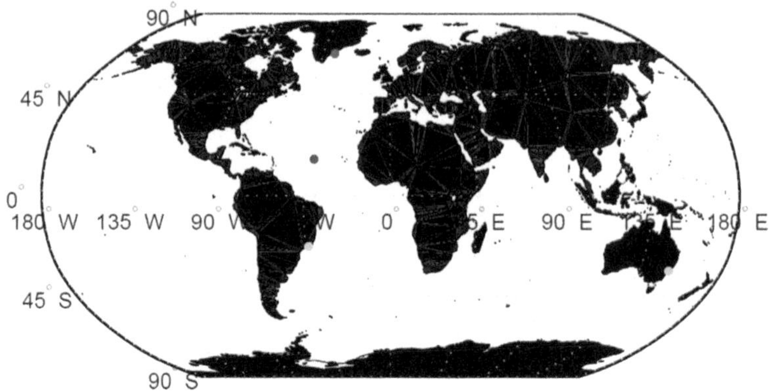

Fig. 5.12 These four points produce a singular matrix, if they are inserted in $\mathrm{matr}_{\{p_1,p_2,p_3,p_4\}}(Y_{0,1},Y_{1,-1},Y_{1,0},Y_{1,1})$ in the case of the fully normalized spherical harmonics. The *red point* and the *blue point* are located on the two curves in Fig. 5.10

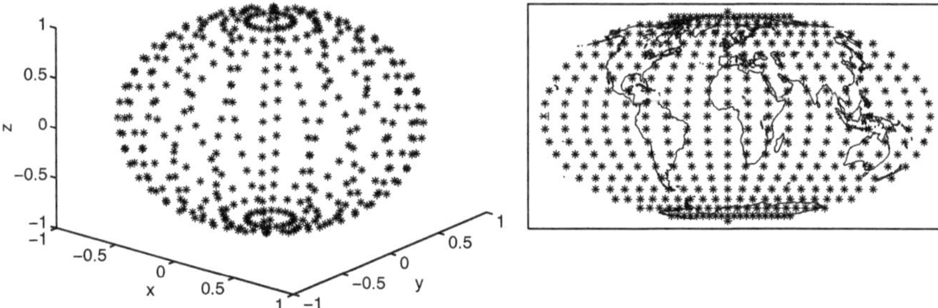

Fig. 5.13 Driscoll–Healy grid for $N = 20$

The associated points

$$\xi_{ij} := \begin{pmatrix} x_{ij} \\ y_{ij} \\ z_{ij} \end{pmatrix} := \begin{pmatrix} \sin\vartheta_j \cos\varphi_i \\ \sin\vartheta_j \sin\varphi_i \\ \cos\vartheta_j \end{pmatrix}, \quad i,j = 0,\dots,N, \tag{5.174}$$

are shown in Fig. 5.13. Note that the points appear equidistributed in the (φ,ϑ)-plane (besides the higher density in the ϑ-direction) but not on the sphere. Moreover, it should be noted that the points ξ_{ij} are not pairwise distinct. More precisely, φ_0 produces the same points as φ_N for periodicity reasons, such that we only need to consider N values for φ. Furthermore, $\xi_{i_1 0} = \xi_{i_2 0}$ and $\xi_{i_1 N} = \xi_{i_2 N}$ for all $i_1, i_2 = 0,\dots,N$. As a consequence, the actual number of pairwise distinct points is

$$N(N-1) + 2 = N^2 - N + 2. \tag{5.175}$$

The naming of the grid is based on a numerical integration method on the sphere which is due to Driscoll and Healy (see [46] and Theorem 7.33 on p. 230).

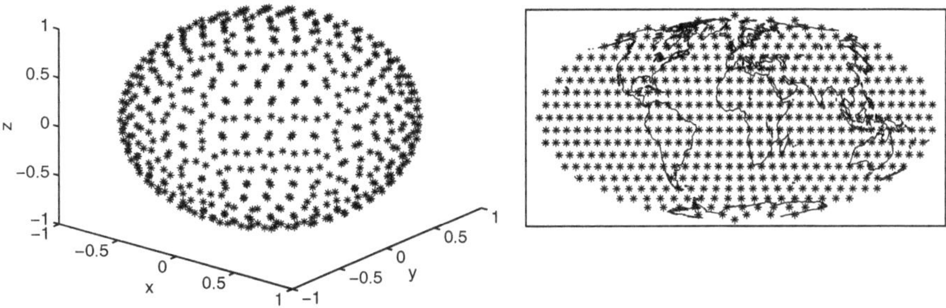

Fig. 5.14 Reuter grid in the case of the parameter 20

Example 5.36 (Reuter Grid). The following point grid, which depends on a parameter $N \in \mathbb{N}$, obviously shows a better equidistribution; see Fig. 5.14:

$$\begin{aligned}
\vartheta_i &:= \frac{\pi}{N}\, i, \qquad i = 0,\ldots,N,\\
\gamma_0 &:= 1,\\
\gamma_i &:= \left\lfloor \frac{2\pi}{\arccos\left(\left(\cos\frac{\pi}{N} - \cos^2\vartheta_i\right) / \sin^2\vartheta_i\right)} \right\rfloor, \quad i = 1,\ldots,N-1,\\
\gamma_N &:= 1,\\
\varphi_{01} &:= 0,\\
\varphi_{ij} &:= \left(j - \frac{1}{2}\right)\frac{2\pi}{\gamma_i}, \qquad i = 1,\ldots,N-1, \quad j = 1,\ldots,\gamma_i,\\
\varphi_{N1} &:= 0,
\end{aligned} \tag{5.176}$$

where $\lfloor\cdot\rfloor$ represents the Gaussian bracket [see (3.165)]. As usual, the points are computed as

$$\xi_{ij} := \begin{pmatrix} x_{ij} \\ y_{ij} \\ z_{ij} \end{pmatrix} := \begin{pmatrix} \sin\vartheta_i \cos\varphi_{ij} \\ \sin\vartheta_i \sin\varphi_{ij} \\ \cos\vartheta_i \end{pmatrix}, \quad i = 0,\ldots,N, \quad j = 1,\ldots,\gamma_i. \tag{5.177}$$

The name of the grid refers to the author of the PhD thesis [157]. There appears to be no explicit formula for the number of points depending on the parameter N. However, [66, Lemma 7.1.10] at least gives an estimate from above: The number of points does not exceed $2 + \frac{4}{\pi} N^2$.

Let us now investigate the following example: We want to expand $F(x_1, x_2, x_3) := x_1 x_2 x_3$ in a truncated Fourier series of degree ≤ 20. This degree is certainly higher than necessary for that particular F. However, in practice, we would not know F and, from this point of view, 20 is a rather low degree.

We now have to solve (5.163), which contains the matrix[7]

$$A := \left(Y_{n,j}\left(\eta_k\right)\right)_{\substack{k=1,\ldots,M \\ n=0,\ldots,N,\, j=-n,\ldots,n}} \tag{5.178}$$

for $N := 20$ and the right-hand side

$$b := \left(F\left(\eta_k\right)\right)_{k=1,\ldots,M}. \tag{5.179}$$

The unknown vector $a := (a_{n,j})_{n=0,\ldots,N,\, j=-n,\ldots,n}$ can (theoretically) be obtained by solving the normal equation

$$A^{\mathrm{T}} A\, a = A^{\mathrm{T}} b. \tag{5.180}$$

We choose a Reuter grid with parameter 20 (see Fig. 5.14), which provides us with 502 points, where we have $21^2 = 441$ unknown coefficients. The condition number of $A^{\mathrm{T}}A$ (with respect to the 2-norm, i.e., the condition number is the ratio of the largest and the smallest singular value of $A^{\mathrm{T}}A$) turns out to be approximately 8.4×10^{17}. The result is accordingly bad (see Fig. 5.15).
A typical way of treating this problem is to add a constant to the diagonal of the matrix. In other words, we now solve

$$\left(A^{\mathrm{T}} A + \lambda I_{(N+1)^2}\right) a = A^{\mathrm{T}} b, \tag{5.181}$$

where $I_{(N+1)^2}$ is the $(N+1)^2 \times (N+1)^2$-identity matrix. The idea behind this "trick" is as follows (see, e.g., [102, pp. 86–90]).

Theorem 5.37 (Tykhonov Regularization). *Let $A \in \mathbb{R}^{m\times n}$, $\lambda \in \mathbb{R}^+$, and $b \in \mathbb{R}^m$ be given. Then $a^* \in \mathbb{R}^n$ is a solution of $(A^{\mathrm{T}}A + \lambda I_n)a = A^{\mathrm{T}}b$ if and only if a^* is a minimizer of*

$$\mathscr{F}(a) := \|Aa - b\|^2_{\mathbb{R}^m} + \lambda \, \|a\|^2_{\mathbb{R}^n}, \quad a \in \mathbb{R}^n. \tag{5.182}$$

Proof. A necessary condition for a minimum of (5.182) is obtained by assuming that the gradient vanishes. This yields

$$2A^{\mathrm{T}}Aa - 2A^{\mathrm{T}}b + 2\lambda a = 0, \tag{5.183}$$

which is equivalent to

$$\left(A^{\mathrm{T}}A + \lambda I_n\right) a = A^{\mathrm{T}} b. \tag{5.184}$$

For a sufficient condition, we check the Hessian of $\mathscr{F}$ in (5.182):

$$\mathrm{Hess}_{\mathscr{F}}(a) = 2A^{\mathrm{T}}A + 2\lambda I_n. \tag{5.185}$$

This matrix is positive definite for $\lambda > 0$, since

---

[7]Note that we use here the fully normalized spherical harmonics and, therefore, change the range for the order to $j = -n, \ldots, n$.

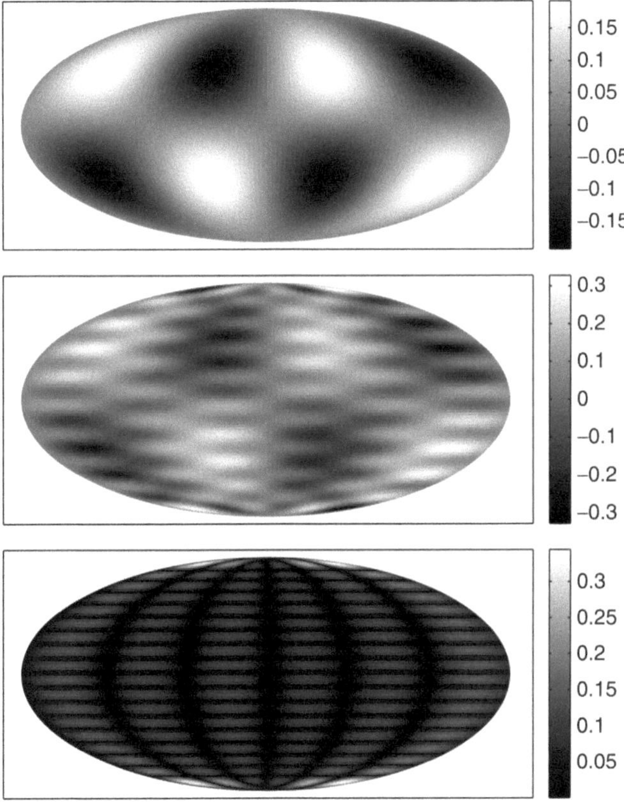

Fig. 5.15 Original function $F(x) := x_1x_2x_3$ (*top*), spherical harmonics expansion of degree ≤ 20 obtained by solving $A^{\mathrm{T}}Aa = A^{\mathrm{T}}b$ (*middle*), and corresponding absolute error (*bottom*); the obtained result is bad due to an ill-conditioned matrix

$$\begin{aligned}\left\langle \left(A^{\mathrm{T}}A + \lambda I_n\right) x, x\right\rangle &= \langle Ax, Ax\rangle + \lambda\langle x, x\rangle \\ &= \|Ax\|^2 + \lambda\,\|x\|^2 \\ &\geq \lambda\,\|x\|^2 \qquad \forall x \in \mathbb{R}^n. \end{aligned} \tag{5.186}$$

Hence, the solutions of (5.184) are *the* minimizers of $\mathscr{F}$. □

The function $\mathscr{F}$ shows that a balance between solving the equation $Aa = b$ and getting a "smooth" approximation by keeping $\|a\|_{\mathbb{R}^n}$ small shall be achieved (note that the bad approximation in Fig. 5.15 had too large extrema). Figure 5.16 shows a logarithmic plot of $\|Aa - b\|$ against $\|a\|$ for our example of an expansion problem. The obtained curve looks (with a bit of fantasy) like an "L." The so-called **L-curve method** (see, for instance, [49, 87, 88]), which is not a mathematically rigorous method but in acceptably many cases a practicable tool, looks for the knickpoint

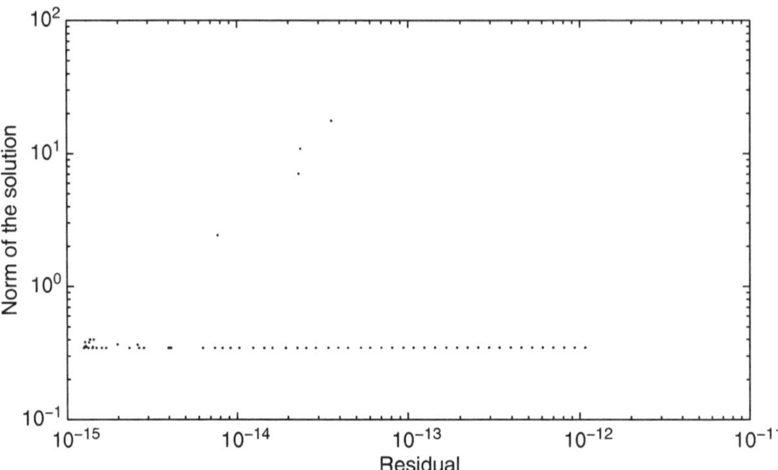

Fig. 5.16 Logarithmic plot of the residual $||Aa-b||$ against the norm $||a||$ of the solution of $(A^{\mathrm{T}}A+\lambda I_{(N+1)^2})a=b$ for different values of λ. An approximate "L-form" is obtained

of the "L." The idea is that, at the knickpoint, the residual and the norm of the solution are both small. In our example, the knickpoint is obtained by choosing $2.3714\times 10^{-16}\max_{i,j}|(A^{\mathrm{T}}A)_{i,j}|$ as the regularization parameter λ, where $(A^{\mathrm{T}}A)_{i,j}$ is the (i,j)-th component of $A^{\mathrm{T}}A$. The result is shown in Fig. 5.17. This looks much better, doesn't it? Note that the effect would be even more evident, if noisy data were taken due to the instability of the inversion of $A^{\mathrm{T}}A$. To illustrate the effect of the regularization, Fig. 5.18 shows the results for different regularization parameters. The smaller value for λ produces a more oscillatory solution, whereas the larger parameter generates a very smooth solution with smaller extrema and a higher approximation error.

In the case of the 1D-polynomial approximation, we saw that irregular point grids cause trouble. Similar problems occur on the sphere. For the purpose of a demonstration, we try to approximate our previous function F by a spherical harmonics expansion of degree ≤ 40. In one test, we use a Reuter grid with $N=40$. This is a grid with 2,013 points, where the number of unknowns is $41^2=1681$. In the second test, we select 1,872 points only out of a Reuter grid with $N=80$, which produces a grid with big gaps. In the third test, we add a Reuter grid with $N=10$ to the grid of the second test, which yields in total 1,995 points. The results are shown in Fig. 5.19. In the case of the uniform grid, we see that the higher degree of the ansatz causes a more serious instability with a worse approximation. For the grid with gaps, we certainly cannot expect to approximate F properly in regions where we do not know anything about F. However, the method produces artefacts. There are oscillatory effects in these areas. It would be more preferable to have a smooth function, something with a "low frequency," in such uncertain areas.

Fig. 5.17 Regularized version of Fig. 5.15 with λ determined via the L-curve method. The maximum of the absolute approximation error function is approximately 0.0186

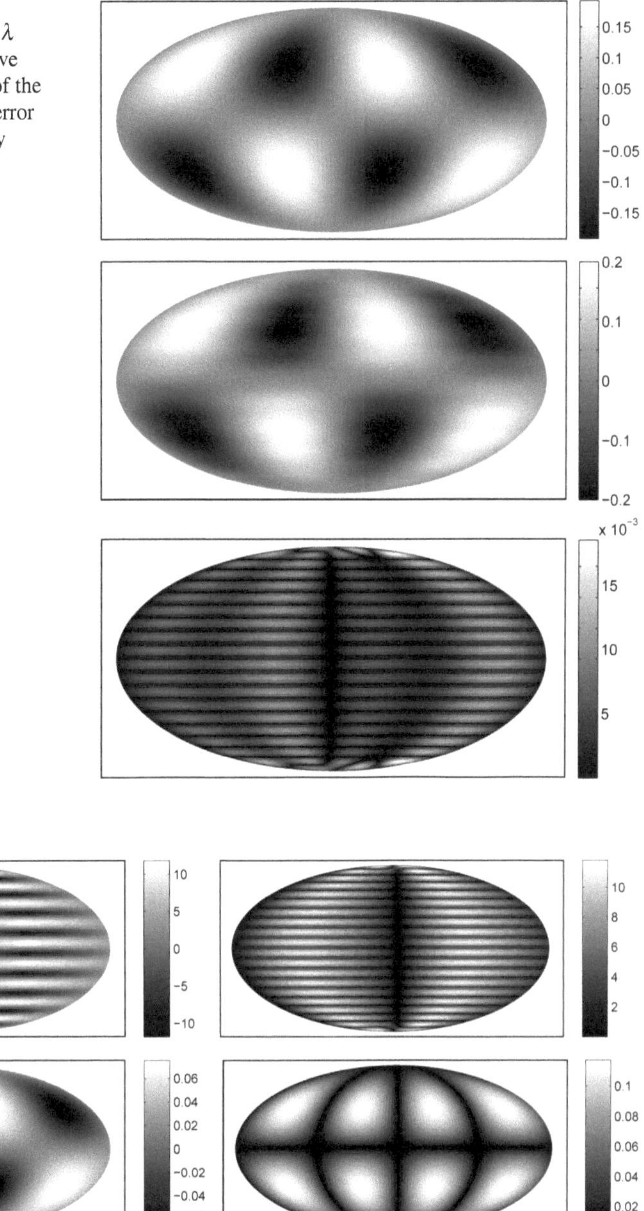

Fig. 5.18 Regularized version of Fig. 5.15, where the *left-hand images* show the approximations and the *right-hand images* show the absolute error. The *top row* corresponds to $\lambda = 7.499 \times 10^{-17} \max_{i,j} |(A^{\mathrm{T}}A)_{i,j}|$, and the *bottom row* corresponds to $\lambda = 0.866 \max_{i,j} |(A^{\mathrm{T}}A)_{i,j}|$

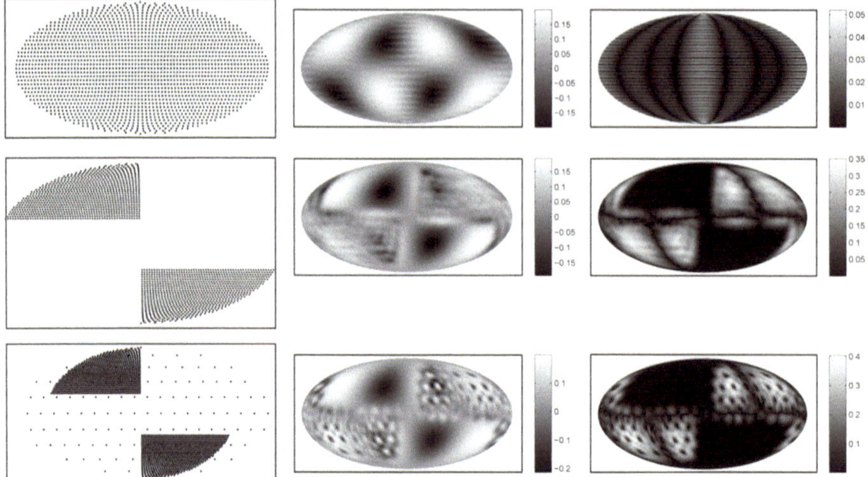

Fig. 5.19 In each row, the *left-hand image* shows the point grid used for the given data $F(\eta_k)$, the *middle image* shows the regularized spherical harmonics expansion of degree ≤ 40, and the *right-hand image* shows the absolute approximation error

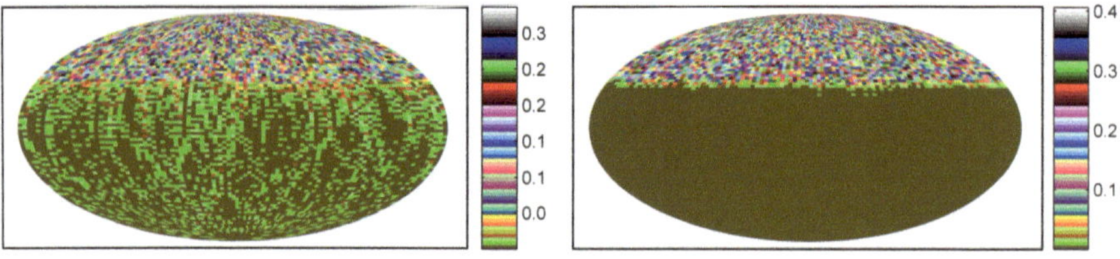

Fig. 5.20 Errors for a spherical harmonics approximation (*left hand*) and a spherical spline approximation (*right hand*) in the case of locally perturbed data: in the case of the global functions (i.e., the spherical harmonics), the error does not remain local

Another problem of global basis functions is that an error does not remain local. In Fig. 5.20, approximately one fourth of the data on a Reuter grid with $N = 80$ is disturbed by random noise, and the approximation error is plotted. The left-hand image shows the result of a regularized spherical harmonics approximation of degree ≤ 60. In comparison to the right-hand image, the noise also has influence on the non-disturbed region. The right-hand image is the result of a spherical spline interpolation. We will learn in the next section how this can be done.

5.4 Questions for Understanding

- How did we define spherical harmonics?
- Which alternatives exist for defining spherical harmonics?
- How did we define the functions $Y_{n,j}$?

- The system $\{Y_{n,j}\}_{n\in\mathbb{N}_0;\, j=1,\dots,2n+1}$ is orthonormal with respect to $\langle\cdot,\cdot\rangle_{\mathrm{L}^2(\Omega)}$. In particular,
$$\left\langle Y_{n,j}, Y_{m,k}\right\rangle_{\mathrm{L}^2(\Omega)} = 0, \quad \text{if } n \neq m \tag{5.187}$$
and
$$\left\langle Y_{n,j}, Y_{m,k}\right\rangle_{\mathrm{L}^2(\Omega)} = 0, \quad \text{if } j \neq k. \tag{5.188}$$
What is the difference between (5.187) and (5.188)?
- Formulate the addition theorem for spherical harmonics!
- We had three simple proofs using the addition theorem. Recapitulate them and the corresponding theorems!
- What is a reproducing kernel?
- Which properties do reproducing kernels have in general?
- Which reproducing kernels do you know (so far)?
- For a Fourier analysis on the sphere, we need a complete orthonormal system in $\mathrm{L}^2(\Omega)$. How can such a system be constructed in general and in particular?
- What were the milestones to show the completeness of the spherical orthonormal basis?
- The Poisson integral formula is (from an abstract point of view) a kind of an analogue of a theorem in Chap. 3. Which theorem is meant?
- In this chapter, we proved further properties of the Legendre polynomials. Name all of them!
- Now, let us have our complete orthonormal system on Ω (more precisely, a finite subset of it). We want to measure a function F on Ω and expand it in this system. What do we have to do? Which problems might occur?
- Are we done? Or are there reasons why spherical harmonics expansions might not always be the first choice?
- Was there a partition of unity again anywhere?
- What is the Abel–Poisson kernel? Which properties does it have?
- What does unisolvence mean?
- Name examples of unisolvent and non-unisolvent systems!
- What are the fundamental theorems concerning (non-)unisolvence? How did we prove them? Which theorem from your undergraduate course in analysis plays an important role in these proofs?
- What is a Tykhonov regularization?
- What is an L-curve?

Chapter 6
Spherical Splines

There are several possibilities of establishing splines on the sphere. Certainly, one can also construct cubic (or other polynomial) splines based on a triangulation of the sphere as a kind of a continuation of the 1D cubic spline concept to the sphere, see, for example, [6, 50, 105, 156, 192].

We will follow here a different way which was established by W. Freeden in [61,62,66] and uses reproducing kernel Hilbert spaces on the sphere. For this reason, we first study reproducing kernels in further details (see also [43, Sect. 12.6]). Note that we only consider scalar real functions here.

6.1 Reproducing Kernel Hilbert Spaces

We have already encountered such spaces in Definition 5.13 and Theorems 5.14, 5.15, and 5.16 (see pp. 109–111). Let us briefly recapitulate what we need to know here: A reproducing kernel Hilbert space H is a Hilbert space of functions on a domain $D \subset \mathbb{R}^n$ with a kernel $K_H : D \times D \to \mathbb{R}$ that satisfies

$$K_H(x,\cdot) \in H \quad \forall x \in D, \tag{6.1}$$

$$\langle K_H(x,\cdot), F\rangle = F(x) \quad \forall x \in D \quad \forall F \in H\,. \tag{6.2}$$

A reproducing kernel is uniquely determined by (6.1) and (6.2) and is symmetric:

$$K_H(x,y) = K_H(y,x) \quad \forall x,y \in D\,. \tag{6.3}$$

Fortunately, there is an equivalent criterion for the existence of reproducing kernels.

Theorem 6.1 (Aronszajn's theorem). *Let $(H, \langle\cdot,\cdot\rangle)$ be a Hilbert space of functions on $D \subset \mathbb{R}^n$. Then H is a reproducing kernel Hilbert space (RKHS) if and only if, for each fixed $x \in D$, the evaluation functional*

$$\begin{aligned} L_x : H &\to \mathbb{R} \\ F &\mapsto F(x) \end{aligned} \tag{6.4}$$

is continuous.

V. Michel, *Lectures on Constructive Approximation*, Applied and Numerical Harmonic Analysis, DOI 10.1007/978-0-8176-8403-7_6,

Proof.

(1) "⇐":
The evaluation functionals are obviously linear. Hence, each L_x is in the dual space H^*. Due to the Riesz representation theorem (Theorem 2.26 on p. 26), there exists $g_x \in H$ such that

$$(F(x) =) L_x F = \langle F, g_x \rangle \qquad \forall F \in H\,. \tag{6.5}$$

Note that the index x refers to the fact that different points $x \in D$ yield different functionals L_x, to which different functions g_x belong.
We set $K_H(x,\cdot) := g_x$ for each $x \in D$. Then the properties of g_x immediately yield (6.1) and (6.2).

(2) "⇒":
Let $x \in D$ be arbitrary. Since L_x is linear, it suffices to show that L_x is bounded (see Theorem 2.25). We get, using (6.2) and the Cauchy–Schwarz inequality,

$$\begin{aligned} |L_x F| &= |F(x)| \\ &= |\langle F, K_H(x,\cdot)\rangle| \\ &\leq \|F\|\,\|K_H(x,\cdot)\| \end{aligned} \tag{6.6}$$

for all $F \in H$, where

$$\begin{aligned} \|K_H(x,\cdot)\|^2 &= \langle K_H(x,\cdot), K_H(x,\cdot)\rangle \\ &\overset{(6.2)}{=} K_H(x,x) \end{aligned} \tag{6.7}$$

is a finite constant. □

Obviously, $\mathrm{L}^2(\Omega)$ is not a RKHS. The fact alone that we do not know what "$F(\xi)$" means for a concrete $\xi \in \Omega$ shows the problem (remember the definition of the L^2-spaces, see Example 2.4 on p. 15). Moreover, we could define a sequence of functions

$$F_n(\eta) := \begin{cases} 0, & \eta \in \Omega \setminus A_n \\ 1, & \eta \in A_n \end{cases}, \quad n \in \mathbb{N}_0,\ \eta \in \Omega\,, \tag{6.8}$$

where $(A_n)_{n\in\mathbb{N}_0}$ is a sequence of subsets of Ω with

$$A_n := \left\{ \eta \in \Omega \,\middle|\, |\xi - \eta| < \frac{1}{n+1} \right\}, \quad n \in \mathbb{N}_0\,, \tag{6.9}$$

for a fixed point $\xi \in \Omega$. Then we get

$$\|F_n\|_{\mathrm{L}^2(\Omega)} = \left(\int_{A_n} 1 \,\mathrm{d}\omega \right)^{1/2} \underset{n\to\infty}{\longrightarrow} 0\,, \tag{6.10}$$

that is, F_n converges to the zero function in the sense of $\mathrm{L}^2(\Omega)$, although

$$L_\xi F_n = F_n(\xi) = 1 \underset{n\to\infty}{\longrightarrow} 1\,. \tag{6.11}$$

Hence, L_ξ is not continuous.

We do not only have a criterion for the existence of a reproducing kernel, but there is also a way of giving a formula for the reproducing kernel (see also [132]).

Theorem 6.2. *Let $(H, \langle\cdot,\cdot\rangle)$ be a reproducing kernel Hilbert space of functions on $D \subset \mathbb{R}^n$. If $\{U_k\}_{k\in\mathbb{N}_0}$ is a complete orthonormal system in H that satisfies*

$$\sum_{k=0}^{\infty} U_k(x)^2 < +\infty \quad \forall x \in D\,, \tag{6.12}$$

then the reproducing kernel of H is represented by

$$K_H(x,y) = \sum_{k=0}^{\infty} U_k(x)\,U_k(y) \quad \forall x,y \in D\,. \tag{6.13}$$

Proof. We will prove that the kernel K_H in (6.13) satisfies (6.1) and (6.2). For (6.1), we observe that, for each fixed $x \in D$, the function

$$K_H(x,\cdot) = \sum_{k=0}^{\infty} U_k(x)\,U_k(\cdot) \tag{6.14}$$

is already expanded in the orthonormal basis of H. Hence, the Parseval identity yields

$$\|K_H(x,\cdot)\|^2 = \sum_{k=0}^{\infty} U_k(x)^2\,, \tag{6.15}$$

which is finite due to (6.12). Thus, $K_H(x,\cdot) \in H$. For (6.2), we use again the Parseval identity: For all $x \in D$ and all $F \in H$, we get [due to (6.14)]

$$\begin{aligned}\langle K_H(x,\cdot), F\rangle &= \sum_{k=0}^{\infty} \langle K_H(x,\cdot), U_k\rangle\,\langle F, U_k\rangle \\ &= \sum_{k=0}^{\infty} U_k(x)\,\langle F, U_k\rangle\,. \end{aligned} \tag{6.16}$$

This is the Fourier series of F, but we are not done, yet, because we need to know that this Fourier series is *pointwise* convergent. However, since H is a RKHS, Aronszajn's theorem yields that

$$\left\| \sum_{k=0}^{\infty} \langle F, U_k\rangle\, U_k - \sum_{k=0}^{N} \langle F, U_k\rangle\, U_k \right\| \underset{N\to\infty}{\longrightarrow} 0 \tag{6.17}$$

implies

$$\left| L_x \left(\sum_{k=0}^{\infty} \langle F, U_k \rangle U_k \right) - L_x \left(\sum_{k=0}^{N} \langle F, U_k \rangle U_k \right) \right| \underset{N \to \infty}{\longrightarrow} 0 . \tag{6.18}$$

Hence,

$$\begin{aligned} \langle K_H(x,\cdot), F \rangle &= \sum_{k=0}^{\infty} \langle F, U_k \rangle U_k(x) \\ &= F(x) \end{aligned} \tag{6.19}$$

for all $x \in D$ and all $F \in H$. □

Remark 6.3. We already know the reproducing kernel of $(\mathrm{Harm}_n(\Omega), \langle \cdot, \cdot \rangle_{\mathrm{L}^2(\Omega)})$ (see Theorem 5.16). Since this space is finite dimensional, the continuity of the evaluation functionals is trivial, and (6.12) is automatically satisfied. Hence, the reproducing kernel of this space is given by

$$\begin{aligned} K_{\mathrm{Harm}_n(\Omega)}(\xi, \eta) &= \sum_{j=1}^{2n+1} Y_{n,j}(\xi) Y_{n,j}(\eta) \\ &= \frac{2n+1}{4\pi} P_n(\xi \cdot \eta) \quad \forall \xi, \eta \in \Omega . \end{aligned} \tag{6.20}$$

In the spherical case, we will only study data of the type $\{F(\eta_k)\}_{k=1,\ldots,N}$. More general data of the form $\{\mathscr{F}^k F\}_{k=1,\ldots,N}$, where each $\mathscr{F}^k$ is a linear and continuous functional,[1] could be considered, see, for example, [63] and Sect. 6.4.1 for the generalization of spherical splines to such data. We will only study such data in the case of the ball since, in practice, functions on the ball are usually solutions of tomographic problems. To make a long story short, for the more general data, we need the following theorem.

Theorem 6.4. *Let $(H, \langle \cdot, \cdot \rangle)$ be a Hilbert space of functions on $D \subset \mathbb{R}^n$ with the reproducing kernel K_H. If $\mathscr{F} : H \to \mathbb{R}$ is a linear and continuous functional, then the function*

$$D \ni y \mapsto \mathscr{F}_x K_H(x,y) \tag{6.21}$$

is an element of H and

$$\mathscr{F} F = \langle F, \mathscr{F}_x K_H(x,\cdot) \rangle \quad \forall F \in H . \tag{6.22}$$

The notation $\mathscr{F}_x K_H(x,y)$ means that y is kept fixed and the function $x \mapsto K_H(x,y)$, which is an element of H, is first considered. We then apply $\mathscr{F}$ to this function and obtain a real number. Since this can be done for each $y \in D$, we, eventually, get a function $D \ni y \mapsto \mathscr{F}_x K_H(x,y) \in \mathbb{R}$.

[1]Note that this is a generalization in the case of a RKHS due to Aronszajn's theorem.

Proof. Due to the Riesz representation theorem (Theorem 2.26 on p. 26), there exists $G \in H$ such that

$$\mathscr{F}F = \langle F, G \rangle \quad \forall F \in H \,. \tag{6.23}$$

In particular,

$$\begin{aligned} \mathscr{F}_x K_H(x,y) &= \langle K_H(\cdot,y), G \rangle \\ &= G(y) \end{aligned} \tag{6.24}$$

for all $y \in D$. Since $G \in H$ and

$$\mathscr{F}F = \langle F, G \rangle = \langle F, \mathscr{F}_x K_H(x,\cdot) \rangle \quad \forall F \in H \,, \tag{6.25}$$

the theorem is proved. □

Note that the particular case $\mathscr{F}^k F := F(x_k)$ corresponds to property (6.2) of a reproducing kernel:

$$F(x_k) = \mathscr{F}^k F = \left\langle F, \mathscr{F}_x^k K_H(x,\cdot) \right\rangle = \langle F, K_H(x_k,\cdot) \rangle \,. \tag{6.26}$$

6.2 Spherical Sobolev Spaces

The reproducing kernel Hilbert spaces we need are called Sobolev spaces. There is a "classical" concept of Sobolev spaces, which you need not know to understand what is done here. However, if you know these spaces, you will hopefully be satisfied by an analogy which is explained later.

Definition 6.5. Let $(A_n)_{n \in \mathbb{N}_0}$ be a given sequence with $A_n \neq 0$ for all $n \in \mathbb{N}_0$. Then $\mathscr{E} := \mathscr{E}((A_n);\Omega)$ represents the subspace of $\mathrm{C}^{(\infty)}(\Omega)$ which consists of all $F \in \mathrm{C}^{(\infty)}(\Omega)$ with

$$\sum_{n=0}^{\infty} \sum_{j=1}^{2n+1} A_n^2 \left\langle F, Y_{n,j} \right\rangle_{\mathrm{L}^2(\Omega)}^2 < +\infty \,. \tag{6.27}$$

This space is equipped with the inner product

$$\langle F, G \rangle_{\mathscr{H}((A_n);\Omega)} := \sum_{n=0}^{\infty} \sum_{j=1}^{2n+1} A_n^2 \left\langle F, Y_{n,j} \right\rangle_{\mathrm{L}^2(\Omega)} \left\langle G, Y_{n,j} \right\rangle_{\mathrm{L}^2(\Omega)}; \quad F, G \in \mathscr{E} \,. \tag{6.28}$$

Moreover, the completion of $\mathscr{E}((A_n);\Omega)$ with respect to $\langle \cdot,\cdot \rangle_{\mathscr{H}((A_n);\Omega)}$ is denoted by $\mathscr{H}((A_n);\Omega)$ and is called a **Sobolev space**. If no confusion is likely to arise, we simply write

$$\mathscr{H} := \mathscr{H}((A_n);\Omega) \text{ and } \langle \cdot,\cdot \rangle_{\mathscr{H}} := \langle \cdot,\cdot \rangle_{\mathscr{H}((A_n);\Omega)} \,. \tag{6.29}$$

Remark 6.6.

(a) The induced norm of $\mathscr{H}$ is, consequently,

$$\|F\|_{\mathscr{H}} := \left(\sum_{n=0}^{\infty} \sum_{j=1}^{2n+1} A_n^2 \langle F, Y_{n,j} \rangle_{\mathrm{L}^2(\Omega)}^2 \right)^{1/2}, \quad F \in \mathscr{H}. \tag{6.30}$$

(b) Due to the Cauchy–Schwarz inequality, the inner product in (6.28) is always finite: For all $F, G \in \mathscr{H}$, we get

$$\sum_{n=0}^{\infty} \sum_{j=1}^{2n+1} A_n^2 \langle F, Y_{n,j} \rangle_{\mathrm{L}^2(\Omega)} \langle G, Y_{n,j} \rangle_{\mathrm{L}^2(\Omega)}$$
$$\leq \left(\sum_{n=0}^{\infty} \sum_{j=1}^{2n+1} A_n^2 \langle F, Y_{n,j} \rangle_{\mathrm{L}^2(\Omega)}^2 \right)^{1/2} \left(\sum_{n=0}^{\infty} \sum_{j=1}^{2n+1} A_n^2 \langle G, Y_{n,j} \rangle_{\mathrm{L}^2(\Omega)}^2 \right)^{1/2}, \tag{6.31}$$

where the right-hand side is finite due to (6.27).

(c) The "bouncer" for $\mathscr{E}$ (or $\mathscr{H}$) is the condition (6.27). It depends on the choice of the sequence (A_n) how strong this condition is. For instance, for $A_n = 1$ $\forall n \in \mathbb{N}_0$, (6.27) simply means that the $\mathrm{L}^2(\Omega)$-norm of F is finite. Accordingly, we get $\mathscr{H}((1);\Omega) = \mathrm{L}^2(\Omega)$. Let us compare this with two sequences (B_n) and (C_n) with $\lim_{n\to\infty} B_n = 0$ and $\lim_{n\to\infty} |C_n| = +\infty$. Then, much more functions than only $\mathrm{L}^2(\Omega)$-functions will be in $\mathscr{H}((B_n);\Omega)$, since the factor B_n causes that there will be functions $F : \Omega \to \mathbb{R}$ with

$$\sum_{n=0}^{\infty} \sum_{j=1}^{2n+1} B_n^2 \langle F, Y_{n,j} \rangle_{\mathrm{L}^2(\Omega)}^2 < +\infty \tag{6.32}$$

but

$$\sum_{n=0}^{\infty} \sum_{j=1}^{2n+1} \langle F, Y_{n,j} \rangle_{\mathrm{L}^2(\Omega)}^2 = +\infty. \tag{6.33}$$

On the other hand, not every $F \in \mathrm{L}^2(\Omega)$ will satisfy

$$\sum_{n=0}^{\infty} \sum_{j=1}^{2n+1} C_n^2 \langle F, Y_{n,j} \rangle_{\mathrm{L}^2(\Omega)}^2 < +\infty, \tag{6.34}$$

since the Fourier coefficients are now equipped with a divergent factor. In particular, functions whose Fourier coefficients do not decay rapidly enough will fail here. Such functions have comparatively high contributions of the components corresponding to large degrees. Briefly, such functions strongly oscillate.

By choosing a sequence (A_n) that diverges "fast enough," we can, therefore, achieve that the functions in $\mathscr{H}((A_n);\Omega)$ are "smooth enough."

The proofs of the following two theorems are easy exercises.

Theorem 6.7. *Let $(A_n)_{n\in\mathbb{N}_0}$ and $(B_n)_{n\in\mathbb{N}_0}$ be two real sequences with $A_n \neq 0 \neq B_n$ for all $n \in \mathbb{N}_0$, $n_0 \in \mathbb{N}_0$ a given integer, and $|A_n| \leq |B_n|$ for all $n \in \mathbb{N}_0$ with $n \geq n_0$. Then*

$$\mathscr{H}((B_n);\Omega) \subset \mathscr{H}((A_n);\Omega)\,. \tag{6.35}$$

Theorem 6.8. *Let $(A_n)_{n\in\mathbb{N}_0}$ be a real sequence with $A_n \neq 0$ for all $n \in \mathbb{N}_0$. If $F \in \mathscr{H}((A_n);\Omega)$ and $G \in \mathscr{H}((A_n^{-1});\Omega)$, then $\langle F,G\rangle_{\mathscr{H}((1);\Omega)}$ (i.e., $\langle F,G\rangle_{\mathrm{L}^2(\Omega)}$) formally exists. In other words, the series*

$$\sum_{n=0}^{\infty}\sum_{j=1}^{2n+1} \left\langle F,Y_{n,j}\right\rangle_{\mathrm{L}^2(\Omega)} \left\langle G,Y_{n,j}\right\rangle_{\mathrm{L}^2(\Omega)} \tag{6.36}$$

converges.

Definition 6.9. The Sobolev spaces $\mathscr{H}_s(\Omega)$, $s \in \mathbb{R}$, are defined by

$$\mathscr{H}_s(\Omega) := \mathscr{H}\left(\left(\left(n+\frac{1}{2}\right)^s\right);\Omega\right)\,. \tag{6.37}$$

From Theorem 6.7, we immediately get that these spaces are nested.

Theorem 6.10. *For all s, $t \in \mathbb{R}$ with $s < t$, we have*

$$\mathscr{H}_t(\Omega) \subset \mathscr{H}_s(\Omega)\,. \tag{6.38}$$

Why do we consider these particular spaces? If we apply Green's second surface identity (see Theorem 4.12 on p. 92) to $F \in \mathrm{C}^{(2)}(\Omega)$ and $Y_{n,j}$ (for arbitrary n and j), we get

$$\int_\Omega F(\xi)\underbrace{\Delta^*_\xi Y_{n,j}(\xi)}_{=-n(n+1)Y_{n,j}(\xi)} - Y_{n,j}(\xi)\Delta^*_\xi F(\xi)\,\mathrm{d}\omega(\xi) = 0\,, \tag{6.39}$$

due to Lemma 5.8 (see p. 101), and consequently,

$$-\,n(n+1)\left\langle F,Y_{n,j}\right\rangle_{\mathrm{L}^2(\Omega)} = \left\langle \Delta^* F,Y_{n,j}\right\rangle_{\mathrm{L}^2(\Omega)}\,. \tag{6.40}$$

Now, let us go on by changing the sign and adding something:

$$\left[n(n+1)+\frac{1}{4}\right]\left\langle F,Y_{n,j}\right\rangle_{\mathrm{L}^2(\Omega)} = -\left\langle \Delta^* F,Y_{n,j}\right\rangle_{\mathrm{L}^2(\Omega)} + \frac{1}{4}\left\langle F,Y_{n,j}\right\rangle_{\mathrm{L}^2(\Omega)}\,. \tag{6.41}$$

Note that $n(n+1)+\frac{1}{4}=n^2+n+\frac{1}{4}=(n+\frac{1}{2})^2$. Based on the right-hand side, we can consider the operator "$-\Delta^*+\frac{1}{4}$," which maps $F\in \mathrm{C}^{(2)}(\Omega)$ to $-\Delta^*F+\frac{1}{4}F$. We know its eigenfunctions (these are again the spherical harmonics) and its eigenvalues: $(n+\frac{1}{2})^2$, $n\in\mathbb{N}_0$. Note that they are all positive.

Formally, we can define this operator via its singular value decomposition:

$$\left(-\Delta^*+\frac{1}{4}\right)F:=\sum_{n=0}^{\infty}\sum_{j=1}^{2n+1}\left(n+\frac{1}{2}\right)^2 F^\wedge(n,j)Y_{n,j}\,, \tag{6.42}$$

that is,

$$\left\langle\left(-\Delta^*+\frac{1}{4}\right)F,Y_{n,j}\right\rangle_{\mathrm{L}^2(\Omega)}=\left(n+\frac{1}{2}\right)^2 F^\wedge(n,j)\ \forall n\in\mathbb{N}_0\ \forall j\in\{1,\ldots,2n+1\}\,. \tag{6.43}$$

This definition avoids the formal requirement of existing second-order derivatives of F. Which space can be used instead? Let $F\in\mathcal{H}_t(\Omega)$ for an arbitrary but fixed $t\in\mathbb{R}$. Then $(-\Delta^*+\frac{1}{4})F$ is defined and contained in $\mathcal{H}_{t-2}(\Omega)$ since

$$\begin{aligned}\left\|\left(-\Delta^*+\frac{1}{4}\right)F\right\|^2_{\mathcal{H}_{t-2}(\Omega)}&=\sum_{n=0}^{\infty}\sum_{j=1}^{2n+1}\left(n+\frac{1}{2}\right)^{2t-4}\left\langle\left(-\Delta^*+\frac{1}{4}\right)F,Y_{n,j}\right\rangle^2_{\mathrm{L}^2(\Omega)}\\&=\sum_{n=0}^{\infty}\sum_{j=1}^{2n+1}\left(n+\frac{1}{2}\right)^{2t-4}\left[\left(n+\frac{1}{2}\right)^2\langle F,Y_{n,j}\rangle_{\mathrm{L}^2(\Omega)}\right]^2\\&=\sum_{n=0}^{\infty}\sum_{j=1}^{2n+1}\left(n+\frac{1}{2}\right)^{2t}\langle F,Y_{n,j}\rangle^2_{\mathrm{L}^2(\Omega)}<+\infty\,.\end{aligned} \tag{6.44}$$

This can be even further generalized.

Definition 6.11. For each $s,t\in\mathbb{R}$, the operator $(-\Delta^*+\frac{1}{4})^{s/2}:\mathcal{H}_t(\Omega)\to\mathcal{H}_{t-s}(\Omega)$ is defined by

$$\left\langle\left(-\Delta^*+\frac{1}{4}\right)^{s/2}F,Y_{n,j}\right\rangle_{\mathrm{L}^2(\Omega)}:=\left(n+\frac{1}{2}\right)^s\langle F,Y_{n,j}\rangle_{\mathrm{L}^2(\Omega)} \tag{6.45}$$

for all $F\in\mathcal{H}_t(\Omega)$, $n\in\mathbb{N}_0$, and $j\in\{1,\ldots,2n+1\}$.

The proof that $(-\Delta^*+\frac{1}{4})^{s/2}F\in\mathcal{H}_{t-s}(\Omega)$ for $F\in\mathcal{H}_t(\Omega)$ is analogous to the previous proof:

$$\begin{aligned}\left\| \left(-\Delta^* + \frac{1}{4}\right)^{s/2} F \right\|^2_{\mathcal{H}_{t-s}(\Omega)} &= \sum_{n=0}^{\infty} \sum_{j=1}^{2n+1} \left(n+\frac{1}{2}\right)^{2t-2s} \left\langle \left(-\Delta^* + \frac{1}{4}\right)^{s/2} F, Y_{n,j} \right\rangle^2_{\mathrm{L}^2(\Omega)} \\ &= \sum_{n=0}^{\infty} \sum_{j=1}^{2n+1} \left(n+\frac{1}{2}\right)^{2t-2s} \left[\left(n+\frac{1}{2}\right)^{s} F^\wedge(n,j)\right]^2 \\ &= \sum_{n=0}^{\infty} \sum_{j=1}^{2n+1} \left(n+\frac{1}{2}\right)^{2t} \left(F^\wedge(n,j)\right)^2 < +\infty \,. \end{aligned} \tag{6.46}$$

We can, therefore, associate the Sobolev spaces $\mathcal{H}_t$ to some kind of generalized differentiability orders (where even fractional derivatives are possible). This corresponds to the concept of the "classical" Sobolev spaces (see, e.g., [103, Sect. 5.4], [184, Sects. 2.3 and 4.2], and [188, Chap. 3]).

Our previous considerations also yield the following result.

Theorem 6.12. *The linear operator* $(-\Delta^* + \frac{1}{4})^{s/2} : \mathcal{H}_t(\Omega) \to \mathcal{H}_{t-s}(\Omega)$ *for some fixed* $s,t \in \mathbb{R}$ *is bounded:*

$$\left\| \left(-\Delta^* + \frac{1}{4}\right)^{s/2} F \right\|_{\mathcal{H}_{t-s}(\Omega)} = \|F\|_{\mathcal{H}_t(\Omega)} \quad \forall F \in \mathcal{H}_t(\Omega) \,. \tag{6.47}$$

In particular,

$$\|F\|_{\mathcal{H}_t(\Omega)} = \left\| \left(-\Delta^* + \frac{1}{4}\right)^{t/2} F \right\|_{\mathrm{L}^2(\Omega)} \qquad \forall F \in \mathcal{H}_t(\Omega) \,. \tag{6.48}$$

Theorem 6.12 provides us with a way to interpret the Sobolev norm as the $\mathrm{L}^2(\Omega)$-norm of a generalized derivative. This is an important feature. When we study the spherical splines, we will compare them to all functions in the same Sobolev space which also satisfy the interpolation conditions. We will find out that the Sobolev norm is minimal for the interpolating spline. Keeping in mind how we interpreted the Sobolev norm, we observe the analogy to Holladay's theorem (Theorem 3.26 on p. 60).

Furthermore, remember Remark 6.6. The weighting of the Fourier coefficients in the Parseval identity (6.27) causes, for the case $|A_n| \xrightarrow[n\to\infty]{} \infty$, that the coefficients associated to large degrees get high weights. Strongly oscillating functions are, therefore, punished by such a Sobolev norm. In other words, if $|A_n| \xrightarrow[n\to\infty]{} \infty$ (which will later always be the case), then the Sobolev norm is a kind of a non-smoothness measure.

We will now investigate further properties of the Sobolev spaces and their elements.

Definition 6.13. Let $(A_n)_{n\in\mathbb{N}_0}$ and $(B_n)_{n\in\mathbb{N}_0}$ be two real sequences with $A_n \neq 0 \neq B_n \; \forall n \in \mathbb{N}_0$. Then $(A_n)_{n\in\mathbb{N}_0}$ is called a $(B_n)_{n\in\mathbb{N}_0}$-**summable** sequence if

$$\sum_{n=0}^{\infty} \frac{2n+1}{4\pi} \frac{B_n^2}{A_n^2} < +\infty \,. \tag{6.49}$$

Moreover, we simply say that $(A_n)_{n\in\mathbb{N}_0}$ is summable, if $(A_n)_{n\in\mathbb{N}_0}$ is $(1)_{n\in\mathbb{N}_0}$-summable, that is, if

$$\sum_{n=0}^{\infty} \frac{2n+1}{4\pi} A_n^{-2} < +\infty \,. \tag{6.50}$$

Note that $|A_n| \underset{n\to\infty}{\longrightarrow} \infty$ is a necessary condition for a summable sequence (A_n).

Lemma 6.14 (Sobolev Lemma). *Let $(A_n)_{n\in\mathbb{N}_0}$ be a $(B_n)_{n\in\mathbb{N}_0}$-summable sequence. Then the following embedding is possible:*

$$\mathscr{H}\left((B_n^{-1}A_n)\,;\Omega\right) \subset \mathrm{C}(\Omega) \,, \tag{6.51}$$

where the Fourier series of each $F \in \mathscr{H}((B_n^{-1}A_n);\Omega)$ is uniformly convergent.

Proof. We show that the Fourier series

$$F = \sum_{n=0}^{\infty} \sum_{j=1}^{2n+1} F^{\wedge}(n,j) Y_{n,j} \tag{6.52}$$

of each $F \in \mathscr{H}((B_n^{-1}A_n);\Omega)$ is uniformly convergent. Since every summand is continuous, Theorem 2.14 on p. 20 eventually allows us to conclude that F is a continuous function.

The uniform convergence can be proved as follows: Let $\xi \in \Omega$ be arbitrary but fixed. Then the Cauchy–Schwarz inequality, the addition theorem (Theorem 5.11), requirement (6.27), and the summability condition (6.49) yield

$$\begin{aligned}
&\left| \sum_{n=0}^{\infty} \sum_{j=1}^{2n+1} F^{\wedge}(n,j) Y_{n,j}(\xi) - \sum_{n=0}^{N-1} \sum_{j=1}^{2n+1} F^{\wedge}(n,j) Y_{n,j}(\xi) \right| \\
&= \left| \sum_{n=N}^{\infty} \sum_{j=1}^{2n+1} F^{\wedge}(n,j) Y_{n,j}(\xi) \right| \\
&= \left| \sum_{n=N}^{\infty} \sum_{j=1}^{2n+1} \frac{A_n}{B_n} F^{\wedge}(n,j) \frac{B_n}{A_n} Y_{n,j}(\xi) \right| \\
&\leq \left(\sum_{n=N}^{\infty} \sum_{j=1}^{2n+1} \frac{A_n^2}{B_n^2} F^{\wedge}(n,j)^2 \right)^{1/2} \left(\sum_{n=N}^{\infty} \sum_{j=1}^{2n+1} \frac{B_n^2}{A_n^2} (Y_{n,j}(\xi))^2 \right)^{1/2} \\
&= \left(\sum_{n=N}^{\infty} \sum_{j=1}^{2n+1} \frac{A_n^2}{B_n^2} F^{\wedge}(n,j)^2 \right)^{1/2} \left(\sum_{n=N}^{\infty} \frac{B_n^2}{A_n^2} \frac{2n+1}{4\pi} \right)^{1/2} \underset{N\to\infty}{\longrightarrow} 0 \,.
\end{aligned} \tag{6.53}$$

Since the right-hand side of the inequality is independent of ξ, the convergence is uniform. □

We will see now that further embeddings can be proved.

Lemma 6.15. *For all $n \in \mathbb{N}_0$, $j \in \{1,\ldots,2n+1\}$, and $\xi \in \Omega$,*

$$\left|\nabla_\xi^* Y_{n,j}(\xi)\right| \le \sqrt{\frac{2n+1}{4\pi} n(n+1)} = \mathcal{O}\left(n^{3/2}\right) \quad \text{as } n \to \infty \tag{6.54}$$

and

$$\sum_{j=1}^{2n+1} \left|\nabla_\xi^* Y_{n,j}(\xi)\right|^2 = \frac{2n+1}{4\pi} n(n+1) = \mathcal{O}\left(n^3\right) \quad \text{as } n \to \infty\,. \tag{6.55}$$

Proof.

(1) Differentiating the Addition Theorem:
We consider the expression

$$\sum_{j=1}^{2n+1} \left(\nabla_\xi^* Y_{n,j}(\xi)\right) \otimes \left(\nabla_\eta^* Y_{n,j}(\eta)\right) = \nabla_\xi^* \otimes \nabla_\eta^* \left(\sum_{j=1}^{2n+1} Y_{n,j}(\xi) Y_{n,j}(\eta)\right), \quad \xi,\eta \in \Omega\,, \tag{6.56}$$

where $a \otimes b := (a_i b_j)_{i,j=1,2,3} \in \mathbb{R}^{3\times 3}$ is the tensor product of $a, b \in \mathbb{R}^3$ and $\nabla_\xi^* \otimes f(\xi)$ for $f \in \mathrm{C}^{(1)}(\Omega,\mathbb{R}^3)$ represents a second-rank tensor with the j-th column $\nabla_\xi^* f_j(\xi)$, if $f = (f_1, f_2, f_3)^{\mathrm{T}}$. The right-hand side of (6.56) equals (see also Theorem 4.15 on p. 93)

$$\begin{aligned}
\nabla_\xi^* \otimes \nabla_\eta^* \left(\frac{2n+1}{4\pi} P_n(\xi\cdot\eta)\right) &= \frac{2n+1}{4\pi} \nabla_\xi^* \otimes \left[P_n'(\xi\cdot\eta)(\xi - (\xi\cdot\eta)\eta)\right] \\
&= \frac{2n+1}{4\pi} \Big[P_n''(\xi\cdot\eta)(\eta - (\xi\cdot\eta)\xi) \otimes (\xi - (\xi\cdot\eta)\eta) \\
&\quad + P_n'(\xi\cdot\eta)\left(\nabla_\xi^* \otimes \xi - (\eta - (\xi\cdot\eta)\xi)\otimes\eta\right)\Big],
\end{aligned} \tag{6.57}$$

where (let $x = r\xi$, $r > 0$) the identity I in $\mathbb{R}^{3\times 3}$ can be represented by (see also Theorem 4.5)

$$I = \nabla_x \otimes x = \left(\xi \frac{\partial}{\partial r} + \frac{1}{r}\nabla_\xi^*\right) \otimes (r\xi) = \xi \otimes \xi + \nabla_\xi^* \otimes \xi \tag{6.58}$$

such that

$$\nabla_\xi^* \otimes \xi = I - \xi \otimes \xi\,. \tag{6.59}$$

Thus, calculating the trace in (6.56) and (6.57) and setting $\xi = \eta$, we get

$$\begin{aligned}\sum_{j=1}^{2n+1} \left|\nabla_\xi^* Y_{n,j}(\xi)\right|^2 &= \frac{2n+1}{4\pi} P_n'(1)\left(3-|\xi|^2\right) \\ &= \frac{2n+1}{4\pi} n(n+1) \end{aligned} \tag{6.60}$$

due to Theorem 3.19 (see p. 49).

(2) Orthogonality of $\nabla^* Y_{n,j}$:
Green's first surface identity (see Theorem 4.12 on p. 92) yields for $F = Y_{n,j}$, $G = Y_{m,k}$ and $\Gamma = \Omega$ the result

$$\int_\Omega \left(\nabla_\xi^* Y_{m,k}(\xi)\right)\cdot\left(\nabla_\xi^* Y_{n,j}(\xi)\right)\, \mathrm{d}\omega(\xi) + \int_\Omega Y_{n,j}(\xi)\, \underbrace{\Delta_\xi^* Y_{m,k}(\xi)}_{=-m(m+1)Y_{m,k}(\xi)}\, \mathrm{d}\omega(\xi) = 0\,. \tag{6.61}$$

Consequently,

$$\int_\Omega \left(\nabla_\xi^* Y_{m,k}(\xi)\right)\cdot\left(\nabla_\xi^* Y_{n,j}(\xi)\right)\, \mathrm{d}\omega(\xi) = m(m+1)\delta_{nm}\delta_{jk} \tag{6.62}$$

for all $n,m \in \mathbb{N}_0$, $j \in \{1,\ldots,2n+1\}$, and $k \in \{1,\ldots,2m+1\}$.

(3) Proceed in analogy to Theorem 5.17 (see p. 111):
The Cauchy–Schwarz inequality yields

$$\begin{aligned}\left|\nabla_\xi^* Y_{n,j}(\xi)\right| &= \left|\sum_{k=1}^{2n+1} \frac{1}{n(n+1)} \left\langle \nabla^* Y_{n,j}, \nabla^* Y_{n,k}\right\rangle_{\mathrm{L}^2(\Omega,\mathbb{R}^3)} \nabla_\xi^* Y_{n,k}(\xi)\right| \\ &\le \left(\sum_{k=1}^{2n+1} \underbrace{\left(\frac{\left\langle \nabla^* Y_{n,j}, \nabla^* Y_{n,k}\right\rangle_{\mathrm{L}^2(\Omega,\mathbb{R}^3)}}{n(n+1)}\right)^2}_{=\delta_{jk}}\right)^{1/2} \left(\sum_{k=1}^{2n+1} \left|\nabla_\xi^* Y_{n,k}(\xi)\right|^2\right)^{1/2} \\ &= \sqrt{\frac{2n+1}{4\pi} n(n+1)}\,. \end{aligned} \tag{6.63}$$

□

Obviously, the Sobolev Lemma (Lemma 6.14) yields that $\mathcal{H}_s(\Omega) \subset \mathrm{C}(\Omega)$ for $s > 1$, since in this case, the sequence $((n+\frac{1}{2})^s)$ is summable:

$$\sum_{n=0}^{\infty} \frac{2n+1}{4\pi}\left(n+\frac{1}{2}\right)^{-2s} = \frac{1}{2\pi}\sum_{n=0}^{\infty}\left(n+\frac{1}{2}\right)^{1-2s} < \infty \Leftrightarrow 1-2s < -1 \Leftrightarrow 1 < s\,. \tag{6.64}$$

Therefore, the Fourier series

$$F(\xi) = \sum_{n=0}^{\infty}\sum_{j=1}^{2n+1} F^{\wedge}(n,j)Y_{n,j}(\xi) \quad \forall \xi \in \Omega \tag{6.65}$$

of every $F \in \mathscr{H}_s(\Omega)$, $s > 1$, is uniformly convergent. Moreover, the following embedding can be proved.

Theorem 6.16. *For all $s \in \mathbb{R}$ with $s > 2$,*

$$\mathscr{H}_s(\Omega) \subset \mathrm{C}^{(1)}(\Omega)\,. \tag{6.66}$$

Proof. We have to show that $\nabla^* F$ exists on Ω and is continuous. Due to the Cauchy–Schwarz inequality and the definition of $\mathscr{H}_s(\Omega)$, we have

$$\begin{aligned}
&\left|\sum_{n=N}^{\infty}\sum_{j=1}^{2n+1} F^{\wedge}(n,j)\nabla^*_\xi Y_{n,j}(\xi)\right| \\
&= \left|\sum_{n=N}^{\infty}\sum_{j=1}^{2n+1} F^{\wedge}(n,j)\left(n+\frac{1}{2}\right)^{s}\left(n+\frac{1}{2}\right)^{-s}\nabla^*_\xi Y_{n,j}(\xi)\right| \\
&\le \left(\sum_{n=N}^{\infty}\sum_{j=1}^{2n+1} F^{\wedge}(n,j)^2\left(n+\frac{1}{2}\right)^{2s}\right)^{1/2}\left(\sum_{n=N}^{\infty}\sum_{j=1}^{2n+1}\left(n+\frac{1}{2}\right)^{-2s}\left|\nabla^*_\xi Y_{n,j}(\xi)\right|^2\right)^{1/2} \\
&\le \left(\sum_{n=N}^{\infty}\sum_{j=1}^{2n+1} F^{\wedge}(n,j)^2\left(n+\frac{1}{2}\right)^{2s}\right)^{1/2} \\
&\quad \times \left(\sum_{n=N}^{\infty}\left(n+\frac{1}{2}\right)^{-2s}\frac{n(n+1)(2n+1)}{4\pi}\right)^{1/2} \underset{N\to\infty}{\longrightarrow} 0
\end{aligned} \tag{6.67}$$

for $s > 2$, where we used Lemma 6.15. Because of this uniform convergence, we may write

$$\nabla^*_\xi F(\xi) = \sum_{n=0}^{\infty}\sum_{j=1}^{2n+1} F^{\wedge}(n,j)\nabla^*_\xi Y_{n,j}(\xi) \quad \forall \xi \in \Omega \tag{6.68}$$

and conclude that $\nabla^* F$ is continuous. □

Theorem 6.17. *Every $F \in \mathscr{H}_s(\Omega)$ for $s > 2$ is Lipschitz continuous on Ω:*

$$|F(\xi) - F(\eta)| \le C_F(s)\,|\xi - \eta| \quad \forall \xi,\eta \in \Omega\,, \tag{6.69}$$

where

$$C_F(s) = \left(\frac{1}{2} \sum_{n=1}^{\infty} \frac{2n+1}{4\pi} \frac{n(n+1)}{\left(n+\frac{1}{2}\right)^{2s}} \right)^{1/2} \|F\|_{\mathcal{H}_s(\Omega)} \,. \tag{6.70}$$

Proof. From Theorem 6.16, we know that $F \in \mathrm{C}^{(1)}(\Omega)$. Since Ω is compact, F is, consequently, Lipschitz continuous. Moreover, (6.64) implies the uniform convergence of the Fourier series of F. We now determine a Lipschitz constant. For arbitrary $\xi, \eta \in \Omega$, we get using the Cauchy–Schwarz inequality and the addition theorem (Theorem 5.11)

$$\begin{aligned}
&|F(\xi) - F(\eta)|^2 \\
&= \left| \sum_{n=0}^{\infty} \sum_{j=1}^{2n+1} F^{\wedge}(n,j) \left(Y_{n,j}(\xi) - Y_{n,j}(\eta)\right) \right|^2 \\
&= \left| \sum_{n=0}^{\infty} \sum_{j=1}^{2n+1} \left(n+\frac{1}{2}\right)^{s} \left(n+\frac{1}{2}\right)^{-s} F^{\wedge}(n,j) \left(Y_{n,j}(\xi) - Y_{n,j}(\eta)\right) \right|^2 \\
&\leq \left(\sum_{n=0}^{\infty} \sum_{j=1}^{2n+1} \left(n+\frac{1}{2}\right)^{2s} \left(F^{\wedge}(n,j)\right)^2 \right) \left(\sum_{n=0}^{\infty} \sum_{j=1}^{2n+1} \left(n+\frac{1}{2}\right)^{-2s} \left(Y_{n,j}(\xi) - Y_{n,j}(\eta)\right)^2 \right) \\
&= \|F\|^2_{\mathcal{H}_s(\Omega)} \sum_{n=0}^{\infty} \left(n+\frac{1}{2}\right)^{-2s} \sum_{j=1}^{2n+1} \left(Y_{n,j}(\xi)^2 - 2Y_{n,j}(\xi)Y_{n,j}(\eta) + Y_{n,j}(\eta)^2\right) \\
&= \|F\|^2_{\mathcal{H}_s(\Omega)} \sum_{n=0}^{\infty} \left(n+\frac{1}{2}\right)^{-2s} \frac{2n+1}{4\pi} 2\left(P_n(1) - P_n(\xi\cdot\eta)\right) .
\end{aligned} \tag{6.71}$$

Due to the mean value theorem of differentiation, there exists $\tau \in [-1,1]$ with

$$P_n(1) - P_n(\xi\cdot\eta) = P_n'(\tau)(1 - \xi\cdot\eta) \,. \tag{6.72}$$

Hence, Theorems 5.18 (see p. 112) and 3.19 (see p. 49) yield

$$P_n(1) - P_n(\xi\cdot\eta) = |P_n(1) - P_n(\xi\cdot\eta)| \leq \frac{n(n+1)}{2}(1 - \xi\cdot\eta) \,. \tag{6.73}$$

Since $|\xi - \eta|^2 = 2(1 - \xi\cdot\eta)$, we, finally, get

$$|F(\xi) - F(\eta)| \leq \|F\|_{\mathcal{H}_s(\Omega)} \left(\sum_{n=1}^{\infty} \frac{2n+1}{4\pi} \frac{n(n+1)}{2\left(n+\frac{1}{2}\right)^{2s}} \right)^{1/2} |\xi - \eta| \,. \tag{6.74}$$

Note that $C_F(s)$ is finite, because

$$(2n+1)\frac{n(n+1)}{\left(n+\frac{1}{2}\right)^{2s}} = \mathscr{O}\left(n^{3-2s}\right) \text{ as } n \to \infty \tag{6.75}$$

and $3-2s < -1$. □

Theorem 6.18. *Let $(A_n)_{n\in\mathbb{N}_0}$ be a summable sequence. Then $\mathscr{H} = \mathscr{H}((A_n);\Omega)$ is a reproducing kernel Hilbert space with the (unique) reproducing kernel*

$$\Omega \times \Omega \ni (\xi,\eta) \mapsto K_{\mathscr{H}}(\xi\cdot\eta) = \sum_{n=0}^{\infty} A_n^{-2}\frac{2n+1}{4\pi}P_n(\xi\cdot\eta). \tag{6.76}$$

Proof.

(1) Existence of the reproducing kernel:
According to Aronszajn's theorem (Theorem 6.1), we have to show that each evaluation functional

$$\begin{aligned} L_\xi : \mathscr{H} &\to \mathbb{R} \\ F &\mapsto F(\xi), \end{aligned} \tag{6.77}$$

$\xi \in \Omega$, is bounded. We can achieve this by using the Sobolev Lemma (Lemma 6.14), the Cauchy–Schwarz inequality, the addition theorem (Theorem 5.11), and the summability condition (6.50):

$$\begin{aligned} \left|L_\xi F\right| &= |F(\xi)| \\ &= \left|\sum_{n=0}^{\infty}\sum_{j=1}^{2n+1} F^\wedge(n,j)Y_{n,j}(\xi)\right| \\ &= \left|\sum_{n=0}^{\infty}\sum_{j=1}^{2n+1} A_nF^\wedge(n,j)A_n^{-1}Y_{n,j}(\xi)\right| \\ &\leq \left(\sum_{n=0}^{\infty}\sum_{j=1}^{2n+1} A_n^2\left(F^\wedge(n,j)\right)^2\right)^{1/2}\left(\sum_{n=0}^{\infty}\sum_{j=1}^{2n+1} A_n^{-2}\left(Y_{n,j}(\xi)\right)^2\right)^{1/2} \\ &= \|F\|_{\mathscr{H}}\underbrace{\left(\sum_{n=0}^{\infty} A_n^{-2}\frac{2n+1}{4\pi}\right)^{1/2}}_{<+\infty}. \end{aligned} \tag{6.78}$$

Note that the evaluation functionals are even uniformly bounded.

(2) Representation of the reproducing kernel:
We will use Theorem 6.2. For this purpose, we need an orthonormal basis of $\mathscr{H}$. Using (6.28), we get

$$\left\langle F, Y_{m,k}\right\rangle_{\mathscr{H}} = \sum_{n=0}^{\infty}\sum_{j=1}^{2n+1} A_n^2 \left\langle F, Y_{n,j}\right\rangle_{\mathrm{L}^2(\Omega)} \underbrace{\left\langle Y_{m,k}, Y_{n,j}\right\rangle_{\mathrm{L}^2(\Omega)}}_{=\delta_{nm}\delta_{jk}}$$

$$= A_m^2 \left\langle F, Y_{m,k}\right\rangle_{\mathrm{L}^2(\Omega)} \tag{6.79}$$

for all $F \in \mathscr{H}$ and all $m \in \mathbb{N}_0$, $k = 1, \ldots, 2m+1$. Hence,

$$\begin{aligned}\left\langle A_p^{-1} Y_{p,q}, A_m^{-1} Y_{m,k}\right\rangle_{\mathscr{H}} &= A_p^{-1} A_m^{-1} A_m^2 \left\langle Y_{p,q}, Y_{m,k}\right\rangle_{\mathrm{L}^2(\Omega)} \\ &= \delta_{pm}\delta_{qk}\end{aligned} \tag{6.80}$$

for all $p, m \in \mathbb{N}_0$, $q \in \{1, \ldots, 2p+1\}$, and all $k \in \{1, \ldots, 2m+1\}$. Moreover, (6.28), that is,

$$\begin{aligned}\langle F, G\rangle_{\mathscr{H}} &= \sum_{n=0}^{\infty}\sum_{j=1}^{2n+1} A_n^2 \left\langle F, Y_{n,j}\right\rangle_{\mathrm{L}^2(\Omega)} \left\langle G, Y_{n,j}\right\rangle_{\mathrm{L}^2(\Omega)} \\ &= \sum_{n=0}^{\infty}\sum_{j=1}^{2n+1} \left\langle F, A_n^{-1} Y_{n,j}\right\rangle_{\mathscr{H}} \left\langle G, A_n^{-1} Y_{n,j}\right\rangle_{\mathscr{H}} ,\end{aligned} \tag{6.81}$$

is the Parseval identity for the orthonormal system $\{A_n^{-1} Y_{n,j}\}_{n\in\mathbb{N}_0; j=1,\ldots,2n+1}$ in $\mathscr{H}$. This identity is equivalent to the completeness (see Theorem 2.18 on p. 21). Finally, we observe that

$$\sum_{n=0}^{\infty}\sum_{j=1}^{2n+1} \left(A_n^{-1} Y_{n,j}(\xi)\right)^2 = \sum_{n=0}^{\infty} A_n^{-2} \frac{2n+1}{4\pi} < +\infty \quad \forall \xi \in \Omega . \tag{6.82}$$

Hence,

$$\begin{aligned}K_{\mathscr{H}}(\xi \cdot \eta) &= \sum_{n=0}^{\infty}\sum_{j=1}^{2n+1} A_n^{-1} Y_{n,j}(\xi) A_n^{-1} Y_{n,j}(\eta) \\ &= \sum_{n=0}^{\infty} A_n^{-2} \frac{2n+1}{4\pi} P_n(\xi \cdot \eta) \quad \forall \xi, \eta \in \Omega .\end{aligned} \tag{6.83}$$

□

Note that we have already used in the notation of $K_{\mathscr{H}}$ that the reproducing kernel is a zonal function.

Corollary 6.19. *For all $F \in \mathscr{H}$ and all $n \in \mathbb{N}_0$, $j = 1, \ldots, 2n+1$,*

$$\left\langle F, Y_{n,j}\right\rangle_{\mathscr{H}} = A_n^2 \left\langle F, Y_{n,j}\right\rangle_{\mathrm{L}^2(\Omega)} . \tag{6.84}$$

Corollary 6.20. *If (A_n) is summable, then*

$$\|F\|_{C(\Omega)} \leq \|F\|_{\mathscr{H}} \left(\sum_{n=0}^{\infty} A_n^{-2} \frac{2n+1}{4\pi} \right)^{1/2} \tag{6.85}$$

for all $F \in \mathscr{H}$.

6.3 Spherical Splines

The reproducing kernels introduced in Sect. 6.2 will now be used to construct spherical splines. For everything that follows, we make the following assumption.

By (A_n), we denote a fixed summable sequence. The corresponding Sobolev space $\mathscr{H} := \mathscr{H}((A_n); \Omega)$ is then equipped with the unique reproducing kernel $K_{\mathscr{H}}$.

A spline is now defined by means of the reproducing kernel.

Definition 6.21. Let $X_N := \{\eta_1, \ldots, \eta_N\} \subset \Omega$ be given.[2] Then every function $S \in \mathscr{H}$ of the form

$$S(\xi) = \sum_{j=1}^{N} a_j K_{\mathscr{H}}(\eta_j \cdot \xi) \quad \forall \xi \in \Omega \tag{6.86}$$

with constants $a_1, \ldots, a_N \in \mathbb{R}$ is called a **spherical spline** in $\mathscr{H}$ relative to X_N. The set of all such splines is denoted by $\mathrm{Spline}((A_n); X_N)$.

The task of a spline interpolation problem is now as follows.

Problem 6.22. Given $y \in \mathbb{R}^N$ and $X_N = \{\eta_1, \ldots, \eta_N\} \subset \Omega$, find $S \in \mathrm{Spline}((A_n); X_N)$ such that $S(\eta_k) = y_k$ for all $k = 1, \ldots, N$.

This is, from the very beginning, a different concept than the one used for spherical harmonics. Though the ansatz in both cases is a linear expansion in a basis system, the basis system is completely different. While spherical harmonics are global functions (see Fig. 5.5 on p. 129 and the numerical experiments in Sect. 5.3), the spline basis function $\xi \mapsto K_{\mathscr{H}}(\eta_j \cdot \xi)$ is localized around the point η_j, which we will call here the **center of the spline basis function**. For an illustration, we consider a particular example.

[2] Note that the usual concept of a set includes the requirement that elements may not occur more than once. Hence, we automatically require that the points are pairwise distinct, that is, $\eta_i \neq \eta_j$ for $i \neq j$.

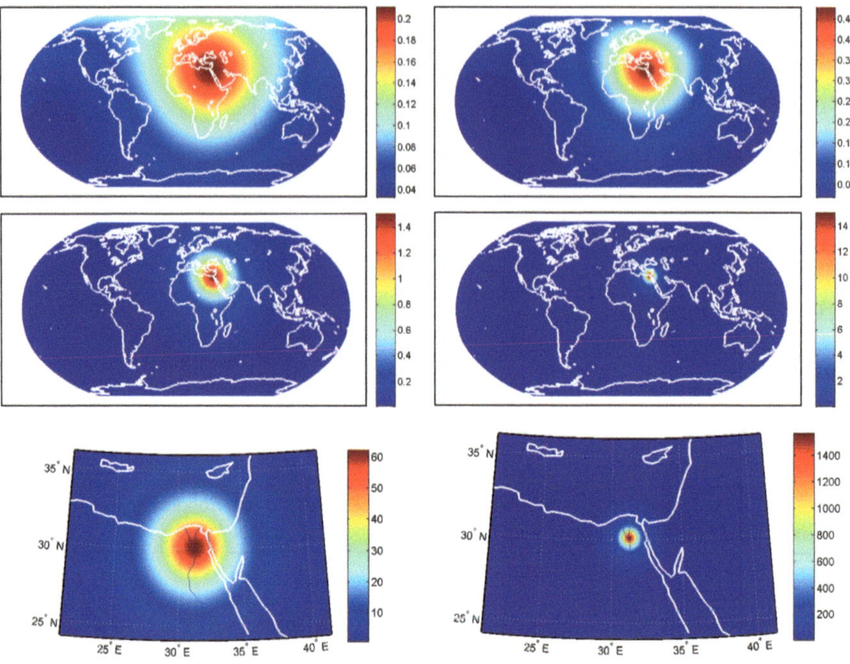

Fig. 6.1 Illustration of the Abel–Poisson kernel as a spline basis function with a fixed center at Cairo for different parameters h: 0.3 (*top left*), 0.5 (*top right*), 0.7 (*middle left*), 0.9 (*middle right*), 0.95 (*bottom left*), and 0.99 (*bottom right*)

Example 6.23. Remember Theorem 3.16 on p. 46. We have

$$\sum_{n=0}^{\infty}(2n+1)h^n P_n(t) = \frac{1-h^2}{(1+h^2-2ht)^{3/2}} \tag{6.87}$$

for all $h \in]-1,1[$ and all $t \in [-1,1]$. Setting $t = \xi \cdot \eta$ with $\xi, \eta \in \Omega$ and applying the addition theorem (Theorem 5.11), we obtain

$$\sum_{n=0}^{\infty} h^n \sum_{j=1}^{2n+1} Y_{n,j}(\xi) Y_{n,j}(\eta) = \frac{1}{4\pi} \frac{1-h^2}{(1+h^2-2h(\xi \cdot \eta))^{3/2}} \tag{6.88}$$

for all $h \in]-1,1[$ and all $\xi, \eta \in \Omega$. This is a reproducing kernel (the **Abel–Poisson kernel**). Theorem 6.18 yields

$$K_{\mathscr{H}((h^{-n/2});\Omega)}(\xi \cdot \eta) = \frac{1}{4\pi} \frac{1-h^2}{(1+h^2-2h(\xi \cdot \eta))^{3/2}} \quad \forall \xi, \eta \in \Omega\,. \tag{6.89}$$

The sequence $(A_n) = (h^{-n/2})$ is obviously summable if $0 < h < 1$. Figure 6.1 shows the corresponding spline basis functions for a fixed center but different parameters h.

We can expect now that the coefficients a_j in (6.86) tell us something about the function in the neighborhood of the center η_j. For instance, regionally limited noise should only influence those coefficients a_j which correspond to basis functions whose centers are located in or near the region of the noise (see also Fig. 5.20 on p. 142). Moreover, if we have a nonuniformly distributed data grid X_N, then more centers are automatically located in areas with more data, since we choose the data points as centers of the spline basis functions (see Problem 6.22). The spline will, consequently, regionally adapt its resolution to the data density. The more we know about the function in a certain area, the better we can resolve it.

Let us now investigate how we can calculate the interpolating spline. The ansatz (6.86) for Problem 6.22 yields the system of linear equations

$$\sum_{j=1}^{N} a_j K_{\mathscr{H}}(\eta_j \cdot \eta_k) = y_k \quad \forall k = 1, \ldots, N . \tag{6.90}$$

Problem 6.22 is, consequently, equivalent to the following problem.

Problem 6.24. Given $y \in \mathbb{R}^N$ and $X_N = \{\eta_1, \ldots, \eta_N\} \subset \Omega$, find $a \in \mathbb{R}^N$ such that

$$\left(K_{\mathscr{H}}\left(\eta_j \cdot \eta_k\right)\right)_{j,k=1,\ldots,N} a = y . \tag{6.91}$$

According to Hadamard, a problem is called **well-posed**, if the following three criteria are satisfied:

- The problem is solvable.
- There is not more than one solution.
- The solution is stable, that is, it continuously depends on the given data.

Otherwise, it is called **ill-posed**.

In our case, we have to verify that the matrix

$$M_{\mathscr{H},X_N} := \left(K_{\mathscr{H}}\left(\eta_j \cdot \eta_k\right)\right)_{j,k=1,\ldots,N} \tag{6.92}$$

is regular, since the finite dimensions of the involved spaces will directly imply the continuity of the inversion. To accomplish this objective, we have to understand the concept of positive definite and strictly positive definite functions.

Definition 6.25. Let $K : [-1,1] \to \mathbb{R}$ be continuous. K is said to be **positive definite**, if, for every system $X_N = \{\eta_1, \ldots, \eta_N\} \subset \Omega$ (i.e., also for every $N \in \mathbb{N}$), the matrix

$$\left(K\left(\eta_j \cdot \eta_k\right)\right)_{j,k=1,\ldots,N} \tag{6.93}$$

is positive semi-definite. If the matrix is additionally positive definite, K is called **strictly positive definite**.

Note the inconsistency in the common usage of the expression "positive definite" for matrices and functions.

We will show now that every reproducing kernel $K_{\mathscr{H}}$ (i.e., no matter which summable sequence (A_n) we have chosen in the very beginning) is strictly positive definite. As a consequence, the spline interpolation problem is always uniquely solvable. Remember that we needed additional conditions for the cubic splines in the 1D case to obtain a unique solution. Furthermore, the matrix is positive definite which allows us to use faster numerical solvers for Problem 6.24. Last but definitely not least, these properties are independent of the chosen point grid X_N. This is a very nice advantage. So, we do not have to face problems like those caused by the Extended Haar's theorem (see Theorem 5.34 on p. 133).

Lemma 6.26. *Let $K:[-1,1]\to\mathbb{R}$ be a continuous function with*

$$\sum_{n=0}^{\infty}\frac{2n+1}{4\pi}K^{\wedge}(n)<+\infty \tag{6.94}$$

and

$$K^{\wedge}(n)>0\quad\forall n\in\mathbb{N}_0\,, \tag{6.95}$$

where $K^{\wedge}(n)$ is the n-th Legendre coefficient of K (see Definition 3.15 on p. 45). Then K is strictly positive definite if and only if, for every $N\in\mathbb{N}$ and every $X_N=\{\eta_1,\ldots,\eta_n\}\subset\Omega$, the functions

$$F_j:\Omega\ni\xi\mapsto K(\eta_j\cdot\xi)\ ;\quad j=1,\ldots,N\,; \tag{6.96}$$

are linearly independent.

Proof.

(1) Investigation of the Legendre series:
Since $\|P_n\|_{C[-1,1]}=1$ for all $n\in\mathbb{N}_0$ (see Theorem 5.18 on p. 112), the requirement (6.94) in combination with (6.95) implies the uniform convergence[3] of the Legendre series

$$K(t)=\sum_{n=0}^{\infty}K^{\wedge}(n)\frac{2n+1}{4\pi}P_n(t)\quad\forall t\in[-1,1]\,. \tag{6.97}$$

Moreover, K is the reproducing kernel of the Sobolev space $\mathscr{H}((K^{\wedge}(n)^{-1/2});\Omega)$, since $A_n:=K^{\wedge}(n)^{-1/2}$ is summable:

$$\sum_{n=0}^{\infty}\frac{2n+1}{4\pi}A_n^{-2}=\sum_{n=0}^{\infty}\frac{2n+1}{4\pi}K^{\wedge}(n)<+\infty\,. \tag{6.98}$$

(2) A Gramian matrix:
Now, since we know that K is a reproducing kernel, we can write the matrix entries as follows:

$$K(\eta_j\cdot\eta_k)=\langle K(\eta_j\cdot),K(\eta_k\cdot)\rangle_{\mathscr{H}}=\langle F_j,F_k\rangle_{\mathscr{H}}\,. \tag{6.99}$$

The notation $K(\eta\cdot)$ represents here the function $\Omega\ni\xi\mapsto K(\eta\cdot\xi)$. The matrix with these entries is, consequently, a Gramian matrix. From Linear Algebra, it is known

[3] By the way, this also implies the continuity of K. For the use of the Legendre coefficients, we have to use, however, an appropriate condition in the theorem.

that a Gramian matrix is positive definite if and only if the vectors which were used to build it are linearly independent.[4] This completes the proof. □

The following theorem is based on more general results in [65, 168, 206].

Theorem 6.27. *$K_{\mathscr{H}}$ is strictly positive definite.*

Proof.

(1) The plan:
Let $N \in \mathbb{N}$ and $X_N = \{\eta_1, \ldots, \eta_N\} \subset \Omega$ be arbitrary. We will use Lemma 6.26 to prove this result. For this purpose, we assume that $a_1, \ldots, a_N \in \mathbb{R}$ are given such that

$$\sum_{j=1}^{N} a_j K_{\mathscr{H}}(\eta_j \cdot \xi) = 0 \quad \forall \xi \in \Omega . \tag{6.100}$$

We have to show that $a_1 = \cdots = a_N = 0$. Note that (6.94) and (6.95) are satisfied, since (A_n) is summable.

(2) The Legendre polynomials again:
If (6.100) holds true, then

$$\left\langle \sum_{j=1}^{N} a_j K_{\mathscr{H}}(\eta_j \cdot), P_n(\zeta \cdot) \right\rangle_{\mathscr{H}} = 0 \tag{6.101}$$

for all $\zeta \in \Omega$ and all $n \in \mathbb{N}_0$. Due to the bilinearity of the inner product and the fact that $K_{\mathscr{H}}$ is the reproducing kernel of $\mathscr{H}$, we obtain

$$\sum_{j=1}^{N} a_j P_n(\eta_j \cdot \zeta) = 0 \quad \forall n \in \mathbb{N}_0 \quad \forall \zeta \in \Omega . \tag{6.102}$$

(3) An auxiliary function:
The separation distance q_{X_N} is defined by

$$q_{X_N} := \min_{\substack{i,j \in \{1,\ldots,N\} \\ i \neq j}} \left| \eta_i - \eta_j \right| . \tag{6.103}$$

We choose now an arbitrary but fixed index $i^* \in \{1, \ldots, N\}$ (we will eventually show that $a_{i^*} = 0$) and constants $0 < \varepsilon < q_{X_N}$, $\sigma := \frac{\varepsilon^2}{2}$, $1 - \sigma \leq h < 1$. Based on these preparations, we define the function $H : \Omega \to \mathbb{R}$ by

$$H(\eta) := \begin{cases} \frac{\xi \cdot \eta - h}{1-h}, & \text{if } \xi \cdot \eta \geq h, \\ 0 & \text{else} \end{cases} , \quad \eta \in \Omega , \tag{6.104}$$

where $\xi \in \Omega$ is chosen as $\xi := \eta_{i^*}$.

---

[4] Otherwise, the matrix is singular.

This function has the following properties:

1. $H(\xi)=1$: Since $\xi\cdot\xi=1>h$, we have $H(\xi)=\frac{\xi\cdot\xi-h}{1-h}=1$.
2. $\operatorname{supp} H\subset\{\eta\in\Omega\,\big|\,|\xi-\eta|\le\varepsilon\}$: The inequality $|\xi-\eta|\le\varepsilon$ holds if and only if

$$\xi\cdot\eta=\frac{1}{2}\left(-|\xi-\eta|^2+2\right)\ge\frac{1}{2}\left(-\varepsilon^2+2\right)=\frac{1}{2}(-2\sigma+2)=1-\sigma\,. \quad (6.105)$$

Moreover, $\xi\in\operatorname{supp} H$ is equivalent to $\xi\cdot\eta\ge h$, where we chose $h\ge 1-\sigma$.

Hence, $H(\eta_i)=\delta_{ii^*}$ for all $i\in\{1,\dots,N\}$, since

$$H(\eta_{i^*})=1 \quad \text{(see above)} \quad (6.106)$$

and

$$|\eta_i-\eta_{i^*}|\ge q_{X_N}>\varepsilon \quad \forall i\ne i^* \Rightarrow \eta_i\notin\operatorname{supp} H \quad \forall i\ne i^*\,. \quad (6.107)$$

Consequently, we can write

$$a_{i^*}=\sum_{i=1}^{N}a_iH(\eta_i)\,. \quad (6.108)$$

(4) The Abel–Poisson kernel again:
Let $Q_r:[-1,1]\to\mathbb{R}$, $r\in]0,1[$, denote the Abel–Poisson kernel

$$Q_r(t):=\frac{1}{4\pi}\frac{1-r^2}{(1+r^2-2rt)^{3/2}} \quad \forall t\in[-1,1]\,. \quad (6.109)$$

Remember that (see Example 6.23 or Theorem 3.16)

$$Q_r(t)=\sum_{n=0}^{\infty}\frac{2n+1}{4\pi}r^nP_n(t) \quad \forall t\in[-1,1]\,. \quad (6.110)$$

If we combine this result with (6.102), we get

$$\sum_{n=0}^{\infty}\frac{2n+1}{4\pi}r^n\sum_{j=1}^{N}a_j\,P_n(\eta_j\cdot)=0 \quad \forall r\in]0,1[\quad (6.111)$$

and, finally,

$$\sum_{j=1}^{N}a_jQ_r(\eta_j\cdot)=0 \quad \forall r\in]0,1[\,. \quad (6.112)$$

Since H is continuous, (6.108) and the Poisson integral formula (Theorem 5.19 on p. 112) yield

$$
\begin{aligned}
a_{i^*} &= \sum_{i=1}^{N} a_i H(\eta_i) \\
&= \sum_{i=1}^{N} a_i \lim_{r\to 1-} \langle Q_r(\eta_i \cdot), H\rangle_{\mathrm{L}^2(\Omega)} \\
&= \lim_{r\to 1-} \left\langle \sum_{i=1}^{N} a_i Q_r(\eta_i \cdot), H \right\rangle_{\mathrm{L}^2(\Omega)} \\
&= 0 \,.
\end{aligned}
\tag{6.113}
$$

This completes the proof. □

We now have our first fundamental result on spherical splines.

Theorem 6.28 (Existence and Uniqueness of the Interpolating Spline). *For every $X_N = \{\eta_1, \ldots, \eta_N\} \subset \Omega$ and every $y \in \mathbb{R}^N$, there exists one and only one $S \in \mathrm{Spline}((A_n); X_N)$ such that $S(\eta_k) = y_k$ for all $k = 1, \ldots, N$.*

Hence, spherical spline interpolation is a well-posed problem (in the sense of Hadamard).

We will now derive further properties of the interpolating spline.

Lemma 6.29. *Let $S = \sum_{k=1}^{N} a_k K_{\mathscr{H}}(\eta_k \cdot) \in \mathrm{Spline}((A_n); X_N)$ and $F \in \mathscr{H}$. Then*

$$
\langle S, F\rangle_{\mathscr{H}} = \sum_{k=1}^{N} a_k \, F(\eta_k) \,. \tag{6.114}
$$

Proof. This lemma is a direct consequence of the defining property of a reproducing kernel, since

$$
\begin{aligned}
\langle S, F\rangle_{\mathscr{H}} &= \left\langle \sum_{k=1}^{N} a_k K_{\mathscr{H}}(\eta_k \cdot), F \right\rangle_{\mathscr{H}} \\
&= \sum_{k=1}^{N} a_k \langle K_{\mathscr{H}}(\eta_k \cdot), F\rangle_{\mathscr{H}} \\
&= \sum_{k=1}^{N} a_k \, F(\eta_k) \,.
\end{aligned}
\tag{6.115}
$$

□

Theorem 6.30 (First Minimum Property). *Let a vector $y \in \mathbb{R}^N$ and a point set $X_N = \{\eta_1, \ldots, \eta_N\} \subset \Omega$ be given. If $S^* \in \mathrm{Spline}((A_n); X_N)$ is the spline given by $S^*(\eta_k) = y_k \ \forall k = 1, \ldots, N$, then*

$$\|S^*\|_{\mathcal{H}} = \min\{\|F\|_{\mathcal{H}} : F \in \mathcal{H} \text{ with } F(\eta_k) = y_k \quad \forall k = 1,\ldots,N\}\,, \tag{6.116}$$

where S^* is the unique minimizer.

Proof. Let $F \in \mathcal{H}$ with $F(\eta_k) = y_k\ \forall k = 1,\ldots,N$. Then the induced norm $\|\cdot\|_{\mathcal{H}}$ satisfies

$$\begin{aligned}\|F\|_{\mathcal{H}}^2 &= \langle F - S^* + S^*, F - S^* + S^*\rangle_{\mathcal{H}}\\ &= \langle F - S^*, F - S^*\rangle_{\mathcal{H}} + 2\langle F - S^*, S^*\rangle_{\mathcal{H}} + \langle S^*, S^*\rangle_{\mathcal{H}}\\ &= \|F - S^*\|_{\mathcal{H}}^2 + 2\langle S^*, F - S^*\rangle_{\mathcal{H}} + \|S^*\|_{\mathcal{H}}^2\,. \end{aligned} \tag{6.117}$$

Let the spline S^* be represented by

$$S^* = \sum_{k=1}^{N} a_k^*\, K_{\mathcal{H}}(\eta_k \cdot)\,. \tag{6.118}$$

Then Lemma 6.29 and the fact that S^* and F satisfy the same interpolation conditions yield

$$\langle S^*, F - S^*\rangle_{\mathcal{H}} = \sum_{k=1}^{N} a_k^* \underbrace{(F - S^*)(\eta_k)}_{=0} = 0\,. \tag{6.119}$$

If we insert this result in (6.117), we obtain

$$\|F\|_{\mathcal{H}}^2 = \|F - S^*\|_{\mathcal{H}}^2 + \|S^*\|_{\mathcal{H}}^2 \geq \|S^*\|_{\mathcal{H}}^2\,, \tag{6.120}$$

where "=" holds if and only if $F = S^*$. □

The first minimum property is an analogue of Holladay's theorem (Theorem 3.26 on p. 60) for the following reasons: In Remark 6.6 and in Theorem 6.12, we saw that the Sobolev norm $\|F\|_{\mathcal{H}}$ can be interpreted as a non-smoothness measure for F and as the $\mathrm{L}^2(\Omega)$-norm of something like a derivative of F. The first minimum property now tells us that the interpolating spline minimizes this norm in comparison to all other interpolating functions (in $\mathcal{H}$). It is, therefore, in some sense the smoothest interpolant—just as the natural cubic spline minimizes the linearized curvature, which is represented by the $\mathrm{L}^2[a,b]$-norm of the second derivative.

Remark 6.31. It is not necessary to require that $A_n \neq 0$ for all $n \in \mathbb{N}_0$. Some exceptions may be allowed. This has, indeed, already been elaborated for the spherical splines in the literature (see, e.g., [66]), and we will see how it works on the ball later, where practical applications require the possibility of vanishing components. So, let us choose $A_n := n(n+1)$ for all $n \in \mathbb{N}_0$. This sequence is obviously summable (if we ignore that $A_0 = 0$). Using Definition 6.5 and Theorem 5.28 [or, alternatively, (6.40)], we find

$$\begin{aligned}\|F\|_{\mathscr{H}}^2 &= \sum_{n=1}^{\infty}\sum_{j=1}^{2n+1} n^2(n+1)^2 \langle F, Y_{n,j}\rangle_{\mathrm{L}^2(\Omega)}^2 \\ &= \sum_{n=1}^{\infty}\sum_{j=1}^{2n+1} \left[-n(n+1)\langle F, Y_{n,j}\rangle_{\mathrm{L}^2(\Omega)}\right]^2 \\ &= \sum_{n=1}^{\infty}\sum_{j=1}^{2n+1} \langle \Delta^* F, Y_{n,j}\rangle_{\mathrm{L}^2(\Omega)}^2 \end{aligned} \tag{6.121}$$

for all $F \in \mathscr{H} := \mathscr{H}((n(n+1));\Omega)$, where Δ^* is defined on $\mathscr{H}$ in analogy to $-\Delta^* + \frac{1}{4}$ on $\mathscr{H}_t(\Omega)$ (see Definition 6.11). As a consequence,

$$\|F\|_{\mathscr{H}} = \|\Delta^* F\|_{\mathrm{L}^2(\Omega)} \tag{6.122}$$

for all $F \in \mathscr{H}$. The term on the right-hand side can be interpreted as the "bending energy" on the sphere (see [75, pp. 188-189]). This stresses the interpretation of the first minimum property as an analogue of Holladay's theorem. Furthermore, the reproducing kernel corresponding to this particular Sobolev space is exactly Green's function[5] for the iterated Beltrami operator $(\Delta^*)^2$, for which an explicit representation is known (see [21]).

Theorem 6.32 (Second Minimum Property). *Let $F \in \mathscr{H}$ and $X_N = \{\eta_1,\ldots,\eta_N\} \subset \Omega$. If $S^* \in \mathrm{Spline}((A_n);X_N)$ is the spline given by $S^*(\eta_k) = F(\eta_k)$ $\forall k = 1,\ldots,N$, then*

$$\|F - S^*\|_{\mathscr{H}} = \min\{\|F - S\|_{\mathscr{H}} : S \in \mathrm{Spline}((A_n);X_N)\}, \tag{6.123}$$

where S^* is the unique minimizer.

Proof. Let $S \in \mathrm{Spline}((A_n);X_N)$ be an arbitrary spline. Then we find

$$\begin{aligned}\|F-S\|_{\mathscr{H}}^2 &= \langle F - S^* + S^* - S, F - S^* + S^* - S\rangle_{\mathscr{H}} \\ &= \langle F - S^*, F - S^*\rangle_{\mathscr{H}} + 2\langle F - S^*, S^* - S\rangle_{\mathscr{H}} + \langle S^* - S, S^* - S\rangle_{\mathscr{H}} \\ &= \|F - S^*\|_{\mathscr{H}}^2 + 2\langle S^* - S, F - S^*\rangle_{\mathscr{H}} + \|S^* - S\|_{\mathscr{H}}^2 . \end{aligned} \tag{6.124}$$

Using Lemma 6.29, the interpolation conditions for S^*, and the representations

$$S = \sum_{k=1}^{N} a_k K_{\mathscr{H}}(\eta_k \cdot), \qquad S^* = \sum_{k=1}^{N} a_k^* K_{\mathscr{H}}(\eta_k \cdot), \tag{6.125}$$

[5] Green's functions represent a tool for solving particular classes of (ordinary or partial) differential equations, which can be combined with boundary-value conditions. They have also been discovered for further applications in potential theory and geomathematics.

we get

$$\langle S^* - S, F - S^* \rangle_{\mathscr{H}} = \sum_{k=1}^{N} (a_k^* - a_k) \underbrace{(F - S^*)(\eta_k)}_{=0} = 0 \,. \tag{6.126}$$

Hence,

$$\|F - S\|_{\mathscr{H}}^2 = \|F - S^*\|_{\mathscr{H}}^2 + \|S^* - S\|_{\mathscr{H}}^2 \geq \|F - S^*\|_{\mathscr{H}}^2 \tag{6.127}$$

has been proved,[6] where "=" holds if and only if $S = S^*$. □

If we regard F as the unknown function, which has been measured on X_N, then the second minimum property says that the interpolating spline is closest to F (in the sense of the Sobolev norm) in comparison to all other splines. Using a common phrase from approximation theory, we can say that the interpolating spline S^* is the **best approximation** in $\mathrm{Spline}((A_n);X_N)$ with respect to the function F.

Theorem 6.33 (Shannon Sampling Theorem in Spline($(A_n);X_N$)). *Let the coefficients*[7] a_l^k*;* $l,k \in \{1,\ldots,N\}$*; be given by the systems of linear equations*

$$\sum_{l=1}^{N} a_l^k \, K_{\mathscr{H}}(\eta_l \cdot \eta_j) = \delta_{jk} \quad \forall j,k = 1,\ldots,N \,. \tag{6.128}$$

Moreover, the functions $L_k \in \mathrm{Spline}((A_n);X_N)$*;* $k = 1,\ldots,N$*; are defined by*

$$L_k(\xi) := \sum_{l=1}^{N} a_l^k \, K_{\mathscr{H}}(\eta_l \cdot \xi) \quad \forall \xi \in \Omega \,. \tag{6.129}$$

Then every spline $S \in \mathrm{Spline}((A_n);X_N)$ *is representable by its samples on* $X_N = \{\eta_1,\ldots,\eta_N\} \subset \Omega$ *as*

$$S(\xi) = \sum_{k=1}^{N} S(\eta_k) L_k(\xi) \quad \forall \xi \in \Omega \,. \tag{6.130}$$

Proof. Comparing (6.128) and (6.129), we observe that the coefficients a_l^k are chosen such that

$$L_k(\eta_j) = \delta_{jk} \quad \forall j,k = 1,\ldots,N \,. \tag{6.131}$$

Hence, the function $G \in \mathrm{Spline}((A_n);X_N)$ defined by (note that $\mathrm{Spline}((A_n);X_N)$ is a linear space)

$$G(\xi) := \sum_{k=1}^{N} S(\eta_k) L_k(\xi) \quad \forall \xi \in \Omega \tag{6.132}$$

[6] Note that the identity $\|F - S\|_{\mathscr{H}}^2 = \|F - S^*\|_{\mathscr{H}}^2 + \|S^* - S\|_{\mathscr{H}}^2$ in (6.127) is a generalization of the identity $\|F\|_{\mathscr{H}}^2 = \|F - S^*\|_{\mathscr{H}}^2 + \|S^*\|_{\mathscr{H}}^2$ in (6.120), since the zero function is also a spline.

[7] Note that "k" in a_l^k is only an upper index and *not* an exponent.

satisfies

$$G(\eta_j) = \sum_{k=1}^{N} S(\eta_k) L_k(\eta_j) = \sum_{k=1}^{N} S(\eta_k)\,\delta_{jk} = S(\eta_j) \quad \forall j = 1,\ldots,N\,. \tag{6.133}$$

Due to the uniqueness of the interpolating spline (see Theorem 6.28), we get $G = S$.
□

For obtaining the coefficients a_l^k we have to solve N linear systems of the size $N \times N$, where previously we solved Problem 6.24, which involved only one $N \times N$ system with exactly the same matrix. So, is there any use of Theorem 6.33? Yes, there is! It is very useful if the same grid X_N is used many times (e.g., for a long time series) or if the time required for computing the spline after obtaining the data has to be kept as short as possible (as it is the case in medical imaging, where almost real-time solutions are required). Before the data are measured for the first time, we can already do our homework and calculate all coefficients a_l^k. Moreover, we can also already calculate the **Lagrange basis** $L_1,\ldots,L_N$ on the point grid that we want to use for plotting the result later and store everything on the computer. As soon as the data on X_N arrive, the spline will very quickly be calculated and plotted.

We will now prove error estimates and convergence results (see also [66]).

Lemma 6.34. *Let $(A_n)_{n\in\mathbb{N}_0}$ be $(n)_{n\in\mathbb{N}_0}$-summable. Then the family of zonal functions*

$$F_\xi : \Omega \ni \eta \mapsto K_{\mathscr{H}}(\xi\cdot\eta), \quad \xi \in \Omega\,, \tag{6.134}$$

is uniformly Lipschitz continuous, that is, there exists a constant $E > 0$, which only depends on (A_n), such that

$$|K_{\mathscr{H}}(\xi\cdot\eta) - K_{\mathscr{H}}(\xi\cdot\zeta)| \le E^2 |\eta - \zeta| \quad \forall \eta,\zeta \in \Omega\,. \tag{6.135}$$

Proof. In analogy to parts of the proof of Lemma 6.17, the mean value theorem of differentiation and Theorem 3.19 (see p. 49) yield, for all $\xi,\eta,\zeta \in \Omega$,

$$\begin{aligned} |K_{\mathscr{H}}(\xi\cdot\eta) - K_{\mathscr{H}}(\xi\cdot\zeta)| &= \left| \sum_{n=0}^{\infty} \frac{2n+1}{4\pi} A_n^{-2} \left(P_n(\xi\cdot\eta) - P_n(\xi\cdot\zeta)\right) \right| \\ &\le \sum_{n=1}^{\infty} \frac{2n+1}{4\pi} A_n^{-2} \frac{n(n+1)}{2} |\xi\cdot\eta - \xi\cdot\zeta| \\ &\le \sum_{n=1}^{\infty} \frac{2n+1}{4\pi} A_n^{-2} \frac{n(n+1)}{2} \underbrace{|\xi|}_{=1} \cdot |\eta - \zeta|\,. \end{aligned} \tag{6.136}$$

Hence,

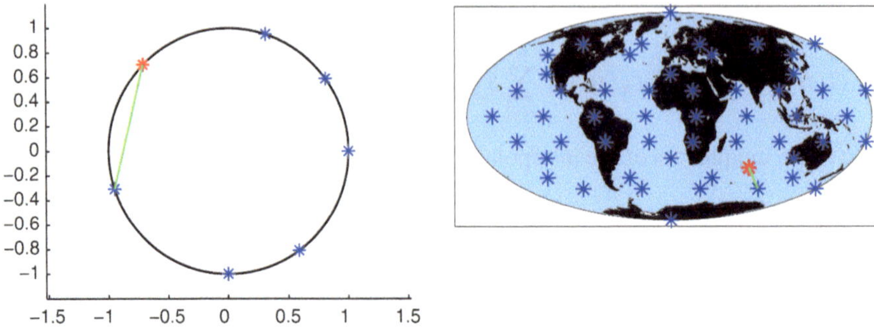

Fig. 6.2 Illustration of the nodal width Θ_{X_N} for the unit circle in $\mathbb{R}^2$ (*left hand*) and the unit sphere in $\mathbb{R}^3$ (*right hand*): The point grid X_N is shown with *blue asterisks* and a point ξ on the 1- or 2-sphere, respectively, with $\Theta_{X_N} = \min_{\eta \in X_N} |\xi - \eta|$ is plotted as a *red asterisk*. The length of the *green line* is Θ_{X_N}

$$E = E_{(A_n)} = \left(\sum_{n=1}^{\infty} \frac{2n+1}{4\pi} A_n^{-2} \frac{n(n+1)}{2} \right)^{1/2}, \tag{6.137}$$

which is finite, since (A_n) is (n)-summable. □

Definition 6.35. For a given finite point system $X_N \subset \Omega$, the **nodal width**[8] Θ_{X_N} is defined by

$$\Theta_{X_N} := \max_{\xi \in \Omega} \min_{\eta \in X_N} |\xi - \eta| . \tag{6.138}$$

The nodal width can be interpreted as the radius of the largest gap in the point grid. Figure 6.2 illustrates the nodal width Θ_{X_N} in $\mathbb{R}^3$ and in $\mathbb{R}^2$ (where no deformation due to a map projection occurs). For a convergence result, we will have to consider the limit $N \to \infty$. If we presumed that $\overline{\lim}_{N\to\infty} \Theta_{X_N} > 0$, there would be an area which is not sufficiently covered by data points. In this case, we cannot expect a satisfactory result. For this reason, it appears to be sensible to assume that $\lim_{N\to\infty} \Theta_{X_N} = 0$.

Theorem 6.36 (Error Estimate and Convergence I). *Let $F \in \mathscr{H}((A_n);\Omega)$, where (A_n) is (n)-summable. For each $N \in \mathbb{N}$, let $X_N = \{\eta_1^{(N)}, \ldots, \eta_N^{(N)}\} \subset \Omega$ be a given point grid. If $S_N \in$ Spline$((A_n);X_N)$ is the spline given by*

$$S_N\left(\eta_j^{(N)}\right) = F\left(\eta_j^{(N)}\right) \quad \forall j = 1, \ldots, N, \tag{6.139}$$

then

$$\|F - S_N\|_{C(\Omega)} \leq 2^{3/2} E_{(A_n)} \Theta_{X_N}^{1/2} \|F\|_{\mathscr{H}} \quad \forall N \in \mathbb{N}, \tag{6.140}$$

where E is given by (6.137). If $\lim\limits_{N\to\infty} \Theta_{X_N} = 0$, then

[8]The maximum exists because Ω is compact, and the minimum exists because X_N is finite.

$$\lim_{N\to\infty} \|F - S_N\|_{C(\Omega)} = 0\,. \tag{6.141}$$

Proof. Let $\xi \in \Omega$ be an arbitrary point. Due to the definition of Θ_{X_N}, there exists $\eta_i^{(N)} \in X_N$, for each $N \in \mathbb{N}$, such that

$$\left|\xi - \eta_i^{(N)}\right| \le \Theta_{X_N}\,. \tag{6.142}$$

Hence, (6.139), the defining property of a reproducing kernel, the Cauchy–Schwarz inequality, the triangle inequality, and Theorem 6.30 (first minimum property) yield

$$\begin{aligned}
|S_N(\xi) - F(\xi)| &= \left|S_N(\xi) - S_N\left(\eta_i^{(N)}\right) + F\left(\eta_i^{(N)}\right) - F(\xi)\right| \\
&= \left|\left\langle K_{\mathcal{H}}(\xi\cdot) - K_{\mathcal{H}}\left(\eta_i^{(N)}\cdot\right), S_N - F\right\rangle_{\mathcal{H}}\right| \\
&\le \left\|K_{\mathcal{H}}(\xi\cdot) - K_{\mathcal{H}}\left(\eta_i^{(N)}\cdot\right)\right\|_{\mathcal{H}} \|S_N - F\|_{\mathcal{H}} \\
&\le \left\|K_{\mathcal{H}}(\xi\cdot) - K_{\mathcal{H}}\left(\eta_i^{(N)}\cdot\right)\right\|_{\mathcal{H}} \left(\|S_N\|_{\mathcal{H}} + \|F\|_{\mathcal{H}}\right) \\
&\le 2\left\|K_{\mathcal{H}}(\xi\cdot) - K_{\mathcal{H}}\left(\eta_i^{(N)}\cdot\right)\right\|_{\mathcal{H}} \|F\|_{\mathcal{H}}\,.
\end{aligned} \tag{6.143}$$

Using the defining property of a reproducing kernel again and Lemma 6.34, we get

$$\begin{aligned}
&\left\|K_{\mathcal{H}}(\xi\cdot) - K_{\mathcal{H}}\left(\eta_i^{(N)}\cdot\right)\right\|_{\mathcal{H}}^2 \\
&= \left\langle K_{\mathcal{H}}(\xi\cdot) - K_{\mathcal{H}}\left(\eta_i^{(N)}\cdot\right), K_{\mathcal{H}}(\xi\cdot) - K_{\mathcal{H}}\left(\eta_i^{(N)}\cdot\right)\right\rangle_{\mathcal{H}} \\
&= K_{\mathcal{H}}(\xi\cdot\xi) - K_{\mathcal{H}}\left(\xi\cdot\eta_i^{(N)}\right) + K_{\mathcal{H}}\left(\eta_i^{(N)}\cdot\eta_i^{(N)}\right) - K_{\mathcal{H}}\left(\eta_i^{(N)}\cdot\xi\right) \\
&\le 2E_{(A_n)}^2\left|\xi - \eta_i^{(N)}\right|\,.
\end{aligned} \tag{6.144}$$

Combining (6.142), (6.143), and (6.144), we obtain

$$\begin{aligned}
|S_N(\xi) - F(\xi)| &\le 2^{3/2} E_{(A_n)} \left|\xi - \eta_i^{(N)}\right|^{1/2} \|F\|_{\mathcal{H}} \\
&\le 2^{3/2} E_{(A_n)} \Theta_{X_N}^{1/2} \|F\|_{\mathcal{H}}\,.
\end{aligned} \tag{6.145}$$

Since the right-hand side is independent of ξ, we are done. □

Lemma 6.37. *For every $n \in \mathbb{N}_0$ and $\xi, \eta \in \Omega$, we have*

$$1 - P_n(\xi\cdot\eta) \le \frac{2n+1}{8}\left[n(n+1)|\xi - \eta|\right]^2\,. \tag{6.146}$$

Note that the left-hand side is nonnegative.

Proof. Let $n \in \mathbb{N}_0$. On the one hand, Theorem 5.16 on p. 111 and the definition of the Legendre polynomials (Definition 3.10 on p. 41) yield

$$\begin{aligned}
&\frac{2n+1}{4\pi}\int_\Omega (P_n(\xi\cdot\zeta)-P_n(\eta\cdot\zeta))^2\,d\omega(\zeta)\\
&=\frac{2n+1}{4\pi}\int_\Omega P_n(\xi\cdot\zeta)^2-2P_n(\xi\cdot\zeta)P_n(\eta\cdot\zeta)+P_n(\eta\cdot\zeta)^2\,d\omega(\zeta)\\
&=P_n(\xi\cdot\xi)-2P_n(\xi\cdot\eta)+P_n(\eta\cdot\eta)\\
&=2(1-P_n(\xi\cdot\eta))
\end{aligned} \tag{6.147}$$

for all $\xi,\eta \in \Omega$. On the other hand, the mean value theorem of differentiation and Theorem 3.19 (see p. 49) yield

$$\begin{aligned}
&\frac{2n+1}{4\pi}\int_\Omega (P_n(\xi\cdot\zeta)-P_n(\eta\cdot\zeta))^2\,d\omega(\zeta)\\
&\le\frac{2n+1}{4\pi}\int_\Omega\left[\frac{n(n+1)}{2}\,|\xi\cdot\zeta-\eta\cdot\zeta|\right]^2 d\omega(\zeta)\\
&\le\frac{2n+1}{4\pi}\,\frac{n^2(n+1)^2}{4}\int_\Omega|\xi-\eta|^2\cdot\underbrace{|\zeta|^2}_{=1}\,d\omega(\zeta)\\
&=(2n+1)\frac{n^2(n+1)^2}{4}\,|\xi-\eta|^2
\end{aligned} \tag{6.148}$$

for all $\xi,\eta \in \Omega$. The combination of (6.147) and (6.148) yields the desired result. □

Theorem 6.38 (Error Estimate and Convergence II). *Let $F \in \mathcal{H}((A_n);\Omega)$, where (A_n) is (n^3)-summable. For each $N \in \mathbb{N}$, let $X_N = \{\eta_1^{(N)},\dots,\eta_N^{(N)}\} \subset \Omega$ be a given point grid. If $S_N \in \mathrm{Spline}((A_n);X_N)$ is the spline given by*

$$S_N\left(\eta_j^{(N)}\right)=F\left(\eta_j^{(N)}\right)\quad \forall j=1,\dots,N\,, \tag{6.149}$$

then

$$\|F-S_N\|_{\mathrm{C}(\Omega)}\le 2\tilde{E}_{(A_n)}\Theta_{X_N}\,\|F\|_{\mathcal{H}}\quad \forall N\in\mathbb{N}\,, \tag{6.150}$$

where

$$\tilde{E}_{(A_n)}=\left(\sum_{n=1}^{\infty}\frac{(2n+1)^2}{4\pi}A_n^{-2}\left(\frac{n(n+1)}{2}\right)^2\right)^{1/2}. \tag{6.151}$$

If $\lim_{N\to\infty}\Theta_{X_N}=0$, then

$$\lim_{N\to\infty}\|F-S_N\|_{\mathrm{C}(\Omega)}=0\,. \tag{6.152}$$

Proof. Let $\xi \in \Omega$ be an arbitrary point. We can proceed in analogy to the proof of Theorem 6.36. We get again (6.142) and (6.143). Instead of (6.144), however, we derive, using Theorem 6.18 and Lemma 6.37,

$$\begin{aligned}
&\left\| K_{\mathscr{H}}(\xi\cdot) - K_{\mathscr{H}}\left(\eta_i^{(N)}\cdot\right)\right\|_{\mathscr{H}}^2 \\
&= K_{\mathscr{H}}(\xi\cdot\xi) - 2K_{\mathscr{H}}\left(\xi\cdot\eta_i^{(N)}\right) + K_{\mathscr{H}}\left(\eta_i^{(N)}\cdot\eta_i^{(N)}\right) \\
&= \sum_{n=0}^{\infty} \frac{2n+1}{4\pi} A_n^{-2}\left(2 - 2P_n\left(\xi\cdot\eta_i^{(N)}\right)\right) \\
&\le \sum_{n=1}^{\infty} \frac{(2n+1)^2}{4\pi} A_n^{-2}\left[\frac{n(n+1)}{2}\left|\xi - \eta_i^{(N)}\right|\right]^2 \\
&= \tilde{E}_{(A_n)}^2 \left|\xi - \eta_i^{(N)}\right|^2 . \qquad (6.153)
\end{aligned}$$

In combination with (6.142) and (6.143), we get

$$\begin{aligned}
|S_N(\xi) - F(\xi)| &\le 2\tilde{E}_{(A_n)}\left|\xi - \eta_i^{(N)}\right| \|F\|_{\mathscr{H}} \\
&\le 2\tilde{E}_{(A_n)}\Theta_{X_N}\|F\|_{\mathscr{H}} , \qquad (6.154)
\end{aligned}$$

where the right-hand side is again independent of $\xi \in \Omega$. □

In analogy to the fact that we know that the global functions $\{Y_{n,j}\}_{n\in\mathbb{N}_0;j=1,\ldots,2n+1}$ are closed systems in $(\mathrm{C}(\Omega), \|\cdot\|_{\mathrm{C}(\Omega)})$ and $(\mathrm{L}^2(\Omega), \|\cdot\|_{\mathrm{L}^2(\Omega)})$, we can now show that there are also localized counterparts, namely, the spline basis functions.

Theorem 6.39. *Let $X \subset \Omega$ be a countable and dense subset and $\mathscr{H} = \mathscr{H}((A_n);\Omega)$ be a Sobolev space corresponding to a summable sequence (A_n). Then the system $\{K_{\mathscr{H}}(\eta\cdot) \,|\, \eta \in X\}$ is closed in the spaces $(\mathscr{H}, \|\cdot\|_{\mathscr{H}})$, $(\mathrm{C}(\Omega), \|\cdot\|_{\mathrm{C}(\Omega)})$, and $(\mathrm{L}^2(\Omega), \|\cdot\|_{\mathrm{L}^2(\Omega)})$.*

Proof.

(1) Closed in $(\mathscr{H}, \|\cdot\|_{\mathscr{H}})$:
Obviously,

$$k := \overline{\operatorname{span}\{K_{\mathscr{H}}(\eta\cdot) \,|\, \eta \in X\}}^{\|\cdot\|_{\mathscr{H}}} \qquad (6.155)$$

is a closed linear subspace of $\mathscr{H}$. Let $F \in \mathscr{H}$ and $F \perp_{\mathscr{H}} k$, that is, for every $G \in k$, we have $\langle F, G\rangle_{\mathscr{H}} = 0$ (see also Theorem 2.23 on p. 25). This implies, in particular,

$$\langle F, K_{\mathscr{H}}(\eta\cdot)\rangle_{\mathscr{H}} = 0 \quad \forall \eta \in X , \qquad (6.156)$$

which is, due to the defining property of a reproducing kernel, equivalent to

$$F(\eta) = 0 \quad \forall \eta \in X\,. \tag{6.157}$$

Since X is dense in Ω and F is continuous, this implies $F \equiv 0$. Hence, $k = \mathscr{H}$.

(2) Closed in $(\mathrm{C}(\Omega), \|\cdot\|_{\mathrm{C}(\Omega)})$:
It is clear that

$$\operatorname{span}\left\{ Y_{n,j} \,\middle|\, n \in \mathbb{N}_0;\, j = 1,\ldots,2n+1 \right\} \subset \mathscr{H}\,, \tag{6.158}$$

since a span only contains finite linear combinations. Furthermore, Theorem 5.22 on p. 120 and Lemma 6.14 (Sobolev Lemma) yield

$$\overline{\operatorname{span}\left\{ Y_{n,j} \,\middle|\, n \in \mathbb{N}_0;\, j = 1,\ldots,2n+1 \right\}}^{\|\cdot\|_{\mathrm{C}(\Omega)}} = \mathrm{C}(\Omega), \quad \mathscr{H}(\Omega) \subset \mathrm{C}(\Omega)\,. \tag{6.159}$$

Combining (6.158) and (6.159), we conclude $\overline{\mathscr{H}}^{\|\cdot\|_{\mathrm{C}(\Omega)}} = \mathrm{C}(\Omega)$. Now, let $F \in \mathrm{C}(\Omega)$ and $\varepsilon > 0$ be arbitrary. Due to the previous result, there exists $G \in \mathscr{H}$ such that

$$\|F - G\|_{\mathrm{C}(\Omega)} < \frac{\varepsilon}{2}\,. \tag{6.160}$$

Due to part 1 of this proof, there exists, associated to $G \in \mathscr{H}$, a function $H \in \operatorname{span}\{K_{\mathscr{H}}(\eta\cdot) \,|\, \eta \in X\}$ such that

$$\|G - H\|_{\mathscr{H}} < \frac{\varepsilon}{2} \left(\sum_{n=0}^{\infty} A_n^{-2} \frac{2n+1}{4\pi} \right)^{-1/2}, \tag{6.161}$$

where the right-hand side exists since (A_n) is summable. We now combine (6.160), Corollary 6.20, and (6.161) and obtain

$$\begin{aligned}
\|F - H\|_{\mathrm{C}(\Omega)} &\leq \|F - G\|_{\mathrm{C}(\Omega)} + \|G - H\|_{\mathrm{C}(\Omega)} \\
&< \frac{\varepsilon}{2} + \|G - H\|_{\mathscr{H}} \left(\sum_{n=0}^{\infty} A_n^{-2} \frac{2n+1}{4\pi} \right)^{1/2} \\
&< \frac{\varepsilon}{2} + \frac{\varepsilon}{2} = \varepsilon\,.
\end{aligned} \tag{6.162}$$

(3) Closed in $(\mathrm{L}^2(\Omega), \|\cdot\|_{\mathrm{L}^2(\Omega)})$:
Let $F \in \mathrm{L}^2(\Omega)$ and $\varepsilon > 0$ be arbitrary. Due to Theorem 4.11 on p. 92, there exists $G \in \mathrm{C}(\Omega)$ such that

$$\|F - G\|_{\mathrm{L}^2(\Omega)} < \frac{\varepsilon}{2}\,. \tag{6.163}$$

Part 2 of this proof now guarantees the existence of $H \in \operatorname{span}\{K_{\mathscr{H}}(\eta \cdot) \mid \eta \in X\}$ with

$$\|G - H\|_{C(\Omega)} < \frac{\varepsilon}{2\sqrt{4\pi}} . \tag{6.164}$$

Finally, we combine (6.163), Theorem 4.10, and (6.164) to get

$$\begin{aligned} \|F - H\|_{L^2(\Omega)} &\leq \|F - G\|_{L^2(\Omega)} + \|G - H\|_{L^2(\Omega)} \\ &< \frac{\varepsilon}{2} + \sqrt{4\pi}\,\|G - H\|_{C(\Omega)} \\ &< \frac{\varepsilon}{2} + \frac{\varepsilon}{2} = \varepsilon . \end{aligned} \tag{6.165}$$

□

6.4 Some Remarks

6.4.1 More General Data

It has already been shown (see, e.g., [63]) that the concept of splines presented here can be extended to the case where the data are of the form

$$y_i = \mathscr{F}^i F, \quad i = 1, \ldots, N ; \tag{6.166}$$

and each $\mathscr{F}^i : \mathscr{H} \to \mathbb{R}$ is a linear and continuous functional. In this case, the spline basis functions are obtained by applying the functionals to one argument of the reproducing kernel, that is, we keep $\xi \in \Omega$ fixed and apply $\mathscr{F}^i$ to $\eta \mapsto K_{\mathscr{H}}(\xi \cdot \eta)$. Since this can be done for all $\xi \in \Omega$, we obtain a function of $\xi \in \Omega$:

$$\Omega \ni \xi \mapsto \mathscr{F}^i_\eta K_{\mathscr{H}}(\xi \cdot \eta) \in \mathbb{R} . \tag{6.167}$$

The ansatz for a spline is now

$$S(\xi) = \sum_{k=1}^{N} a_k \mathscr{F}^k_\eta K_{\mathscr{H}}(\xi \cdot \eta), \quad \xi \in \Omega . \tag{6.168}$$

Note that the splines in the previous section correspond to the particular case where $\mathscr{F}^k F = F(\eta_k)$. These functionals are linear and continuous due to the existence of $K_{\mathscr{H}}$ [see Aronszajn's theorem (Theorem 6.1)].

Due to Theorem 6.4, the basic properties of the splines can be transferred to this more general setting.

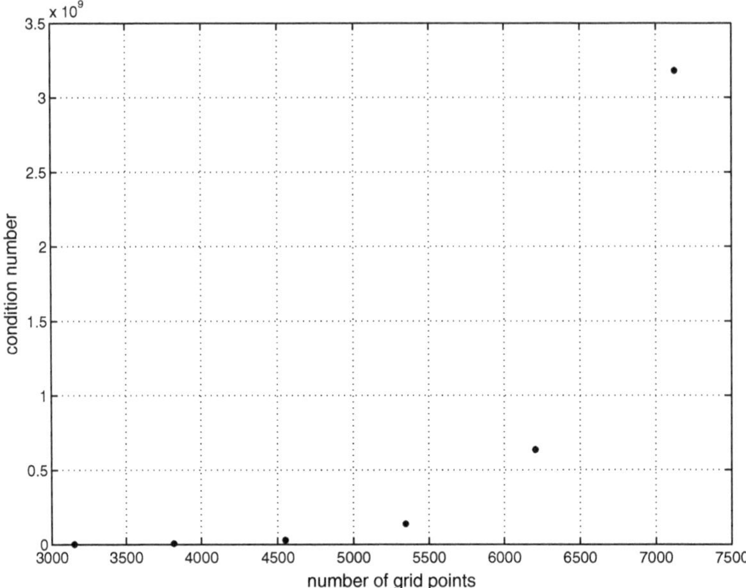

Fig. 6.3 Condition number of the Abel–Poisson spline matrix for different Reuter grids: the *horizontal axis* corresponds to the number of grid points and the *vertical axis* represents the condition number

6.4.2 The Spline Matrix

For cubic splines, it is known that the matrix associated to the system of linear equations has a very low condition number. More precisely, in the case of Hermite boundary conditions and an equidistant grid, the condition number does not exceed 6. With minor stabilizing multiplications of the equations, the upper bound of the condition number can (also for non-equidistant grids and all three kinds of boundary conditions) be reduced to 3 (where different norms define here the condition number, see [170, pp. 131–132], [195, pp. 171–172] for further details).

However, transferring a method from the Euclidean setting to the sphere is often connected to the loss of certain pleasant properties. This is the case for the condition number of the spline matrix. In Fig. 6.3, we show the condition number of the (Abel–Poisson) spline matrix

$$\left(\frac{1}{4\pi}\frac{1-h^2}{(1+h^2-2h\eta_i\cdot\eta_j)^{3/2}}\right)_{i,j=1,\ldots,N} \tag{6.169}$$

for Reuter grids of different sizes N and $h=0.8$. For ill-conditioned matrices, a regularization as in Theorem 5.37 (see p. 138) is necessary. This can be justified by the following theoretical result.

Theorem 6.40 (Spline Approximation). *Let $y \in \mathbb{R}^N$, $X_N = \{\eta_1, \ldots, \eta_N\} \subset \Omega$, $\lambda > 0$, and a summable sequence (A_n) be given. Then the spline*

$$S(\xi) = \sum_{k=1}^{N} a_k K_{\mathscr{H}}(\eta_k \cdot \xi), \quad \xi \in \Omega, \tag{6.170}$$

where the coefficient vector $a = (a_1, \ldots, a_N)^{\mathrm{T}}$ is the solution of

$$\left((K_{\mathscr{H}}(\eta_i \cdot \eta_j))_{i,j=1,\ldots,N} + \lambda I_N \right) a = y \tag{6.171}$$

and I_N is the $N \times N$-identity matrix, is the unique minimizer of the functional

$$\mathscr{H} \ni F \mapsto \sum_{i=1}^{N} (y_i - F(\eta_i))^2 + \lambda \|F\|_{\mathscr{H}}^2 . \tag{6.172}$$

Proof. Since the first matrix in (6.171) is positive definite (see Theorem 6.27) and $\lambda > 0$, a is uniquely determined by (6.171). We now have a closer look at the functional in (6.172). For every $F \in \mathscr{H}$,

$$\begin{aligned} &\sum_{i=1}^{N} (y_i - F(\eta_i))^2 + \lambda \|F\|_{\mathscr{H}}^2 \\ &= \sum_{i=1}^{N} (y_i - S(\eta_i))^2 + \lambda \|S\|_{\mathscr{H}}^2 + \sum_{i=1}^{N} (S(\eta_i) - F(\eta_i))^2 \\ &\quad +2\sum_{i=1}^{N} (F(\eta_i) - S(\eta_i))(S(\eta_i) - y_i) + \lambda \|F - S\|_{\mathscr{H}}^2 + 2\lambda \langle S, F - S \rangle_{\mathscr{H}} . \end{aligned} \tag{6.173}$$

Moreover, the equations in (6.171) obviously require that

$$\sum_{j=1}^{N} a_j K_{\mathscr{H}}(\eta_i \cdot \eta_j) + \lambda a_i = y_i \quad \forall i = 1, \ldots, N, \tag{6.174}$$

that is,

$$S(\eta_i) + \lambda a_i = y_i \quad \forall i = 1, \ldots, N . \tag{6.175}$$

As a consequence,

$$a_i = \frac{1}{\lambda}(y_i - S(\eta_i)) \quad \forall i = 1, \ldots, N \tag{6.176}$$

and, thus,

$$\sum_{i=1}^{N} a_i (F - S)(\eta_i) = \frac{1}{\lambda} \sum_{i=1}^{N} (F - S)(\eta_i)(y_i - S(\eta_i)) . \tag{6.177}$$

Hence, Lemma 6.29 yields

$$\langle S, F - S\rangle_{\mathscr{H}} = \frac{1}{\lambda} \sum_{i=1}^{N} (F - S)(\eta_i)(y_i - S(\eta_i)) \ . \tag{6.178}$$

We insert this result in (6.173) and obtain

$$\begin{aligned}
&\sum_{i=1}^{N} (y_i - F(\eta_i))^2 + \lambda \|F\|^2_{\mathscr{H}} \\
&\quad = \sum_{i=1}^{N} (y_i - S(\eta_i))^2 + \lambda \|S\|^2_{\mathscr{H}} + \sum_{i=1}^{N} (S(\eta_i) - F(\eta_i))^2 + \lambda \|F - S\|^2_{\mathscr{H}} \\
&\quad \geq \sum_{i=1}^{N} (y_i - S(\eta_i))^2 + \lambda \|S\|^2_{\mathscr{H}} \,,
\end{aligned} \tag{6.179}$$

where "=" holds if and only if $F = S$. □

The regularization parameter λ controls a balance between the accuracy of the approximation and the smoothness of the solution as we can see in (6.172).

Remark 6.41. There is more than the analogue of Holladay's theorem and the first minimum property which justifies the name "spline" in the spherical case. The best approximation property of natural cubic splines based on Schoenberg's theorem (see Remark 3.27 on p. 61) also has its analogue for the spherical splines presented here. As a consequence, numerical integration on the sphere can also be performed by integrating the interpolating spline. For further details, see [66, Sects. 5.4, 7.3, and 7.4].

6.4.3 *Another Remark on Point Grids*

As we have already noticed, the choice of the sequence (A_n) influences the localization, that is, the hat-width, of the spline basis functions. In the case of the Abel–Poisson kernel, the parameter h controls the hat-width. We now calculate the interpolating spline for a Reuter grid with 786 points (see Fig. 6.4) and equal data $y_1 = \cdots = y_N = 1$. However, the parameter h of the used Abel–Poisson kernel is varied. The result is shown in Fig. 6.5. We observe the following phenomenon: If h is too large, then the hats are too narrow and "gaps" occur in between the centers. For this reason, the sequence (A_n) always has to be chosen with care, and some preliminary experiments with different sequences are sometimes necessary.

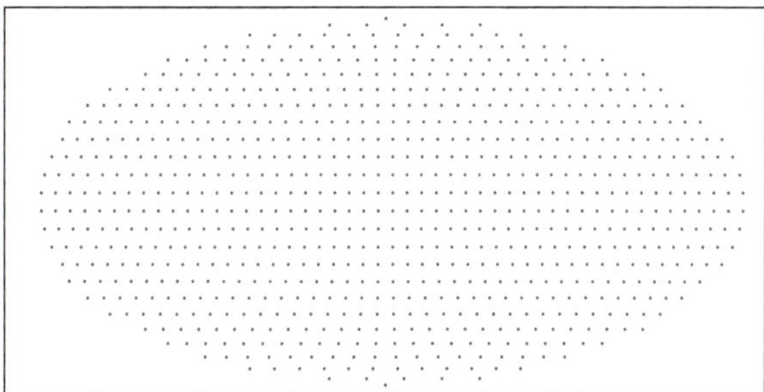

Fig. 6.4 A Reuter grid (see Example 5.36 on p. 137) with parameter $N = 25$ and 786 points used for the spline interpolation experiments in Fig. 6.5

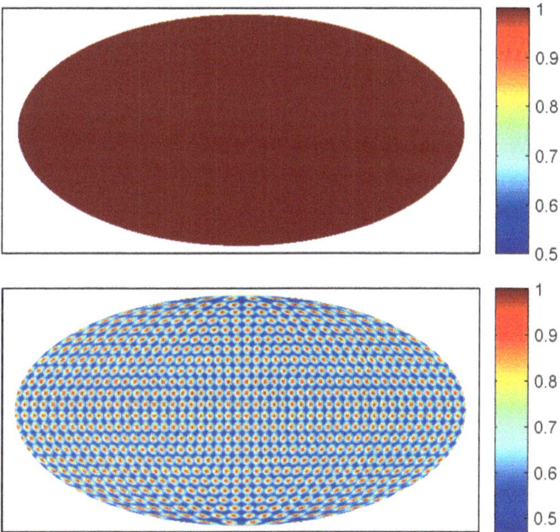

Fig. 6.5 Interpolation of the constant function $F \equiv 1$ on a Reuter grid with 786 points (see Fig. 6.4) by an Abel–Poisson spline for $h = 0.75$ (*top*) and $h = 0.95$ (*bottom*): Obviously, artefacts occur, if the hat-width of the basis functions is too small in relation to the gaps in the point grid. This has to be taken into account, when the sequence (A_n) is chosen. Note that equal colorbars were chosen here for a better comparison

6.5 Questions for Understanding

- What is a reproducing kernel?
- When does it exist?
- If it exists, is it unique?
- Is there a representation for a reproducing kernel?
- Does $\mathrm{L}^2(\Omega)$ have a reproducing kernel?
- We constructed so-called Sobolev spaces $\mathscr{H}(\Omega)$. Why do we need them?
- How is $\mathscr{H}(\Omega)$ defined? What is $\mathscr{H}_s(\Omega)$?
- Under particular conditions, the Sobolev spaces can be embedded into other classical function spaces. What do you know about this?
- What does summability mean? Can you give a summable and a non-summable example?

- Reproducing kernels are represented in terms of a series. This is a problem for numerical implementations. Do you know an example of a reproducing kernel where such a problem does not occur?
- How did we define spherical splines?
- What is a well-posed problem? What is an ill-posed problem?
- Is spherical spline interpolation well-posed or ill-posed? Why?
- In the context of spherical Fourier analysis, we found out that not every point grid can be used (by the way, do you remember more about this?). What is the situation for spherical spline interpolation?
- The concept of spherical splines is very different in comparison to the 1D cubic splines we investigated before. What are nevertheless the main features, that both have in common (and that justify the name "spline")?
- We proved another "nice" property of spherical splines in this context. What is it?
- What does the Shannon sampling theorem say? In which cases is it useful?
- What is the difference between spline interpolation and spline approximation? When should we prefer spline approximation?

Chapter 7
Spherical Wavelet Analysis

In this section, we will study the spherical wavelets introduced by W. Freeden, M. Schreiner, and U. Windheuser in [66,74,77,200], which have been applied to numerous geodetic and geophysical problems and have been adapted for several other classes of tasks (e.g., the approximation of vectorial and tensorial functions) since their introduction. These wavelets have similar features in comparison to Euclidean wavelets, although several aspects become more complicated in connection with the change of the geometry and the underlying orthonormal basis. An essential tool for an Approximate Identity is again a convolution.

Before we continue, we should acknowledge that there exist numerous other multiscale methods on the sphere, where some of them also proved to be applicable to geophysical problems. For instance, some principles from image processing motivated the spherical wavelet technique in [169]. These biorthogonal wavelets are based on a so-called lifting scheme and a hierarchical subdivision of the sphere and turned out to be a useful tool for topographic and other maps. A similar application is possible for the, however, essentially different concept of spherical wavelets which were developed in [10, 11] by tracing a group-theoretic approach. Moreover, the Poisson wavelets on the sphere have been applied to geomagnetic and gravity field modeling; see, for example, [24, 95–97]. Another alternative wavelet approach where more general surfaces than a sphere can be handled (this includes the real Earth's surface or subsurfaces of it) is based on potential theoretic concepts such as layer potentials or Green's functions. These wavelets, which were developed, for example, in [64, 67, 79, 80], can be used for the analysis of different kinds of geophysical data, such as gravity and magnetic field data. Wavelets on nonspherical surfaces for the modeling or downward continuation of gravity data can also be constructed based on the Runge–Walsh theorem from potential theory. For further details, see, for example, [69, 72].

An essential feature of wavelets is their localization (in contrast to polynomial bases). In this line of thought, several approaches to construct spherical trial functions have been developed so far. In [106], optimally localized wavelets on the sphere are constructed by means of an uncertainty principle. A different localization measure is used in [133] to compute Approximate Identities on the sphere, which

V. Michel, *Lectures on Constructive Approximation*, Applied and Numerical Harmonic Analysis, DOI 10.1007/978-0-8176-8403-7_7,

use kernels with an optimal localization. Furthermore, the Slepian functions in Chap. 8 are optimally localized on a region that one may choose almost arbitrarily. Moreover, the aspect of fast computations is particularly treated, for example, in [23, 100].

There are by far more concepts of multiscale methods on the sphere. For instance, in [31] a survey on several multiresolution analyses on the sphere is given. Further examples of spherical multiscale techniques, where some of them are also listed in [31], are investigated in [36, 38, 39, 70, 82, 111, 142, 148, 164, 194].

This list of concepts and references, however, is most probably not complete.

7.1 Convolutions

In Definition 3.31 on p. 67, we defined a general convolution of a kernel $\Phi \in \mathrm{L}^2(D\times D)$ and a function $F \in \mathrm{L}^2(D)$. In the spherical case $D = \Omega$, the particular case where the kernel can be associated to a zonal function is relevant.

Definition 7.1. Let $G \in \mathrm{L}^q[-1,1]$ and $F \in \mathrm{L}^p(\Omega)$ be given, where $p,q \in]1,+\infty[$ with $\frac{1}{p}+\frac{1}{q}=1$. Then the **spherical convolution** $G*F : \Omega \to \mathbb{R}$ is defined by

$$(G*F)(\xi) := \int_\Omega G(\xi\cdot\eta)F(\eta)\,\mathrm{d}\omega(\eta), \quad \xi \in \Omega\,. \tag{7.1}$$

From (3.156), we know that $G*F \in \mathrm{L}^2(\Omega)$ for $p=q=2$. A more general result can be derived by using the Hölder inequality.

Theorem 7.2. *Let $G \in \mathrm{L}^q[-1,1]$ and $F \in \mathrm{L}^p(\Omega)$, where $p,q \in]1,+\infty[$ with $\frac{1}{p}+\frac{1}{q}=1$. Then the spherical convolution $G*F$ exists almost everywhere on Ω and $G*F \in \mathrm{L}^r(\Omega)$ for all $r \in [1,+\infty[$. Moreover,*

$$\|G*F\|_{\mathrm{L}^r(\Omega)} \le (2\pi)^{1/q+1/r}\,2^{1/r}\|G\|_{\mathrm{L}^q[-1,1]}\|F\|_{\mathrm{L}^p(\Omega)}. \tag{7.2}$$

Proof. For every $r \in [1,+\infty[$, we get by using Definition 7.1, the triangle inequality for integrals, the Hölder inequality (Theorem 2.5 on p. 17), and Theorem 4.16 on p. 93 (note that we use different colors for the different integrals for the sake of clearness):

$$\begin{aligned}
&\|G*F\|_{\mathrm{L}^r(\Omega)}\\
&= \left(\int_\Omega |(G*F)(\xi)|^r\,\mathrm{d}\omega(\xi)\right)^{1/r}\\
&= \left(\int_\Omega \left|\int_\Omega G(\xi\cdot\eta)F(\eta)\,\mathrm{d}\omega(\eta)\right|^r\,\mathrm{d}\omega(\xi)\right)^{1/r}\\
&\le \left[\int_\Omega \left(\int_\Omega |G(\xi\cdot\eta)F(\eta)|\,\mathrm{d}\omega(\eta)\right)^r\,\mathrm{d}\omega(\xi)\right]^{1/r}
\end{aligned}$$

$$\leq \left[\int_\Omega \left(\int_\Omega |G(\xi\cdot\eta)|^q \,\mathrm{d}\omega(\eta)\right)^{r/q} \left(\int_\Omega |F(\eta)|^p \,\mathrm{d}\omega(\eta)\right)^{r/p} \mathrm{d}\omega(\xi)\right]^{1/r}$$

$$= \left[\int_\Omega \left(2\pi\int_{-1}^1 |G(t)|^q \,\mathrm{d}t\right)^{r/q} \left(\int_\Omega |F(\eta)|^p \,\mathrm{d}\omega(\eta)\right)^{r/p} \mathrm{d}\omega(\xi)\right]^{1/r}$$

$$= (2\pi)^{1/q} \|G\|_{\mathrm{L}^q[-1,1]} \|F\|_{\mathrm{L}^p(\Omega)} \left(\int_\Omega \mathrm{d}\omega(\xi)\right)^{1/r}$$

$$= (2\pi)^{1/q} \|G\|_{\mathrm{L}^q[-1,1]} \|F\|_{\mathrm{L}^p(\Omega)} (4\pi)^{1/r}. \tag{7.3}$$

Since the right-hand side is finite, $G*F$ is an element of $\mathrm{L}^r(\Omega)$, and due to the definition of this function space, $G*F$ exists (as a function) almost everywhere on Ω. □

Further properties can be proved by using an important theorem of spherical analysis: the Funk–Hecke formula.

Theorem 7.3 (Funk-Hecke Formula). *Let $G \in \mathrm{L}^1[-1,1]$ and $n \in \mathbb{N}_0$ be arbitrary. Then*

$$\int_\Omega G(\xi\cdot\zeta) P_n(\eta\cdot\zeta)\,\mathrm{d}\omega(\zeta) = G^\wedge(n) P_n(\xi\cdot\eta) \tag{7.4}$$

for all $\xi, \eta \in \Omega$.

Note that $G^\wedge(n)$ is the n-th Legendre coefficient of G (see Definition 3.15 on p. 45) and P_n is the Legendre polynomial of degree n (see Definition 3.10).

Proof.

(1) Properties of the left-hand side:
We define the function $H : \Omega \times \Omega \to \mathbb{R}$ by

$$H(\xi,\eta) := \int_\Omega G(\xi\cdot\zeta) P_n(\eta\cdot\zeta)\,\mathrm{d}\omega(\zeta), \quad \xi,\eta \in \Omega. \tag{7.5}$$

If $A \in \mathrm{O}(3)$, that is, A is a 3×3 orthogonal matrix ($A^\mathrm{T}A = I$), we get, in analogy to the considerations in the proof of Theorem 4.16 on p. 93, that

$$\begin{aligned} H(A\xi, A\eta) &= \int_\Omega G(A\xi\cdot\zeta) P_n(A\eta\cdot\zeta)\,\mathrm{d}\omega(\zeta) \\ &= \int_\Omega G\left(\xi\cdot(A^\mathrm{T}\zeta)\right) P_n\left(\eta\cdot(A^\mathrm{T}\zeta)\right)\mathrm{d}\omega(\zeta) \\ &= \int_\Omega G(\xi\cdot\zeta) P_n(\eta\cdot\zeta)\,\mathrm{d}\omega(\zeta) \\ &= H(\xi,\eta). \end{aligned} \tag{7.6}$$

Hence, H is invariant with respect to orthogonal transformations.

(2) The function $H(\xi,\cdot)$:
Keep an arbitrary unit vector $\xi \in \Omega$ fixed. Then $H(\xi,\cdot)$ is a spherical harmonic of degree n since the addition theorem (Theorem 5.11 on p. 103) yields

$$H(\xi,\cdot) = \frac{4\pi}{2n+1} \sum_{j=1}^{2n+1} \int_\Omega G(\xi\cdot\zeta)\, Y_{n,j}(\zeta)\, \mathrm{d}\omega(\zeta)\, Y_{n,j}(\cdot) \in \mathrm{Harm}_n(\Omega). \tag{7.7}$$

Note that the occurring integrals are finite since $Y_{n,j}$ and P_n are bounded on Ω and $[-1,1]$, respectively, and $G \in \mathrm{L}^1[-1,1]$.

(3) A zonal function:
In analogy to parts 2 and 3 of the proof of the addition theorem (Theorem 5.11), we can now conclude that $H(\xi,\eta)$ depends on $\xi\cdot\eta$ only and is a 1D-polynomial of degree n.

(4) Orthogonality:
For each $n \in \mathbb{N}_0$, we set

$$H_n(\xi\cdot\eta) := \int_\Omega G(\xi\cdot\zeta)\, P_n(\eta\cdot\zeta)\, \mathrm{d}\omega(\zeta), \quad \xi,\eta \in \Omega. \tag{7.8}$$

Using Theorem 4.16 on p. 93, we get for $n \neq m$, due to part 2 of this proof and Theorem 5.9 on p. 102,

$$\begin{aligned} \langle H_n, H_m\rangle_{\mathrm{L}^2[-1,1]} &= \int_{-1}^{1} H_n(t)\, H_m(t)\, \mathrm{d}t \\ &= \frac{1}{2\pi} \int_\Omega H_n(\xi\cdot\eta)\, H_m(\xi\cdot\eta)\, \mathrm{d}\omega(\eta) \\ &= 0\,, \end{aligned} \tag{7.9}$$

where $\xi \in \Omega$ is arbitrary.

Hence, Theorem 3.9 on p. 40 and Definition 3.10 imply that there exist constants γ_n, $n \in \mathbb{N}_0$, such that

$$H_n(t) = \gamma_n P_n(t) \quad \forall t \in [-1,1]. \tag{7.10}$$

(5) Determination of the constants:
If $\xi = \eta$, then Theorem 4.16 yields

$$\begin{aligned} \gamma_n &= \gamma_n P_n(1) \\ &= H_n(\xi\cdot\xi) \\ &= \int_\Omega G(\xi\cdot\zeta)\, P_n(\xi\cdot\zeta)\, \mathrm{d}\omega(\zeta) \\ &= 2\pi \int_{-1}^{1} G(t)\, P_n(t)\, \mathrm{d}t \\ &= G^\wedge(n) \end{aligned} \tag{7.11}$$

for all $n \in \mathbb{N}_0$. Hence,

$$H_n(\xi \cdot \eta) = \int_\Omega G(\xi \cdot \zeta)\, P_n(\eta \cdot \zeta)\, \mathrm{d}\omega(\zeta) = G^\wedge(n)\, P_n(\xi \cdot \eta) \tag{7.12}$$

for all $n \in \mathbb{N}_0$ and all $\xi, \eta \in \Omega$. □

Note that $\mathrm{L}^p[-1,1] \subset \mathrm{L}^1[-1,1]$ for all $p \in [1,+\infty[$ due to Theorem 2.6 on p. 17. Hence, the Funk–Hecke formula is applicable to all functions on $[-1,1]$ that we might want to use for spherical convolutions.

Corollary 7.4. *Let $G \in \mathrm{L}^q[-1,1]$ with $q \in]1,+\infty[$ and $Y_n \in \mathrm{Harm}_n(\Omega)$ for a given $n \in \mathbb{N}_0$. Then*

$$(G * Y_n)(\xi) = G^\wedge(n)\, Y_n(\xi) \tag{7.13}$$

for all $\xi \in \Omega$.

Proof. We use the reproducing kernel of $\mathrm{Harm}_n(\Omega)$ (see Theorem 5.16 on p. 111), Fubini's theorem, the Funk–Hecke formula, and again Theorem 5.16 to derive (note again the use of different colors for the different integrals)

$$\begin{aligned}(G * Y_n)(\xi) &= \int_\Omega G(\xi \cdot \eta)\, Y_n(\eta)\, \mathrm{d}\omega(\eta)\\ &= \int_\Omega G(\xi \cdot \eta) \int_\Omega Y_n(\zeta) \frac{2n+1}{4\pi} P_n(\zeta \cdot \eta)\, \mathrm{d}\omega(\zeta)\, \mathrm{d}\omega(\eta)\\ &= \frac{2n+1}{4\pi} \int_\Omega Y_n(\zeta) \int_\Omega G(\xi \cdot \eta)\, P_n(\zeta \cdot \eta)\, \mathrm{d}\omega(\eta)\, \mathrm{d}\omega(\zeta)\\ &= \frac{2n+1}{4\pi} \int_\Omega Y_n(\zeta)\, G^\wedge(n)\, P_n(\xi \cdot \zeta)\, \mathrm{d}\omega(\zeta)\\ &= G^\wedge(n)\, Y_n(\xi) \end{aligned} \tag{7.14}$$

for all $n \in \mathbb{N}_0$ and all $\xi \in \Omega$.

This corollary now allows us to derive a main feature of the spherical convolution, which actually justifies the name "convolution," since the 1D-convolution has an analogous property. □

Theorem 7.5. *Let $G \in \mathrm{L}^q[-1,1]$ and $F \in \mathrm{L}^p(\Omega)$ with $p, q \in]1,+\infty[$ and $\frac{1}{p} + \frac{1}{q} = 1$ be given, and let $Y_n \in \mathrm{Harm}_n(\Omega)$, $n \in \mathbb{N}_0$, be an arbitrary spherical harmonic. Then*

$$\langle G * F, Y_n \rangle_{\mathrm{L}^2(\Omega)} = G^\wedge(n)\, \langle F, Y_n \rangle_{\mathrm{L}^2(\Omega)}\,. \tag{7.15}$$

Proof. This result is a consequence of the definition of the spherical convolution (Definition 7.1), Fubini's theorem, and Corollary 7.4:

$$\begin{aligned}\langle G * F, Y_n \rangle_{\mathrm{L}^2(\Omega)} &= \int_\Omega (G * F)(\xi)\, Y_n(\xi)\, \mathrm{d}\omega(\xi)\\ &= \int_\Omega \int_\Omega G(\xi \cdot \eta)\, F(\eta)\, \mathrm{d}\omega(\eta)\, Y_n(\xi)\, \mathrm{d}\omega(\xi)\end{aligned}$$

$$
\begin{aligned}
&= \int_\Omega F(\eta) \int_\Omega G(\xi\cdot\eta) Y_n(\xi)\, d\omega(\xi)\, d\omega(\eta) \\
&= \int_\Omega F(\eta) G^\wedge(n) Y_n(\eta)\, d\omega(\eta) \\
&= G^\wedge(n) \langle F, Y_n \rangle_{L^2(\Omega)} \,. \qquad (7.16)
\end{aligned}
$$

□

Corollary 7.6. *Let $G \in L^q[-1,1]$ and $F \in L^p(\Omega)$ with $p,q \in]1,+\infty[$ and $\frac{1}{p}+\frac{1}{q}=1$. Then the Fourier series of $G*F \in L^2(\Omega)$ is given (in the sense of $L^2(\Omega)$) by*

$$
G*F = \sum_{n=0}^{\infty} \sum_{j=1}^{2n+1} G^\wedge(n) F^\wedge(n,j) Y_{n,j}. \qquad (7.17)
$$

Equation (7.17) is a key to the introduction of Approximate Identities and wavelets on the sphere Ω. If $F \in L^2(\Omega)$ is our signal, we can manipulate its Fourier coefficients by convolving it with a function $G \in L^2[-1,1]$ whose Legendre coefficients we previously designed to fulfill our purposes. For instance, if we want to suppress the Fourier coefficients of F corresponding to high degrees (e.g., because they are too noisy), we choose a monotonically decreasing sequence $(G^\wedge(n))_{n\in\mathbb{N}_0}$.[1] By varying the speed of the decay of this sequence, we get low-pass filters of different "strengths." In some limit, that we will have to discuss further, we could get the whole signal F again. This provides us with the prerequisites for an Approximate Identity (which itself is a starting point for wavelets here).

Before we can proceed, we also have to define how to convolve two zonal functions.

Definition 7.7. Let $G,H \in L^2[-1,1]$ be given. Then the **convolution** $G*H : [-1,1] \to \mathbb{R}$ is defined by

$$
(G*H)(\xi\cdot\eta) = \int_\Omega G(\xi\cdot\zeta) H(\zeta\cdot\eta)\, d\omega(\zeta), \quad \xi,\eta \in \Omega. \qquad (7.18)
$$

Theorem 7.8. *For all $G,H \in L^2[-1,1]$, the convolution $G*H$ is, indeed, a zonal function in $L^2[-1,1]$ and the Legendre coefficients satisfy*

$$
(G*H)^\wedge(n) = G^\wedge(n) H^\wedge(n) \quad \forall n \in \mathbb{N}_0. \qquad (7.19)
$$

Proof.

(1) A zonal function:
We consider the function $K : \Omega\times\Omega \to \mathbb{R}$ defined by

$$
K(\xi,\eta) := \int_\Omega G(\xi\cdot\zeta) H(\zeta\cdot\eta)\, d\omega(\zeta), \quad \xi,\eta \in \Omega. \qquad (7.20)
$$

---

[1] Note that $\lim_{n\to\infty} G^\wedge(n) = 0$ is necessary for the convergence of the Legendre series expansion of G. This also contributes to the attenuation of the signal. Furthermore, we can choose G such that there exists $N \in \mathbb{N}$ with $G^\wedge(n) = 0$ for all $n \geq N$.

In analogy to part 1 of the proof of the Funk–Hecke formula (Theorem 7.3), we can show that $K(A\xi, A\eta) = K(\xi, \eta)$ for all $\xi, \eta \in \Omega$ and every orthogonal 3×3-matrix A. Then part 2 of the proof of the addition theorem (Theorem 5.11 on p. 103) implies that $K(\xi, \eta)$ depends on the inner product $\xi \cdot \eta$ only.

(2) $G * H \in \mathrm{L}^2[-1,1]$:
Once again, we use Theorem 4.16 (see p. 93) and once more the Cauchy–Schwarz inequality:

$$\begin{aligned}
\|G*H\|^2_{\mathrm{L}^2[-1,1]} &= \frac{1}{2\pi}\int_\Omega [(G*H)(\xi\cdot\eta)]^2 \,\mathrm{d}\omega(\eta) \\
&= \frac{1}{2\pi}\int_\Omega \left[\int_\Omega G(\xi\cdot\zeta)H(\zeta\cdot\eta)\,\mathrm{d}\omega(\zeta)\right]^2 \mathrm{d}\omega(\eta) \\
&\leq \frac{1}{2\pi}\int_\Omega \int_\Omega G(\xi\cdot\zeta)^2\,\mathrm{d}\omega(\zeta) \int_\Omega H(\zeta\cdot\eta)^2\,\mathrm{d}\omega(\zeta)\,\mathrm{d}\omega(\eta) \\
&= 2\pi \int_\Omega \int_{-1}^1 G(t)^2\,\mathrm{d}t \int_{-1}^1 H(t)^2\,\mathrm{d}t\,\mathrm{d}\omega(\eta) \\
&= 8\pi^2 \|G\|^2_{\mathrm{L}^2[-1,1]} \|H\|^2_{\mathrm{L}^2[-1,1]} < +\infty\,,
\end{aligned} \tag{7.21}$$

where $\xi \in \Omega$ is arbitrary.

(3) The Legendre coefficients:
Equation (7.19) can be derived from the definition of the Legendre coefficients (Definition 3.15 on p. 45), Theorem 4.16 (with an arbitrary $\xi \in \Omega$), Definition 7.7, Fubini's theorem, and the Funk–Hecke formula (Theorem 7.3, applied twice) and the definition of the Legendre polynomials (Definition 3.10 on p. 41) as follows:

$$\begin{aligned}
(G*H)^\wedge(n) &= 2\pi \int_{-1}^1 (G*H)(t)\,P_n(t)\,\mathrm{d}t \\
&= \int_\Omega (G*H)(\xi\cdot\eta)\,P_n(\xi\cdot\eta)\,\mathrm{d}\omega(\eta) \\
&= \int_\Omega \int_\Omega G(\xi\cdot\zeta)H(\zeta\cdot\eta)\,\mathrm{d}\omega(\zeta)\,P_n(\xi\cdot\eta)\,\mathrm{d}\omega(\eta) \\
&= \int_\Omega G(\xi\cdot\zeta)\int_\Omega H(\zeta\cdot\eta)\,P_n(\xi\cdot\eta)\,\mathrm{d}\omega(\eta)\,\mathrm{d}\omega(\zeta) \\
&= \int_\Omega G(\xi\cdot\zeta)\,H^\wedge(n)\,P_n(\zeta\cdot\xi)\,\mathrm{d}\omega(\zeta) \\
&= G^\wedge(n)\,H^\wedge(n)\,P_n(\xi\cdot\xi) \\
&= G^\wedge(n)\,H^\wedge(n)
\end{aligned} \tag{7.22}$$

for all $n \in \mathbb{N}_0$. □

We can now summarize Corollary 7.6 and Theorem 7.8 as follows: If $G,H \in \mathrm{L}^2[-1,1]$ and $F \in \mathrm{L}^2(\Omega)$, then

$$\begin{aligned} G*(H*F) &= \sum_{n=0}^{\infty}\sum_{j=1}^{2n+1} G^\wedge(n)H^\wedge(n)F^\wedge(n,j)Y_{n,j} \\ &= (G*H)*F \\ &= (H*G)*F \\ &= H*(G*F) \end{aligned} \tag{7.23}$$

in the sense of $\mathrm{L}^2(\Omega)$.

Furthermore, iterated kernels can be defined as well.

Definition 7.9. If $G \in \mathrm{L}^2[-1,1]$, then the k-th **iterated kernel** $G^{(k)}$ for $k \in \mathbb{N}$ is defined recursively by

$$\begin{aligned} G^{(1)} &:= G\,, \\ G^{(k+1)} &:= G^{(k)} * G \quad \forall k \in \mathbb{N}\,. \end{aligned} \tag{7.24}$$

Theorem 7.10. *If $G \in \mathrm{L}^2[-1,1]$ and $k \in \mathbb{N}$, then the Legendre coefficients of the k-th iterated kernel $G^{(k)}$ satisfy*

$$\left(G^{(k)}\right)^\wedge(n) = \left(G^\wedge(n)\right)^k \quad \forall n \in \mathbb{N}_0. \tag{7.25}$$

Theorem 7.10 is an immediate consequence of Theorem 7.8.

7.2 Scaling Functions

Scaling functions on the sphere are used to construct an Approximate Identity. We first define a scaling function (see also [128] for a generalized approach).

Definition 7.11. A family $\{\Phi_J\}_{J\in\mathbb{N}_0} \subset \mathrm{L}^2[-1,1]$ is called a (spherical) **scaling function** if the Legendre coefficients of the functions Φ_J satisfy the following conditions:

(S1) For all $n,J \in \mathbb{N}_0$, we have $\Phi_J^\wedge(n) \leq \Phi_{J+1}^\wedge(n)$. In other words, for all $n \in \mathbb{N}_0$, the sequence $(\Phi_J^\wedge(n))_{J\in\mathbb{N}_0}$ is monotonically increasing.

(S2) $\lim_{J\to\infty} \Phi_J^\wedge(n) = 1$ for all $n \in \mathbb{N}_0$.

(S3) $\Phi_J^\wedge(n) \geq 0$ for all $n,J \in \mathbb{N}_0$.

Remark 7.12. Some comments should be made here:

(a) It depends on your particular preference if you call each Φ_J a scaling function or if you use this name for the whole family $\{\Phi_J\}_{J\in\mathbb{N}_0}$. You will find both in the literature.

(b) Based on many years of experience in teaching scaling functions, wavelets etc., let me remind you that (S1) is a condition for the monotonicity of the *Legendre coefficients*. In no way does it include any monotonicity property of the scaling functions themselves.
(c) Note that there are actually four conditions on the scaling function, since the requirement $\Phi_J \in \mathrm{L}^2[-1,1]$ is not trivial, as you will see below.
(d) More precisely, these are scale-discrete scaling functions since we make definite steps "$+1$" from one scale $J \in \mathbb{N}_0$ to the next one $J+1$. There is also a scale-continuous theory (see, e.g., [76–78, 128]), which is, however, not useful for numerical purposes.
(e) Note that, though the spherical scaling functions correspond to functions on the interval $[-1,1]$, the reference to the sphere makes sense, as you will see soon. We will namely use them as zonal functions $\Omega \ni \eta \mapsto \Phi_J(\xi \cdot \eta)$ with (fixed) $\xi \in \Omega$.

We will now investigate a way of constructing a scaling function out of a so-called generator. This corresponds to the mentioned ambiguity in the use of the name "scaling function" since, in this case, one single function (e.g., Φ_0) defines the whole family.

Definition 7.13. A function $\gamma_0 : \mathbb{R}_0^+ \to \mathbb{R}$ is called **admissible** if it satisfies the **admissibility condition**:

$$\sum_{n=0}^{\infty} \frac{2n+1}{4\pi} \left(\sup_{x \in [n,n+1[} |\gamma_0(x)| \right)^2 < +\infty. \tag{7.26}$$

We will later define the scaling function by using a family of functions $\varphi_J : \mathbb{R}_0^+ \to \mathbb{R}$, $J \in \mathbb{N}_0$, and setting $\Phi_J^\wedge(n) := \varphi_J(n)$ for all $J, n \in \mathbb{N}_0$. We will see first that the requirement $\Phi_J \in \mathrm{L}^2[-1,1]$ corresponds to the admissibility of φ_J.

Theorem 7.14. *If $\gamma_0 : \mathbb{R}_0^+ \to \mathbb{R}$ is admissible, then a function $\Gamma_0 \in \mathrm{L}^2[-1,1]$ is defined by setting the Legendre coefficients as follows:*

$$\Gamma_0^\wedge(n) := \gamma_0(n) \quad \forall n \in \mathbb{N}_0. \tag{7.27}$$

Proof. We have to verify that $\|\Gamma_0\|_{\mathrm{L}^2[-1,1]} < +\infty$. Using the Parseval identity (see Theorem 2.18 on p. 21), Theorem 3.14 on p. 44, and Definition 3.15, we obtain

$$\begin{aligned}
\|\Gamma_0\|_{\mathrm{L}^2[-1,1]}^2 &= \sum_{n=0}^{\infty} \left\langle \Gamma_0, \|P_n\|_{\mathrm{L}^2[-1,1]}^{-1} P_n \right\rangle_{\mathrm{L}^2[-1,1]}^2 \\
&= \sum_{n=0}^{\infty} \frac{2n+1}{2} \langle \Gamma_0, P_n \rangle_{\mathrm{L}^2[-1,1]}^2
\end{aligned}$$

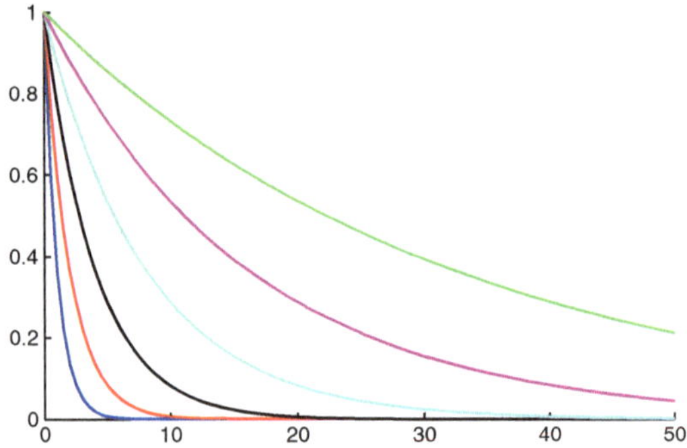

Fig. 7.1 Dilation of the function $\gamma_0(x) := \mathrm{e}^{-x}$, $x \in \mathbb{R}_0^+$, for $J = 0$ (*blue*), $J = 1$ (*red*), $J = 2$ (*black*), $J = 3$ (*cyan*), $J = 4$ (*purple*), and $J = 5$ (*green*)

$$= \frac{1}{2\pi} \sum_{n=0}^{\infty} \frac{2n+1}{4\pi} \left(\Gamma_0^\wedge(n)\right)^2$$

$$\leq \frac{1}{2\pi} \sum_{n=0}^{\infty} \frac{2n+1}{4\pi} \left(\sup_{x \in [n,n+1[} |\gamma_0(x)| \right)^2 < +\infty . \tag{7.28}$$

□

The trick to construct a whole family of admissible functions out of one admissible function is the so-called dilation (see also Fig. 7.1).

Definition 7.15. The **dilation operator** D_J, where $J \in \mathbb{N}_0$, maps a given function $\gamma_0 : \mathbb{R}_0^+ \to \mathbb{R}$ to a function $\mathrm{D}_J \gamma_0 := \gamma_J$ with

$$\gamma_J(x) := \gamma_0\left(2^{-J} x\right) \quad \forall x \in \mathbb{R}_0^+ . \tag{7.29}$$

γ_J is called a **dilated function**.

Theorem 7.16. *If $\gamma_0 : \mathbb{R}_0^+ \to \mathbb{R}$ is admissible, then every corresponding dilated function γ_J, $J \in \mathbb{N}_0$, is admissible.*

Proof. We prove the theorem by induction. By the assumption of the theorem, γ_0 is admissible. Now let γ_J be admissible for an arbitrary but fixed $J \in \mathbb{N}_0$. Then we proceed as follows:

$$A_{J+1,N} := \sum_{n=0}^{N} \frac{2n+1}{4\pi} \left(\sup_{y \in [n,n+1[} |\gamma_{J+1}(y)| \right)^2$$

$$= \sum_{n=0}^{N} \frac{2n+1}{4\pi} \sup_{y \in [n,n+1[} \left(\gamma_J\left(\frac{y}{2}\right)\right)^2$$

$$\overset{x:=y/2}{=} \sum_{n=0}^{N} \frac{2n+1}{4\pi} \sup_{x \in [n/2,(n+1)/2[} (\gamma_J(x))^2$$

$$= \sum_{\substack{n=0\\ n \text{ even}}}^{N} \frac{2n+1}{4\pi} \sup_{x\in[n/2,(n+1)/2[} (\gamma_J(x))^2$$

$$+ \sum_{\substack{n=0\\ n \text{ odd}}}^{N} \frac{2n+1}{4\pi} \sup_{x\in[n/2,(n+1)/2[} (\gamma_J(x))^2 . \tag{7.30}$$

We now consider, as we indicated in the previous line, all even n separately from all odd n:

- Every even n is represented by $n = 2k$, where $n = 0,\dots,N$ corresponds (for even n) to $k = 0,\dots,\lfloor\frac{N}{2}\rfloor$ with $\lfloor\cdot\rfloor$ representing the Gaussian bracket [see (3.165) on p. 71].
- Every odd n is representable by $n = 2k+1$.
  If N is odd, then $k = \lfloor\frac{N}{2}\rfloor$ yields $n = 2\lfloor\frac{N}{2}\rfloor + 1 = (N-1)+1 = N$.
  However, if N is even, then $k = \lfloor\frac{N}{2}\rfloor$ yields $n = 2\lfloor\frac{N}{2}\rfloor + 1 = N+1$. Hence, in this case, we get one more summand such that we obtain an inequality.

Thus,

$$\begin{aligned} A_{J+1,N} &\le \sum_{k=0}^{\lfloor N/2\rfloor} \frac{4k+1}{4\pi} \sup_{x\in[k,k+1/2[} (\gamma_J(x))^2 + \sum_{k=0}^{\lfloor N/2\rfloor} \frac{4k+3}{4\pi} \sup_{x\in[k+1/2,k+1[} (\gamma_J(x))^2 \\ &\le \sum_{k=0}^{\lfloor N/2\rfloor} \underbrace{\frac{4k+3}{4\pi}}_{\le\frac{6k+3}{4\pi}} \underbrace{\left(\sup_{x\in[k,k+1/2[} (\gamma_J(x))^2 + \sup_{x\in[k+1/2,k+1[} (\gamma_J(x))^2 \right)}_{\le 2\sup_{x\in[k,k+1[}(\gamma_J(x))^2} \\ &\le 6 \sum_{k=0}^{\infty} \frac{2k+1}{4\pi} \sup_{x\in[k,k+1[} (\gamma_J(x))^2 , \end{aligned} \tag{7.31}$$

where the right-hand side is finite due to the assumption of the induction. Hence, $\lim_{N\to\infty} A_{J+1,N}$ exists and is finite. □

Up to now, we have managed to construct a family $\{\Gamma_J\}_{J\in\mathbb{N}_0} \subset \mathrm{L}^2[-1,1]$ out of a function $\gamma_0 : \mathbb{R}_0^+ \to \mathbb{R}$ by requiring the admissibility condition for γ_0 and setting $\Gamma_J^\wedge(n) := \gamma_J(n) = \gamma_0\left(2^{-J}n\right)$ for all $J,n \in \mathbb{N}_0$. We will also use this for the wavelets later, which is why we used the somewhat "neutral" notation with the letters γ and Γ up to this point.

We need further conditions to make sure that (S1)–(S3) are satisfied too.

Definition 7.17. A function $\varphi_0 : \mathbb{R}_0^+ \to \mathbb{R}$ is called a **generator of a scaling function** if it is admissible and additionally satisfies the following conditions:

(GS1) $\varphi_0(0) = 1$.
(GS2) φ_0 is continuous at 0.
(GS3) φ_0 is monotonically decreasing.

Theorem 7.18. *If $\varphi_0 : \mathbb{R}_0^+ \to \mathbb{R}$ is a generator of a scaling function, then a scaling function $\{\Phi_J\}_{J\in\mathbb{N}_0} \subset \mathrm{L}^2[-1,1]$ is defined by setting*

$$\Phi_J^\wedge(n) := \varphi_J(n) = \varphi_0\left(2^{-J}n\right) \quad \forall n, J \in \mathbb{N}_0. \tag{7.32}$$

Proof. From Theorems 7.14 and 7.16, we already know that $\Phi_J \in \mathrm{L}^2[-1,1]$ for all $J \in \mathbb{N}_0$. Now let us verify the remaining conditions of Definition 7.11:

(S1) Let $n, J \in \mathbb{N}_0$ be arbitrary. Then

$$\Phi_J^\wedge(n) = \varphi_0\left(2^{-J}n\right) \overset{\text{(GS3)}}{\leq} \varphi_0\left(2^{-J-1}n\right) = \Phi_{J+1}^\wedge(n). \tag{7.33}$$

(S2) For every $n \in \mathbb{N}_0$, we obtain

$$\lim_{J\to\infty} \Phi_J^\wedge(n) = \lim_{J\to\infty} \varphi_0\left(2^{-J}n\right) \overset{\text{(GS2)}}{=} \varphi_0(0) \overset{\text{(GS1)}}{=} 1. \tag{7.34}$$

(S3) The admissibility condition for φ_0 says that

$$\sum_{n=0}^{\infty} \frac{2n+1}{4\pi} \left(\sup_{x\in[n,n+1[} |\varphi_0(x)| \right)^2 < +\infty. \tag{7.35}$$

It is an elementary conclusion from real analysis that

$$\lim_{n\to\infty} \left(\sup_{x\in[n,n+1[} |\varphi_0(x)| \right)^2 = 0, \tag{7.36}$$

where we can, certainly, skip the power of 2 here. This implies that

$$\lim_{x\to\infty} \varphi_0(x) = 0. \tag{7.37}$$

Referring to (GS3), we observe that $\varphi_0(x) \geq 0$ for all $x \in \mathbb{R}_0^+$. Hence, $\Phi_J^\wedge(n) = \varphi_0(2^{-J}n) \geq 0$ for all $n, J \in \mathbb{N}_0$. □

We will now study several common scaling functions and try to classify them. Note that the Clenshaw algorithm (Theorem 3.17 on p. 47) can be used to compute the scaling functions.

Definition 7.19. A scaling function $\{\Phi_J\}_{J\in\mathbb{N}_0} \subset \mathrm{L}^2[-1,1]$ is called **bandlimited** if, for every $J \in \mathbb{N}_0$, the coefficients $\Phi_J^\wedge(n)$, $n \in \mathbb{N}_0$, are nonvanishing for only a finite number of degrees n. Otherwise, $\{\Phi_J\}_{J\in\mathbb{N}_0}$ is called **non-bandlimited**.

All bandlimited scaling functions $\{\Phi_J\}_{J\in\mathbb{N}_0}$ have something in common: Each Φ_J is a 1D-polynomial, and every $\Phi_J * F$ with $F \in \mathrm{L}^2(\Omega)$ is a polynomial on Ω.

Example 7.20. We now classify the following scaling functions depending on their existing or nonexisting origin from a generator and their bandlimited or non-bandlimited character.

(A) **Scaling Functions Constructed by a Generator**
(A1) **Bandlimited**
(A1.1) **Shannon Scaling Function**

In some sense, the generator of the Shannon scaling function represents the easiest way of defining a generator of a scaling function:

$$\varphi_0(x) := \begin{cases} 1, 0 \leq x < 1, \\ 0, 1 \leq x. \end{cases} \tag{7.38}$$

Consequently,

$$\Phi_J(t) = \sum_{n=0}^{2^J-1} \frac{2n+1}{4\pi} P_n(t) \quad \forall t \in [-1,1]. \tag{7.39}$$

The generator and its scaling function are illustrated in Fig. 7.2. Note the oscillatory behavior in comparison to the scaling functions discussed below. Note also that $\Phi_J * F$ yields a truncated Fourier series of $F \in \mathrm{L}^2(\Omega)$, as we can derive from Corollary 7.6:

$$(\Phi_J * F)(\xi) = \sum_{n=0}^{2^J-1} \sum_{j=1}^{2n+1} F^\wedge(n,j) Y_{n,j}(\xi) \quad \forall \xi \in \Omega. \tag{7.40}$$

(A1.2) **Modified Shannon Scaling Function**

This generator is now continuous but not everywhere differentiable. For a fixed parameter $h \in]0,1[$, we set

$$\varphi_0(x) := \begin{cases} 1, & 0 \leq x < h \\ \frac{1-x}{1-h}, & h \leq x < 1 \\ 0, & 1 \leq x \end{cases} ; \tag{7.41}$$

see Fig. 7.3.

(A1.3) **Cubic Polynomial Scaling Function**

Note that the name of this scaling function, which is sometimes abbreviated as cp scaling function or CuP scaling function, refers to the generator and not to the obtained scaling function. It is given by

$$\varphi_0(x) := \begin{cases} (1-x)^2(1+2x), & 0 \leq x < 1, \\ 0, & 1 \leq x. \end{cases} \tag{7.42}$$

Note that, if we set $\varphi_0(x) = 1$ for $x < 0$, then $\varphi_0 \in \mathrm{C}^{(1)}(\mathbb{R})$ with $\varphi_0(0) = 1$, $\varphi_0(1) = 0$, and $\varphi_0'(0) = \varphi_0'(1) = 0$. The generator and the scaling function are shown in Fig. 7.4.

(A1.4) **Blackman Scaling Function**

Another example of a bandlimited scaling function was introduced in [165]. It represents in some sense a compromise of the modified Shannon scaling function and the cp scaling function, since its generator starts with a constant part, but has a smoother way down to 0. It is defined by

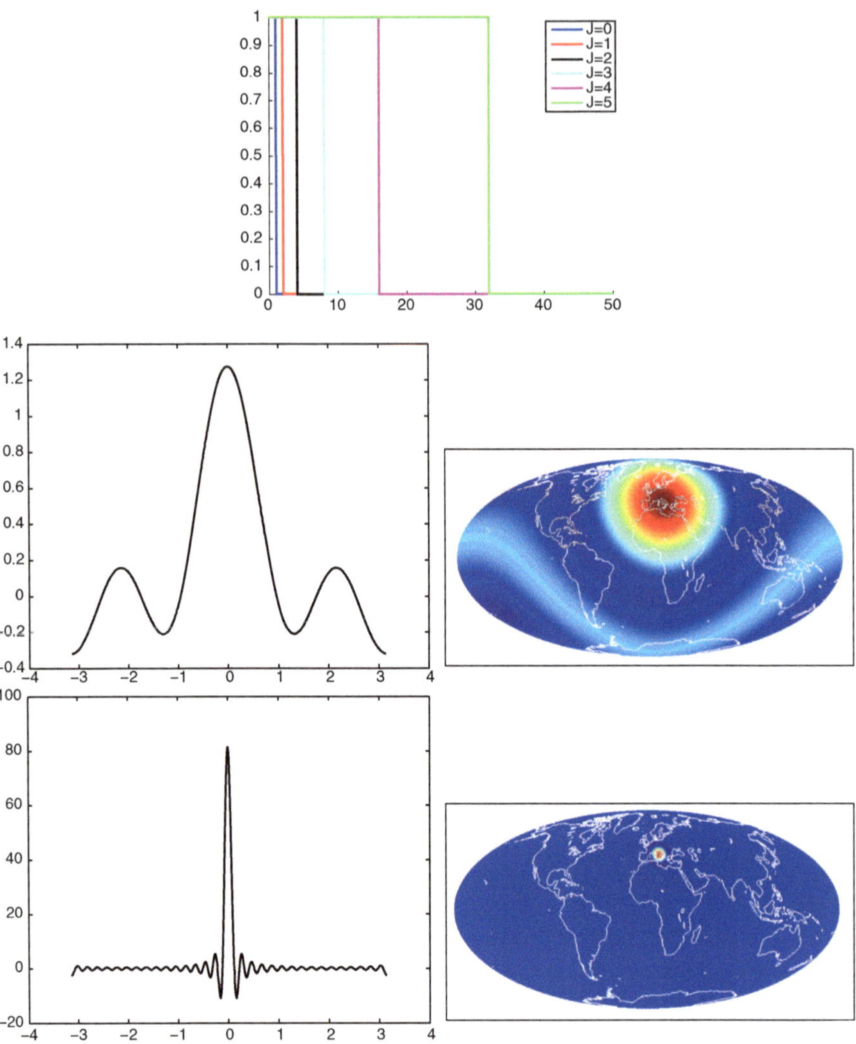

Fig. 7.2 Dilated generator of the Shannon scaling function (*top*) as well as spatial plots of the scaling functions Φ_2 (*second row*) and Φ_5 (*third row*) as 1D-function $[-\pi,\pi] \ni \vartheta \mapsto \Phi_J(\cos\vartheta)$ (*left hand*) and zonal function $\Omega \ni \eta \mapsto \Phi_J(\xi \cdot \eta)$, where ξ is located in Rome

$$\varphi_0(x) := \begin{cases} 1, & 0 \le x < \frac{1}{2}, \\ \frac{21}{50} - \frac{1}{2}\cos(2\pi x) + \frac{2}{25}\cos(4\pi x), & \frac{1}{2} \le x < 1, \\ 0, & 1 \le x. \end{cases} \tag{7.43}$$

Note that

$$\lim_{x\to 1/2+} \varphi_0(x) = \frac{21}{50} - \frac{1}{2}\cos\pi + \frac{2}{25}\cos(2\pi) = \frac{21}{50} + \frac{1}{2} + \frac{2}{25} = 1 \tag{7.44}$$

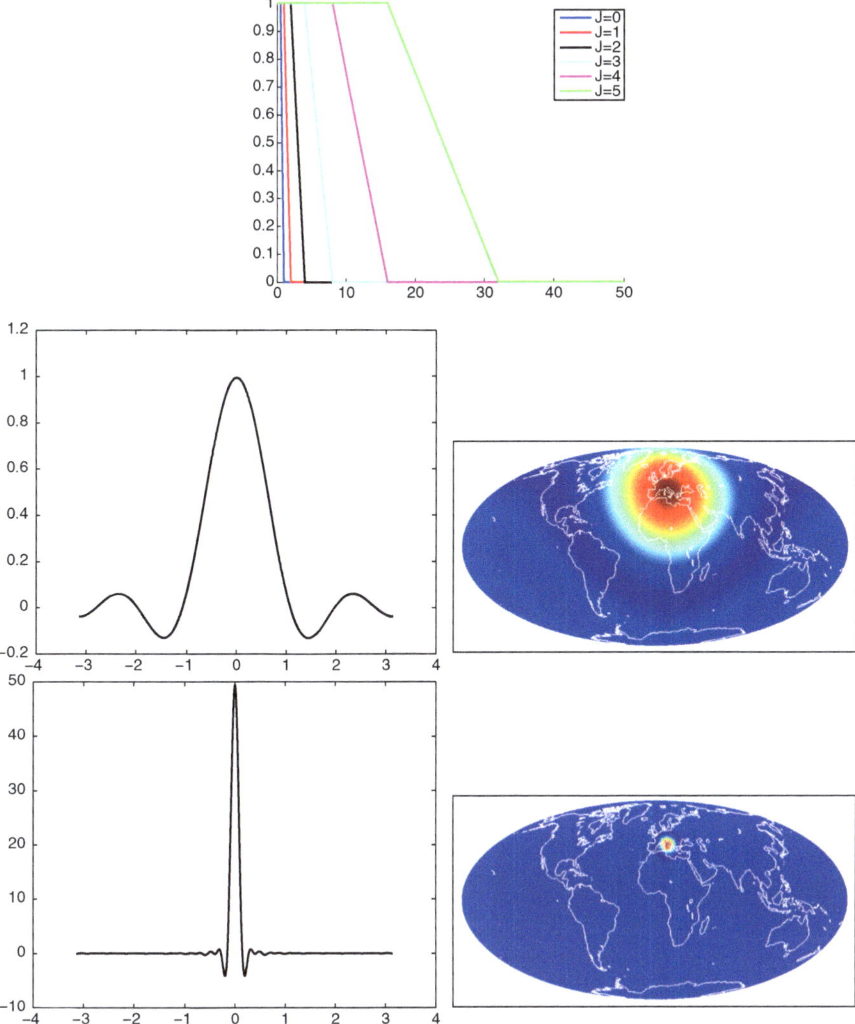

Fig. 7.3 Dilated generator of the modified Shannon scaling function with the chosen parameter $h = 0.5$ (*top*) as well as spatial plots of the scaling functions Φ_2 (*second row*) and Φ_5 (*third row*) as 1D-function $[-\pi, \pi] \ni \vartheta \mapsto \Phi_J(\cos\vartheta)$ (*left hand*) and zonal function $\Omega \ni \eta \mapsto \Phi_J(\xi \cdot \eta)$, where ξ is located in Rome

and

$$\lim_{x \to 1-} \varphi_0(x) = \frac{21}{50} - \frac{1}{2}\cos(2\pi) + \frac{2}{25}\cos(4\pi) = \frac{21}{50} - \frac{1}{2} + \frac{2}{25} = 0, \tag{7.45}$$

that is, φ_0 is continuous. Moreover, for $\frac{1}{2} < x < 1$, we get

$$\varphi_0'(x) = \pi \sin(2\pi x) - \frac{8\pi}{25} \sin(4\pi x). \tag{7.46}$$

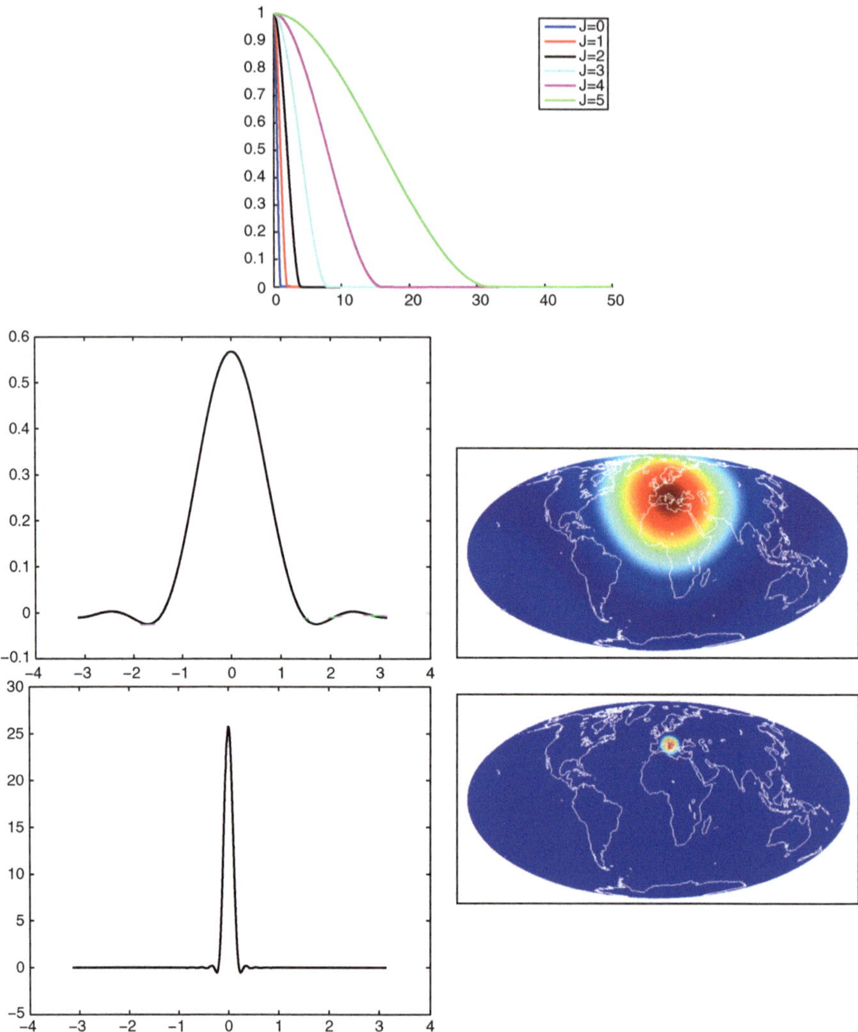

Fig. 7.4 Dilated generator of the cp scaling function (*top*) as well as spatial plots of the scaling functions Φ_2 (*second row*) and Φ_5 (*third row*) as 1D-function $[-\pi, \pi] \ni \vartheta \mapsto \Phi_J(\cos\vartheta)$ (*left hand*) and zonal function $\Omega \ni \eta \mapsto \Phi_J(\xi \cdot \eta)$, where ξ is located in Rome

As a consequence, we have

$$\lim_{x \to 1/2+} \varphi_0'(x) = 0, \quad \lim_{x \to 1-} \varphi_0'(x) = 0, \tag{7.47}$$

that is, φ_0 is continuously differentiable. Furthermore, φ_0 is, indeed, a generator of a scaling function. To verify this, we observe that (GS1) and (GS2) in

Definition 7.17 are trivially satisfied. In order to prove the validity of (GS3), we first show that φ_0' has no zeros in $]\frac{1}{2},1[$. If x were such a zero, then it would have to satisfy

$$\pi\sin(2\pi x) = \frac{8\pi}{25}\sin(4\pi x). \tag{7.48}$$

Since $\sin(2\pi x) \neq 0$ for $\frac{1}{2} < x < 1$, (7.48) is equivalent to

$$\frac{25}{8} = \frac{\sin(4\pi x)}{\sin(2\pi x)}. \tag{7.49}$$

Using the well-known identity $\sin(2y) = 2\sin y\cos y$, $y \in \mathbb{R}$, we arrive at

$$\frac{25}{8} = \frac{2\sin(2\pi x)\cos(2\pi x)}{\sin(2\pi x)} \Leftrightarrow \frac{25}{16} = \cos(2\pi x), \tag{7.50}$$

which has no real solution. Furthermore, we have

$$\varphi_0'\left(\frac{3}{4}\right) = \pi\sin\left(\frac{3}{2}\pi\right) - \frac{8\pi}{25}\sin(3\pi) = -\pi < 0, \tag{7.51}$$

and φ_0' is continuous. Hence, φ_0' is negative on $]\frac{1}{2},1[$, and we can, eventually, conclude that φ_0 is monotonically decreasing.

The generator and the scaling function are shown in Fig. 7.5.

(A2) **Non-bandlimited**

(A2.1) **Rational Scaling Function**

Like the cubic polynomial scaling function, the rational scaling function refers in its name to its generator. For a fixed parameter $s > 1$, we set

$$\varphi_0(x) := (1+x)^{-s} \quad \forall x \in \mathbb{R}_0^+. \tag{7.52}$$

For generators of non-bandlimited scaling functions, we have to verify the admissibility condition, that is, we have to investigate the term

$$\sum_{n=0}^{\infty} \frac{2n+1}{4\pi}\left(\sup_{x\in[n,n+1[}|\varphi_0(x)|\right)^2 = \sum_{n=0}^{\infty}\frac{2n+1}{4\pi}(1+n)^{-2s}. \tag{7.53}$$

This series converges if and only if $1-2s < -1$, which is equivalent to $s > 1$. Note that the series [see also (3.60) on p. 45]

$$\Phi_J = \sum_{n=0}^{\infty}\frac{2n+1}{4\pi}\Phi_J^{\wedge}(n)P_n = \sum_{n=0}^{\infty}\frac{2n+1}{4\pi}\left(1+2^{-J}n\right)^{-s}P_n \tag{7.54}$$

has to be truncated for numerical implementations.

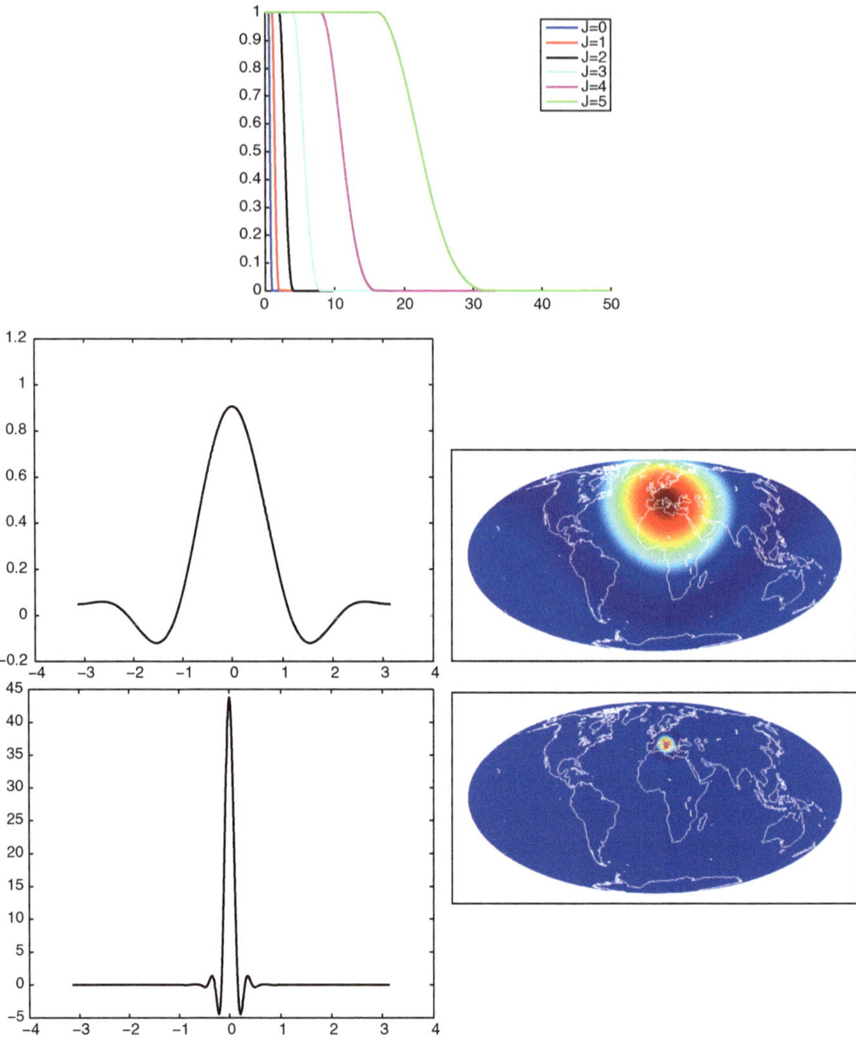

Fig. 7.5 Dilated generator of the Blackman scaling function (*top*) as well as spatial plots of the scaling functions Φ_2 (*second row*) and Φ_5 (*third row*) as 1D-function $[-\pi,\pi] \ni \vartheta \mapsto \Phi_J(\cos\vartheta)$ (*left hand*) and zonal function $\Omega \ni \eta \mapsto \Phi_J(\xi \cdot \eta)$, where ξ is located in Rome

For an illustration of φ_J and Φ_J for $s = 2$, see Fig. 7.6. One can already see here a typical feature of non-bandlimited scaling functions: They are more localized.

(A2.2) **Modified Rational Scaling Function**

For a fixed parameter $s > \frac{1}{2}$, we set

$$\varphi_0(x) := \left(1 + x^2\right)^{-s} \quad \forall x \in \mathbb{R}_0^+; \tag{7.55}$$

see Fig. 7.7, where $s = 1$ was chosen. The proof of the admissibility is analogous to the proof for the rational scaling function.

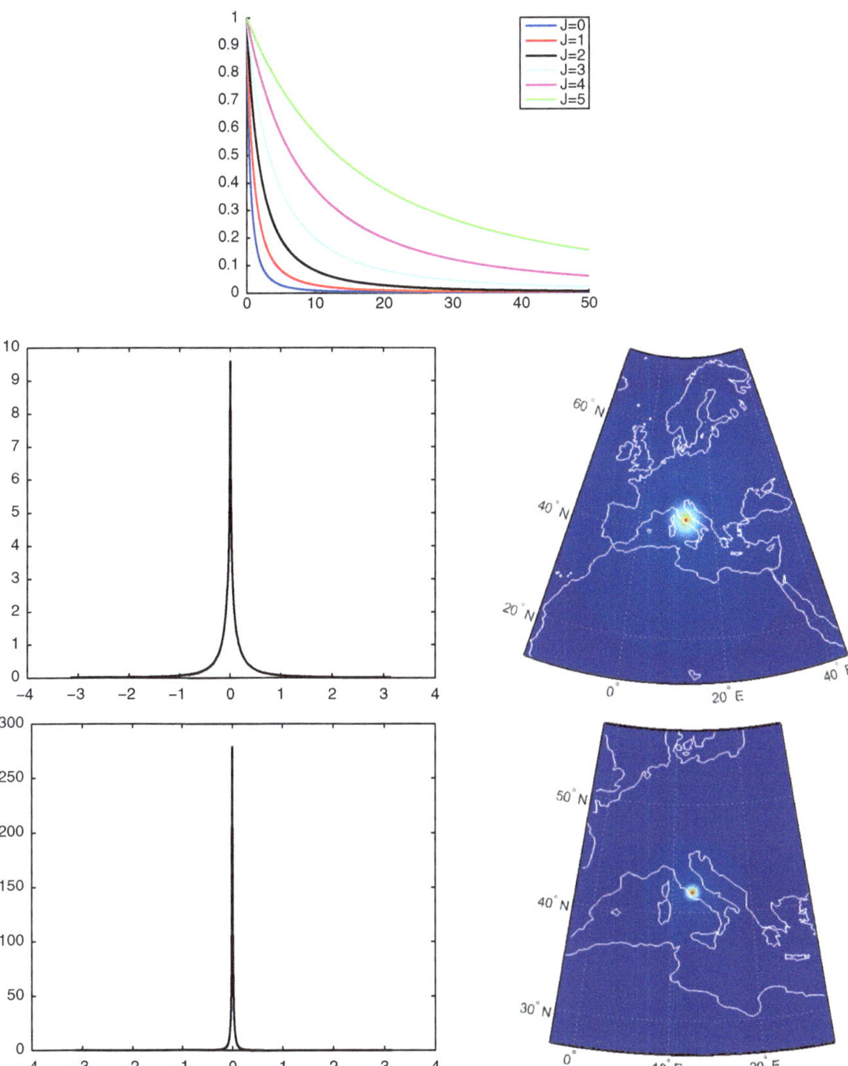

Fig. 7.6 Dilated generator of the rational scaling function with the chosen parameter $s = 2$ (*top*) as well as spatial plots of the scaling functions Φ_2 (*second row*) and Φ_5 (*third row*) as 1D-function $[-\pi, \pi] \ni \vartheta \mapsto \Phi_J(\cos\vartheta)$ (*left hand*) and zonal function $\Omega \ni \eta \mapsto \Phi_J(\xi \cdot \eta)$, where ξ is located in Rome. The series expansion of Φ_J was truncated after degree 1,000

(A2.3) **Abel–Poisson Scaling Function**

We already know this scaling function. It is generated by

$$\varphi_0(x) := \mathrm{e}^{-Rx} \quad \forall x \in \mathbb{R}_0^+ \tag{7.56}$$

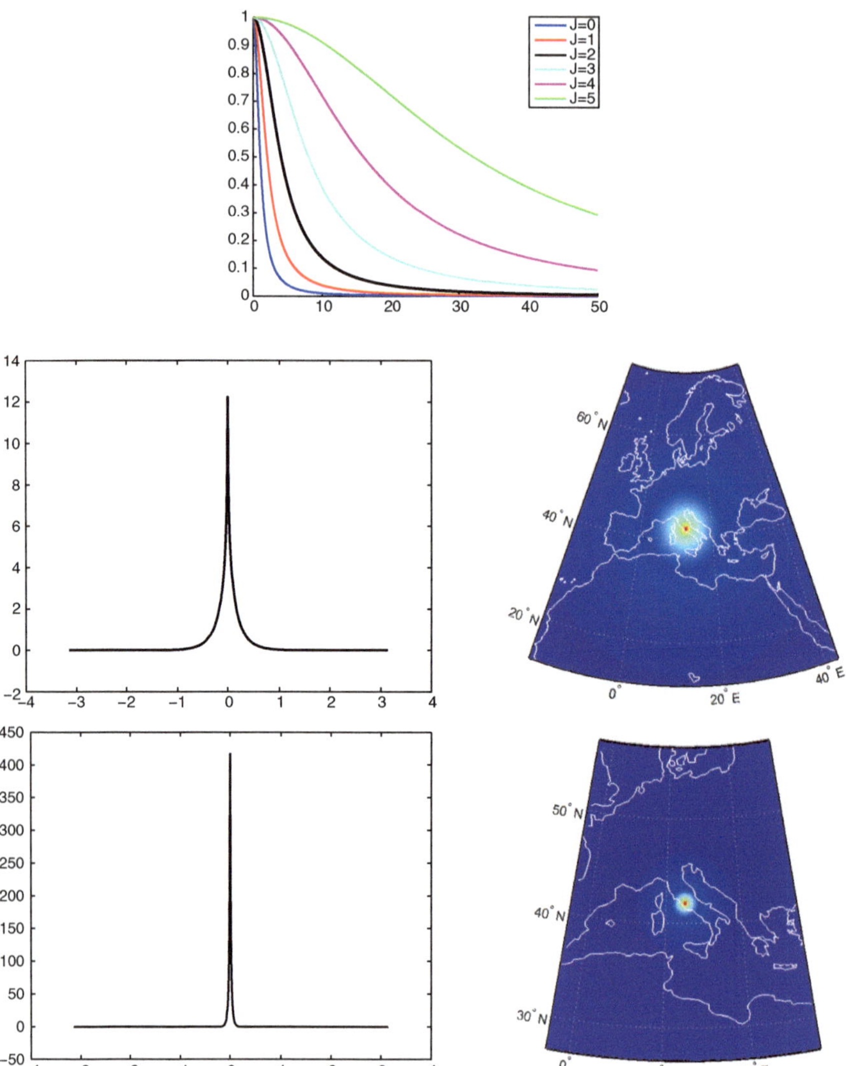

Fig. 7.7 Dilated generator of the modified rational scaling function with the chosen parameter $s = 1$ (*top*) as well as spatial plots of the scaling functions Φ_2 (*second row*) and Φ_5 (*third row*) as 1D-function $[-\pi, \pi] \ni \vartheta \mapsto \Phi_J(\cos\vartheta)$ (*left hand*) and zonal function $\Omega \ni \eta \mapsto \Phi_J(\xi \cdot \eta)$, where ξ is located in Rome. The series expansion of Φ_J was truncated after degree 1,000

for a fixed parameter $R > 0$. It has an essential advantage in comparison to other non-bandlimited scaling functions: It is not necessary to truncate the series, because there is a closed representation (see Theorem 3.16 on p. 46):

$$\begin{aligned}\Phi_J(t) &= \sum_{n=0}^{\infty} \frac{2n+1}{4\pi} \mathrm{e}^{-R2^{-J}n} P_n(t) \\ &= \frac{1}{4\pi} \frac{1-\exp\left(-R2^{1-J}\right)}{\left(1+\exp(-R2^{1-J})-2\exp(-R2^{-J})t\right)^{3/2}} \quad \forall t \in [-1,1]\,. \quad (7.57)\end{aligned}$$

The admissibility is given due to the exponential decay of φ_0. We even get a uniform convergence of the series in (7.57). For an illustration, see Fig. 7.8.

(A2.4) **Gauß–Weierstraß Scaling Function**

The Gauß–Weierstraß scaling function has faster decaying Legendre coefficients than the Abel–Poisson scaling function. It is, therefore, appropriate if a stronger attenuation of the signal is desired. However, there is no closed representation for the functions Φ_J. The generator of the scaling function is defined as follows:

$$\varphi_0(x) := \mathrm{e}^{-Rx(x+1)} \quad \forall x \in \mathbb{R}_0^+, \tag{7.58}$$

where $R > 0$ is again a fixed parameter. For an illustration, see Fig. 7.9.

(B) **Another Scaling Function**

The (non-bandlimited) **Tykhonov–Philips scaling function** is of particular importance in a more general setting which involves inverse problems (see [128]). Its Legendre coefficients are defined as follows:

$$\Phi_J^\wedge(n) := \frac{1}{1+\gamma_{J,n}^2} \quad \forall n,J \in \mathbb{N}_0, \tag{7.59}$$

where $(\gamma_{J,n})_{J,n\in\mathbb{N}_0}$ has to satisfy the following properties:

(a) For every $J \in \mathbb{N}_0$,

$$\begin{aligned}\|\Phi_J\|^2_{\mathrm{L}^2[-1,1]} &= \frac{1}{2\pi} \sum_{n=0}^{\infty} \frac{2n+1}{4\pi} \left(\Phi_J^\wedge(n)\right)^2 \\ &= \frac{1}{2\pi} \sum_{n=0}^{\infty} \frac{2n+1}{4\pi} \left(\frac{1}{1+\gamma_{J,n}^2}\right)^2 < +\infty\,, \quad (7.60)\end{aligned}$$

see also (7.28).

(b) For every $n \in \mathbb{N}_0$, the sequence $(\gamma_{J,n})_{J\in\mathbb{N}_0}$ is monotonically decreasing. This guarantees that (S1) is satisfied.

(c) For all $n \in \mathbb{N}_0$, the convergence $\lim_{J\to\infty} \gamma_{J,n} = 0$ is needed to obtain (S2).

Note that (S3) is always satisfied.

Several choices for $(\gamma_{J,n})_{J,n\in\mathbb{N}_0}$ are possible. If $\gamma_{J,n} = 2^{-J}n$, one obtains, for instance, the modified rational scaling function with $s = 1$. However, also non-dilated versions are possible such as $\gamma_{J,n} = \frac{n}{J+1}$; see Fig. 7.10.

Our next step is to prove one of the fundamental properties of a spherical scaling function.

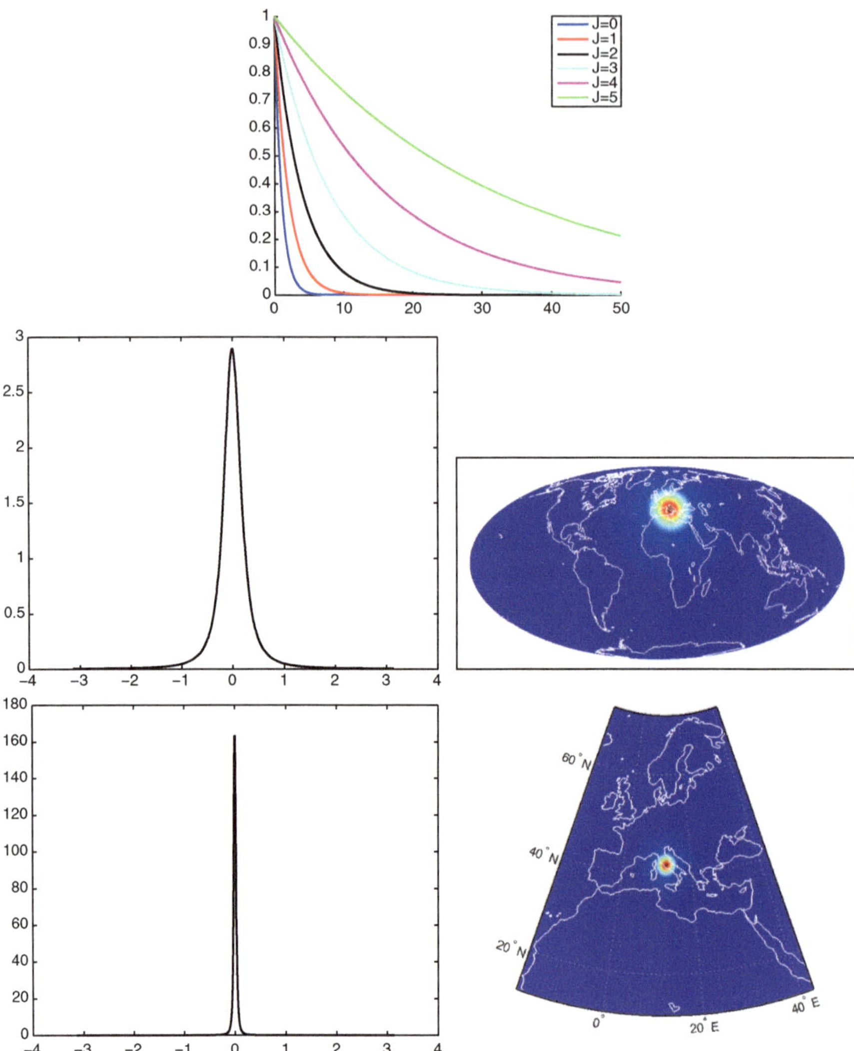

Fig. 7.8 Dilated generator of the Abel–Poisson scaling function with the chosen parameter $R = 1$ (*top*) as well as spatial plots of the scaling functions Φ_2 (*second row*) and Φ_5 (*third row*) as 1D-function $[-\pi, \pi] \ni \vartheta \mapsto \Phi_J(\cos\vartheta)$ (*left hand*) and zonal function $\Omega \ni \eta \mapsto \Phi_J(\xi \cdot \eta)$, where ξ is located in Rome

Theorem 7.21 (Spherical Approximate Identity). *Let $\{\Phi_J\}_{J\in\mathbb{N}_0}$ be an arbitrary scaling function and $F \in \mathrm{L}^2(\Omega)$ be an arbitrary function. Then*

$$\lim_{J\to\infty} \left\| F - \Phi_J^{(k)} * F \right\|_{\mathrm{L}^2(\Omega)} = 0 \tag{7.61}$$

for all levels of iteration $k \in \mathbb{N}$.

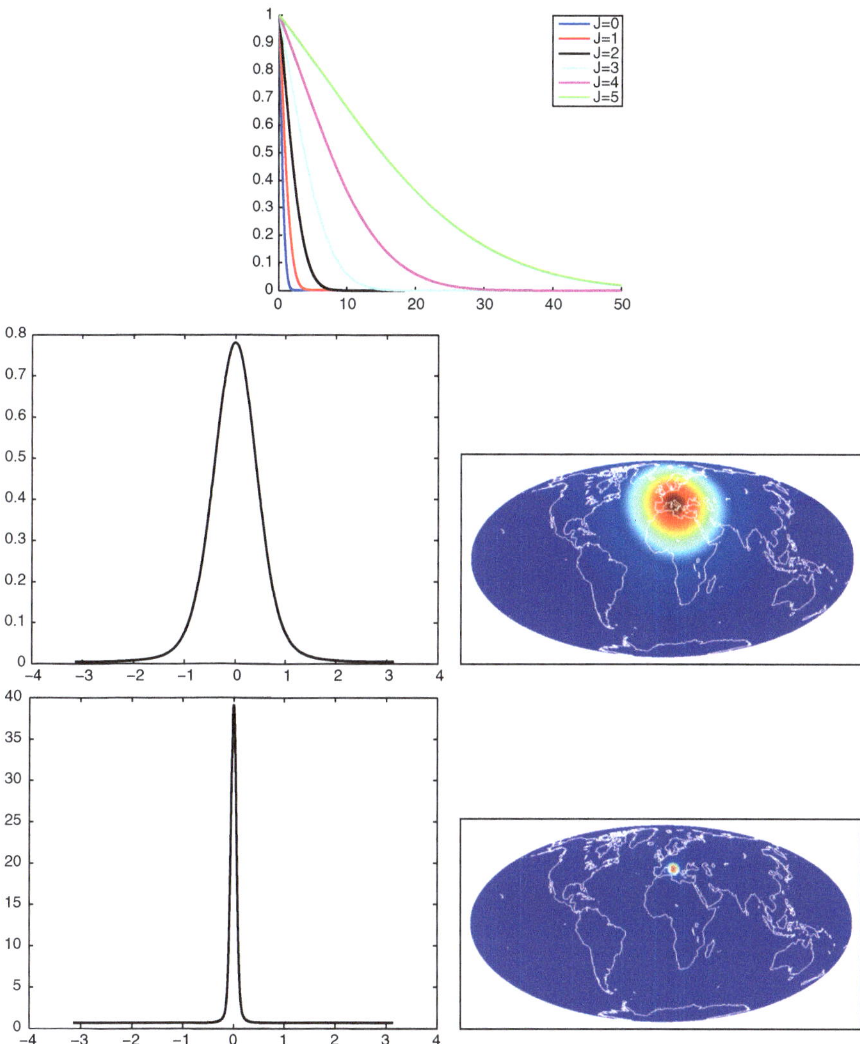

Fig. 7.9 Dilated generator of the Gauß–Weierstraß scaling function with the chosen parameter $R = 1$ (*top*) as well as spatial plots of the scaling functions Φ_2 (*second row*) and Φ_5 (*third row*) as 1D-function $[-\pi,\pi] \ni \vartheta \mapsto \Phi_J(\cos\vartheta)$ (*left hand*) and zonal function $\Omega \ni \eta \mapsto \Phi_J(\xi\cdot\eta)$, where ξ is located in Rome. The series expansion of the scaling function was truncated after degree 1,000

We will use, in particular, the case $k = 2$ here.

Proof. From Corollary 7.6 and Theorem 7.10, we already know that

$$\Phi_J^{(k)} * F = \sum_{n=0}^{\infty} \sum_{j=1}^{2n+1} \left(\Phi_J^{\wedge}(n)\right)^k F^{\wedge}(n,j) Y_{n,j} \tag{7.62}$$

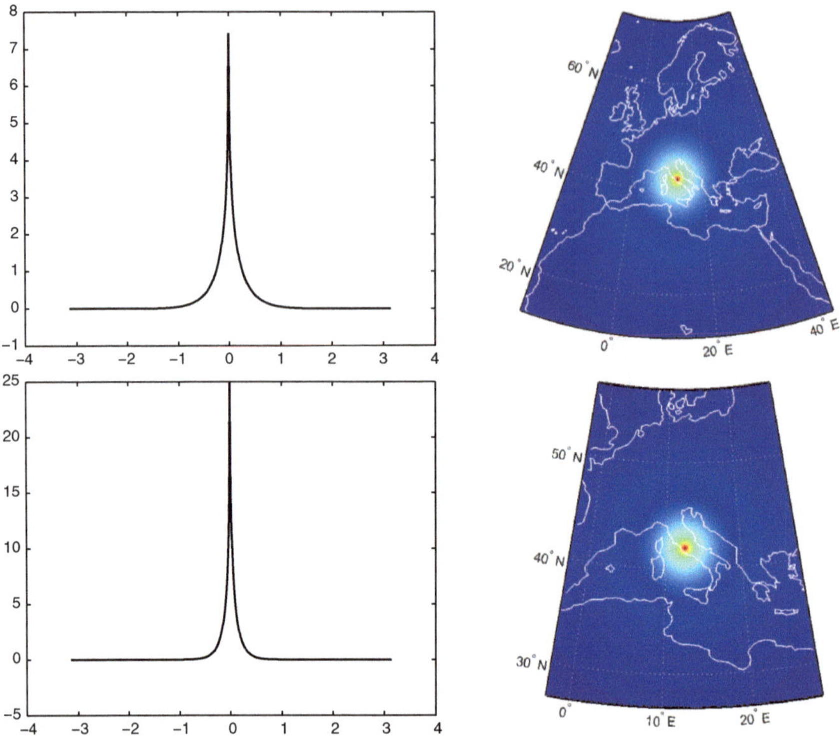

Fig. 7.10 Tykhonov–Philips scaling function Φ_J with the chosen sequence $\gamma_{J,n} = \frac{n}{J+1}$: The *left-hand* plots show $[-\pi,\pi] \ni \vartheta \mapsto \Phi_J(\cos\vartheta)$, whereas the *right-hand* plots show the corresponding zonal functions $\Omega \ni \eta \mapsto \Phi_J(\xi \cdot \eta)$ with ξ located in Rome. The *first row* corresponds to $J = 2$ and the *second one* to $J = 5$. The series expansion was truncated after degree 1,000

and, consequently,

$$F - \Phi_J^{(k)} * F = \sum_{n=0}^{\infty} \sum_{j=1}^{2n+1} \left[1 - \left(\Phi_J^\wedge(n)\right)^k\right] F^\wedge(n,j) Y_{n,j} \tag{7.63}$$

in the sense of $\mathrm{L}^2(\Omega)$. Hence, the Parseval identity (see Theorem 5.25 on p. 122) yields

$$\left\| F - \Phi_J^{(k)} * F \right\|_{\mathrm{L}^2(\Omega)}^2 = \sum_{n=0}^{\infty} \sum_{j=1}^{2n+1} \left[1 - \left(\Phi_J^\wedge(n)\right)^k\right]^2 \left(F^\wedge(n,j)\right)^2. \tag{7.64}$$

We now need to interchange the limit $\lim_{J\to\infty}$ with the series $\lim_{N\to\infty} \sum_{n=0}^{N}$. For this purpose, we recall the definition of a scaling function (see Definition 7.11). Combining (S1)–(S3), we conclude that

$$0 \leq \Phi_J^\wedge(n) \leq 1 \quad \forall n, J \in \mathbb{N}_0. \tag{7.65}$$

Therefore,

$$0 \leq \left[1 - \left(\Phi_J^\wedge(n)\right)^k\right]^2 \leq 1 \quad \forall n, J \in \mathbb{N}_0. \tag{7.66}$$

Since $F \in \mathrm{L}^2(\Omega)$ and, consequently,

$$\|F\|_{\mathrm{L}^2(\Omega)}^2 = \sum_{n=0}^{\infty} \sum_{j=1}^{2n+1} \left(F^\wedge(n,j)\right)^2 < +\infty \tag{7.67}$$

(which is again a consequence of the Parseval identity in Theorem 5.25), the series in (7.64) is uniformly convergent with respect to $J \in \mathbb{N}_0$. Thus, we get

$$\lim_{J\to\infty} \left\| F - \Phi_J^{(k)} * F \right\|_{\mathrm{L}^2(\Omega)}^2 = \sum_{n=0}^{\infty} \sum_{j=1}^{2n+1} \underbrace{\lim_{J\to\infty} \left[1 - \left(\Phi_J^\wedge(n)\right)^k\right]^2}_{\overset{(S2)}{=} 0} \left(F^\wedge(n,j)\right)^2 = 0. \tag{7.68}$$

□

Moreover, the spherical scaling functions can be used to construct a multiresolution analysis (see Theorem 3.35 on p. 74 for the Euclidean case).

Theorem 7.22 (Spherical Multiresolution Analysis). *Let $\{\Phi_J\}_{J\in\mathbb{N}_0} \subset \mathrm{L}^2[-1,1]$ be a scaling function. Then the* ***scale spaces*** *$\{V_J\}_{J\in\mathbb{N}_0}$ defined by*

$$V_J := \left\{ \Phi_J^{(2)} * F \,\middle|\, F \in \mathrm{L}^2(\Omega) \right\} \tag{7.69}$$

represent a multiresolution analysis, that is,

(i) $V_0 \subset \cdots \subset V_J \subset V_{J+1} \subset \cdots \subset \mathrm{L}^2(\Omega)$ *for all* $J \in \mathbb{N}_0$.
(ii) $\overline{\bigcup_{J\in\mathbb{N}_0} V_J}^{\|\cdot\|_{\mathrm{L}^2(\Omega)}} = \mathrm{L}^2(\Omega)$.

Proof.

(1) $V_J \subset \mathrm{L}^2(\Omega)$:
Theorem 7.8 yields that $\Phi_J^{(2)} \in \mathrm{L}^2[-1,1]$. In combination with Theorem 7.2, we get that $\Phi_J^{(2)} * F \in \mathrm{L}^2(\Omega)$ for all $F \in \mathrm{L}^2(\Omega)$.

(2) $V_J \subset V_{J+1}$:
Let $F \in \mathrm{L}^2(\Omega)$ be arbitrary such that $\Phi_J^{(2)} * F$ represents an arbitrary element of V_J. We have to show that $\Phi_J^{(2)} * F \in V_{J+1}$. In other words, we have to find $G \in \mathrm{L}^2(\Omega)$ such that $\Phi_J^{(2)} * F = \Phi_{J+1}^{(2)} * G \in V_{J+1}$. We will show here that

$$G^\wedge(n,j) := \begin{cases} \left(\frac{\Phi_J^\wedge(n)}{\Phi_{J+1}^\wedge(n)}\right)^2 F^\wedge(n,j), & \text{if } \Phi_{J+1}^\wedge(n) \neq 0 \\ 0, & \text{if } \Phi_{J+1}^\wedge(n) = 0 \end{cases}, \tag{7.70}$$

$n \in \mathbb{N}_0$, $j \in \{1,\ldots,2n+1\}$, yields the desired result. First, we verify that $G \in \mathrm{L}^2(\Omega)$ by using the Parseval identity (see Theorem 5.25 on p. 122):

$$\begin{aligned} \|G\|^2_{\mathrm{L}^2(\Omega)} &= \sum_{\substack{n=0 \\ \Phi_{J+1}^\wedge(n)\neq 0}}^{\infty} \sum_{j=1}^{2n+1} \left(\frac{\Phi_J^\wedge(n)}{\Phi_{J+1}^\wedge(n)}\right)^4 \left(F^\wedge(n,j)\right)^2 \\ &\overset{(\mathrm{S1}),(\mathrm{S3})}{\leq} \sum_{\substack{n=0 \\ \Phi_{J+1}^\wedge(n)\neq 0}}^{\infty} \sum_{j=1}^{2n+1} \left(F^\wedge(n,j)\right)^2 \\ &\leq \|F\|^2_{\mathrm{L}^2(\Omega)} < +\infty\,. \end{aligned} \tag{7.71}$$

Furthermore, (S1) in combination with (S3) yields that $0 \leq \Phi_J^\wedge(n) \leq \Phi_{J+1}^\wedge(n)$ for all $n, J \in \mathbb{N}_0$. Hence, if $\Phi_{J+1}^\wedge(n) = 0$, then $\Phi_J^\wedge(n) = 0$. In other words,

$$\left\{n \in \mathbb{N}_0 \,\middle|\, \Phi_{J+1}^\wedge(n) = 0\right\} \subset \left\{n \in \mathbb{N}_0 \,\middle|\, \Phi_J^\wedge(n) = 0\right\}. \tag{7.72}$$

Now, we can apply (7.23) to get

$$\begin{aligned} \Phi_{J+1}^{(2)} * G &= \sum_{n=0}^{\infty} \sum_{j=1}^{2n+1} \left(\Phi_{J+1}^\wedge(n)\right)^2 G^\wedge(n,j)\, Y_{n,j} \\ &= \sum_{\substack{n=0 \\ \Phi_{J+1}^\wedge(n)\neq 0}}^{\infty} \sum_{j=1}^{2n+1} \left(\Phi_{J+1}^\wedge(n)\right)^2 \left(\frac{\Phi_J^\wedge(n)}{\Phi_{J+1}^\wedge(n)}\right)^2 F^\wedge(n,j)\, Y_{n,j} \\ &= \sum_{\substack{n=0 \\ \Phi_{J+1}^\wedge(n)\neq 0}}^{\infty} \sum_{j=1}^{2n+1} \left(\Phi_J^\wedge(n)\right)^2 F^\wedge(n,j)\, Y_{n,j} \\ &= \sum_{n=0}^{\infty} \sum_{j=1}^{2n+1} \left(\Phi_J^\wedge(n)\right)^2 F^\wedge(n,j)\, Y_{n,j} \\ &= \Phi_J^{(2)} * F \end{aligned} \tag{7.73}$$

in the sense of $\mathrm{L}^2(\Omega)$.

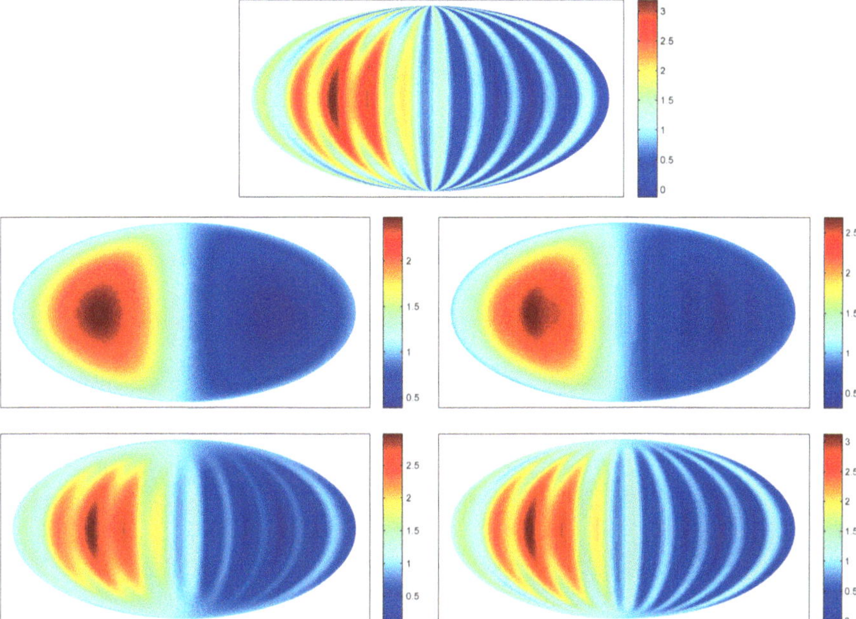

Fig. 7.11 Multiresolution analysis $\Phi_J^{(2)} * F$ for $F(\xi(\varphi,\vartheta)) = \mathrm{e}^{-\xi_2(\varphi,\vartheta)} + 0.5\sin(10\varphi)$ (see *top*) and the cp scaling function. The figure shows $\Phi_J^{(2)} * F$ for $J=3$ (*middle left hand*), $J=4$ (*middle right hand*), $J=5$ (*bottom left hand*), and $J=6$ (*bottom right hand*)

(3) The density property:
Part (ii) of the theorem is an immediate consequence of the fact that the scaling function represents an Approximate Identity (see Theorem 7.21): Let $F \in \mathrm{L}^2(\Omega)$ be arbitrary. Then $\Phi_J * F \in V_J$ for each $J \in \mathbb{N}_0$, that is, the sequence $(\Phi_J * F)_{J\in\mathbb{N}_0}$ belongs to $\bigcup_{J\in\mathbb{N}_0} V_J$. Since this sequence converges to F in the sense of $\|\cdot\|_{\mathrm{L}^2(\Omega)}$, the theorem is proved. □

We demonstrate the method in Fig. 7.11, where $\Phi_J^{(2)} * F$ is calculated for the function $F(\xi(\varphi,\vartheta)) = \mathrm{e}^{-\xi_2(\varphi,\vartheta)} + 0.5\sin(10\varphi)$ (φ and ϑ are the polar coordinates). The chosen scaling function is the cp scaling function. Obviously, the method is able to separate parts of different spatial size. Moreover, for increasing scales, the approximation improves. Furthermore, Fig. 7.12 shows the corresponding functions for a noisy F. The results show that the low-pass filters $\Phi_J^{(2)}$ are also appropriate for denoising. For the numerical integrations, the Driscoll–Healy method (see Theorem 7.33) with polynomial accuracy up to degree 129 was used.

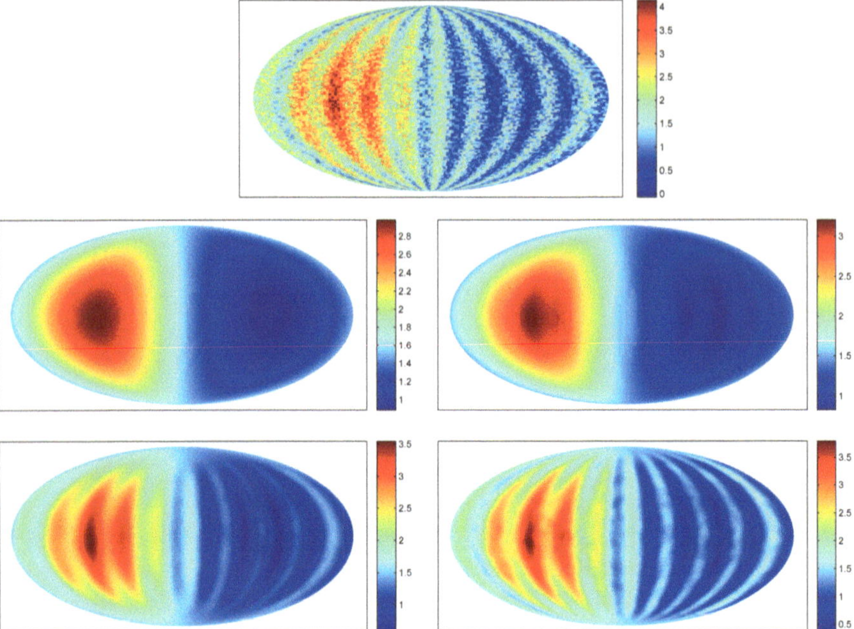

Fig. 7.12 The function in Fig. 7.11 is now replaced by $F+\varepsilon$, where the samples of ε are pseudorandom numbers on the interval $]0,1[$. The convolutions $\Phi_J^{(2)} * (F+\varepsilon)$ reduce, in particular for low scales, the noise in the data

7.3 Wavelets

Based on a given scaling function, wavelets can be constructed.

Definition 7.23. Let $\{\Phi_J\}_{J\in\mathbb{N}_0} \subset \mathrm{L}^2[-1,1]$ be a given scaling function. Then families $\{\Psi_J\}_{J\in\mathbb{N}_0\cup\{-1\}}, \{\tilde{\Psi}_J\}_{J\in\mathbb{N}_0\cup\{-1\}} \subset \mathrm{L}^2[-1,1]$ are called a **(spherical) primal** and a **(spherical) dual wavelet**, respectively, if their Legendre coefficients satisfy the following **refinement equation**:

$$\Psi_J^\wedge(n)\,\tilde{\Psi}_J^\wedge(n) = \left(\Phi_{J+1}^\wedge(n)\right)^2 - \left(\Phi_J^\wedge(n)\right)^2 \quad \forall n,J \in \mathbb{N}_0. \tag{7.74}$$

The functions Ψ_0 and $\tilde{\Psi}_0$ are called the primal and dual **mother wavelets**, respectively. Moreover, we set $\Psi_{-1} := \tilde{\Psi}_{-1} := \Phi_0$.

The meaning of the name "mother wavelet" becomes clear when we study how generators can be used again to construct such a family of kernels by dilation. In this sense, Φ_0 could also be called a mother scaling function.

Theorem 7.24. *Let φ_0 be a generator of a scaling function. Then all functions $\psi_0, \tilde{\psi}_0 : \mathbb{R}_0^+ \to \mathbb{R}$ which are admissible and satisfy the* ***refinement equation***

$$\psi_0(x)\tilde{\psi}_0(x) = \left(\varphi_0\left(\frac{x}{2}\right)\right)^2 - (\varphi_0(x))^2 \quad \forall x \in \mathbb{R}_0^+ \tag{7.75}$$

are ***generators of a primal and a dual mother wavelet****, respectively, that is, the families* $\{\Psi_J\}_{J\in\mathbb{N}_0\cup\{-1\}}$, $\{\tilde{\Psi}_J\}_{J\in\mathbb{N}_0\cup\{-1\}} \subset \mathrm{L}^2[-1,1]$ *defined via their Legendre coefficients by dilating* ψ_0 *and* $\tilde{\psi}_0$ *as*

$$\Psi_J^\wedge(n) := \psi_J(n) = \psi_0\left(2^{-J}n\right), \quad \tilde{\Psi}_J^\wedge(n) := \tilde{\psi}_J(n) = \tilde{\psi}_0\left(2^{-J}n\right) \quad \forall n,J \in \mathbb{N}_0 \tag{7.76}$$

and by

$$\Psi_{-1}^\wedge(n) := \tilde{\Psi}_{-1}^\wedge(n) := \varphi_0(n) \quad \forall n \in \mathbb{N}_0 \tag{7.77}$$

are a primal and a dual wavelet, respectively.

Proof. Due to Theorems 7.14 and 7.16, we have $\Psi_J, \tilde{\Psi}_J \in \mathrm{L}^2[-1,1]$ for all $J \in \mathbb{N}_0 \cup \{-1\}$. Moreover, the refinement equation (7.74) is an immediate consequence of the refinement equation (7.75) and (7.76), since

$$\begin{aligned}\Psi_J^\wedge(n)\tilde{\Psi}_J^\wedge(n) &= \psi_0\left(2^{-J}n\right)\,\tilde{\psi}_0\left(2^{-J}n\right)\\ &= \left(\varphi_0\left(2^{-J-1}n\right)\right)^2 - \left(\varphi_0\left(2^{-J}n\right)\right)^2\\ &= \left(\Phi_{J+1}^\wedge(n)\right)^2 - \left(\Phi_J^\wedge(n)\right)^2\end{aligned} \tag{7.78}$$

for all $n,J \in \mathbb{N}_0$. □

Note that the wavelets are *not* uniquely determined by the scaling function. In particular, two choices are common.

Definition 7.25. Let $\{\Phi_J\}_{J\in\mathbb{N}_0} \subset \mathrm{L}^2[-1,1]$ be a scaling function.

(a) The primal and dual wavelets defined by

$$\Psi_J^\wedge(n) := \tilde{\Psi}_J^\wedge(n) := \sqrt{\left(\Phi_{J+1}^\wedge(n)\right)^2 - \left(\Phi_J^\wedge(n)\right)^2} \quad \forall n,J \in \mathbb{N}_0 \tag{7.79}$$

and $\Psi_{-1} := \tilde{\Psi}_{-1} := \Phi_0$ are called the **P-wavelets**.

(b) The primal and dual wavelets defined by

$$\begin{aligned}\Psi_J^\wedge(n) &:= \Phi_{J+1}^\wedge(n) - \Phi_J^\wedge(n) \quad \forall n,J \in \mathbb{N}_0\\ \tilde{\Psi}_J^\wedge(n) &:= \Phi_{J+1}^\wedge(n) + \Phi_J^\wedge(n) \quad \forall n,J \in \mathbb{N}_0\\ \Psi_{-1} &:= \tilde{\Psi}_{-1} := \Phi_0\end{aligned} \tag{7.80}$$

are called the **M-wavelets**.

Remark 7.26. Here are some comments on these types of wavelets:

(a) As we mentioned above, there is also a scale-continuous wavelet theory. In this context, wavelet scales are not summed up but integrated. Since this is not practical, one introduces scale-discretized wavelets which integrate the scales in small steps. The results are scale-discrete wavelets, which are called wavelet packets. These wavelets coincide with the P-wavelets. This is the reason for the abbreviation P. For further details, see [66, Sect. 10.3] and [128, Sect. 2.3]. P-wavelets are most commonly used, since the identity $\Psi_J = \tilde{\Psi}_J$ reduces the computational costs.

(b) The radicand in (7.79) is nonnegative due to condition (S1) in the definition of a scaling function (Definition 7.11).

(c) "M" is an abbreviation for "modified." The M-wavelets can also be derived from the scale-continuous theory (see [66, Sect. 10.3]). Note that the presented theory is bilinear. A linear theory would use $\Phi_J * F$ instead of $\Phi_J^{(2)} * F$ for the scale spaces and have only one wavelet instead of a primal and a dual wavelet. In this context, the primal M-wavelet from the bilinear approach is the unique wavelet corresponding to a given scaling function in the linear theory.

(d) In the case of the use of generators, the P-wavelets are defined by

$$\psi_0(x) := \tilde{\psi}_0(x) := \sqrt{(\varphi_1(x))^2 - (\varphi_0(x))^2} \quad \forall x \in \mathbb{R}_0^+, \tag{7.81}$$

and the M-wavelets are defined by

$$\begin{aligned} \psi_0(x) &:= \varphi_1(x) - \varphi_0(x) \quad \forall x \in \mathbb{R}_0^+ \\ \tilde{\psi}_0(x) &:= \varphi_1(x) + \varphi_0(x) \quad \forall x \in \mathbb{R}_0^+ . \end{aligned} \tag{7.82}$$

A dilation obviously yields (7.79) and (7.80), respectively.

In order to use the P- and M-wavelets, we have to verify that they satisfy the requirements of Definition 7.23.

Theorem 7.27. *The P- and M-wavelets defined in Definition 7.25 are, indeed, wavelets associated to a scaling function* $\{\Phi_J\}_{J\in\mathbb{N}_0}$.

Proof. The verification of the refinement equation (7.74) is trivial. It remains to show that $\Psi_J, \tilde{\Psi}_J \in \mathrm{L}^2[-1,1]$ for all $J \in \mathbb{N}_0$ in both cases. For the P-wavelets, we obtain [see (7.28)]

$$\begin{aligned} \|\Psi_J\|^2_{\mathrm{L}^2[-1,1]} = \|\tilde{\Psi}_J\|^2_{\mathrm{L}^2[-1,1]} &= \frac{1}{2\pi}\sum_{n=0}^{\infty}\frac{2n+1}{4\pi}\left(\Psi_J^\wedge(n)\right)^2 \\ &= \frac{1}{2\pi}\sum_{n=0}^{\infty}\frac{2n+1}{4\pi}\left(\Phi_{J+1}^\wedge(n)\right)^2 - \frac{1}{2\pi}\sum_{n=0}^{\infty}\frac{2n+1}{4\pi}\left(\Phi_J^\wedge(n)\right)^2 \end{aligned}$$

for all $J \in \mathbb{N}_0$, where both series in the last line are convergent, because $\Phi_J, \Phi_{J+1} \in \mathrm{L}^2[-1,1]$ by definition. For the M-wavelets, we use (S1) and (S3) in Definition 7.11 to derive

$$\begin{aligned} \left|\Psi_J^\wedge(n)\right| &\le \left|\Phi_{J+1}^\wedge(n)\right| + \left|\Phi_J^\wedge(n)\right| \le 2\Phi_{J+1}^\wedge(n) \quad \forall n, J \in \mathbb{N}_0 \\ \left|\tilde{\Psi}_J^\wedge(n)\right| &\le \left|\Phi_{J+1}^\wedge(n)\right| + \left|\Phi_J^\wedge(n)\right| \le 2\Phi_{J+1}^\wedge(n) \quad \forall n, J \in \mathbb{N}_0 . \end{aligned} \tag{7.83}$$

Hence,

$$\max\left(\|\Psi_J\|^2_{\mathrm{L}^2[-1,1]},\|\tilde{\Psi}_J\|^2_{\mathrm{L}^2[-1,1]}\right)\le\frac{2}{\pi}\sum_{n=0}^{\infty}\frac{2n+1}{4\pi}\left(\Phi^{\wedge}_{J+1}(n)\right)^2<+\infty. \tag{7.84}$$

□

Since the scaling functions can be interpreted as low-pass filters, the refinement equation (7.74) shows that the wavelets represent differences of low-pass filters. Hence, they are considered to be band-pass filters.

We will now study the P-wavelets corresponding to the scaling functions listed in Example 7.20.

Example 7.28. We start again with the bandlimited kernels which are derived from a generator. Note that we sometimes use the representation of the generator and sometimes the Legendre symbols.

(A1.1) **Shannon P-Wavelet**

In this case, we have

$$\varphi_J(x)=\begin{cases}1, & 0\le x<2^J\\ 0, & 2^J\le x\end{cases}\Rightarrow \Psi_J^{\wedge}(n)=\tilde{\Psi}_J^{\wedge}(n)=\begin{cases}0, & 0\le n<2^J,\\ 1, & 2^J\le n<2^{J+1},\\ 0, & 2^{J+1}\le n.\end{cases} \tag{7.85}$$

Consequently,

$$\Psi_J(t)=\tilde{\Psi}_J(t)=\sum_{n=2^J}^{2^{J+1}-1}\frac{2n+1}{4\pi}P_n(t)\quad\forall t\in[-1,1]. \tag{7.86}$$

For plots of the generator and the wavelet, see Fig. 7.13.

(A1.2) **Modified Shannon P-Wavelet**

We choose again a fixed parameter $h\in]0,1[$. Since

$$\varphi_0(x)=\begin{cases}1, & 0\le x<h\\ \frac{1-x}{1-h}, & h\le x<1\\ 0, & 1\le x\end{cases},\quad \varphi_1(x)=\begin{cases}1, & 0\le x<2h\\ \frac{1-x/2}{1-h}, & 2h\le x<2\\ 0, & 2\le x\end{cases}\quad\forall x\in\mathbb{R}_0^+, \tag{7.87}$$

we have to distinguish two cases.

Case 1, $h\le\frac{1}{2}$:

$$\psi_0(x)=\tilde{\psi}_0(x)=\begin{cases}0, & 0\le x<h\\ \sqrt{1-\left(\frac{1-x}{1-h}\right)^2}=\frac{\sqrt{-2h+h^2+2x-x^2}}{1-h}, & h\le x<2h\\ \sqrt{\left(\frac{1-x/2}{1-h}\right)^2-\left(\frac{1-x}{1-h}\right)^2}=\frac{\sqrt{x-3x^2/4}}{1-h}, & 2h\le x<1\\ \frac{1-x/2}{1-h}, & 1\le x<2\\ 0, & 2\le x\end{cases}\quad\forall x\in\mathbb{R}_0^+. \tag{7.88}$$

Case 2, $h>\frac{1}{2}$:

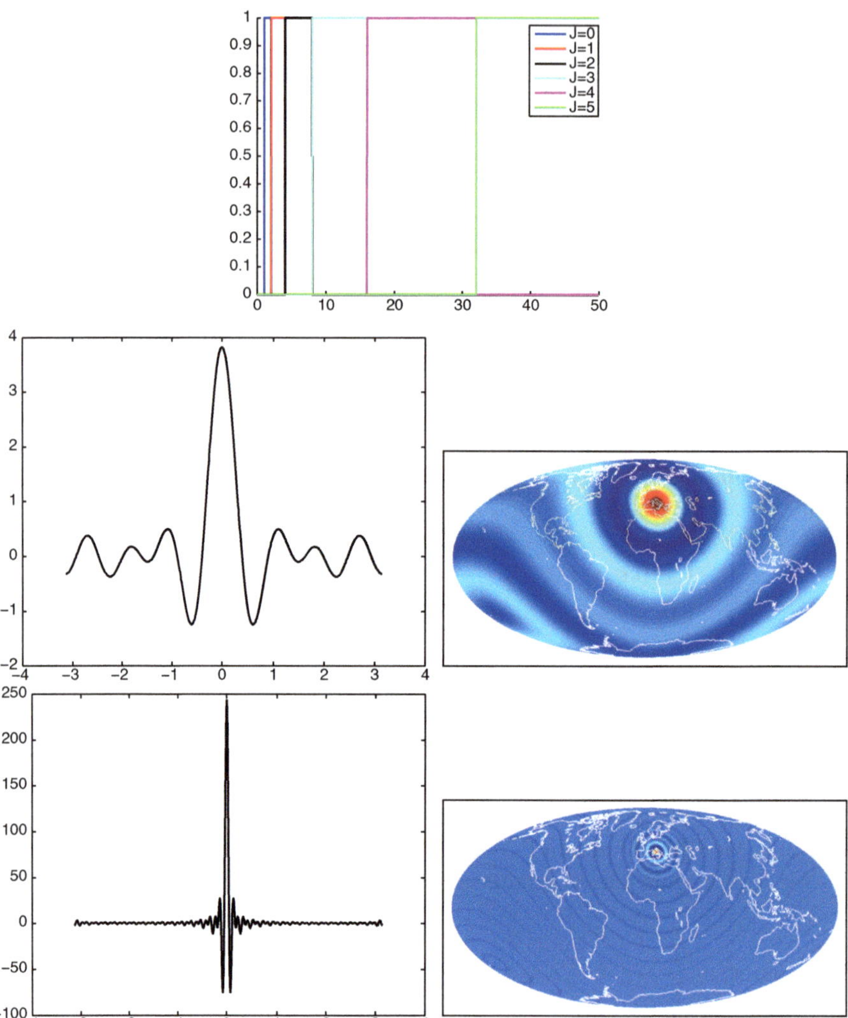

Fig. 7.13 Dilated generator of the Shannon P-wavelet (*top*) as well as spatial plots of the wavelets Ψ_2 (*second row*) and Ψ_5 (*third row*) as 1D-function $[-\pi,\pi] \ni \vartheta \mapsto \Psi_J(\cos\vartheta)$ (*second and third row, left hand*) and zonal function $\Omega \ni \eta \mapsto \Psi_J(\xi\cdot\eta)$, where ξ is located in Rome (*second and third row, right hand*)

$$\psi_0(x) = \tilde{\psi}_0(x) = \begin{cases} 0, & 0 \leq x < h \\ \sqrt{1-\left(\frac{1-x}{1-h}\right)^2} = \frac{\sqrt{-2h+h^2+2x-x^2}}{1-h}, & h \leq x < 1 \\ 1, & 1 \leq x < 2h \\ \frac{1-x/2}{1-h}, & 2h \leq x < 2 \\ 0, & 2 \leq x \end{cases} \quad \forall x \in \mathbb{R}_0^+ . \tag{7.89}$$

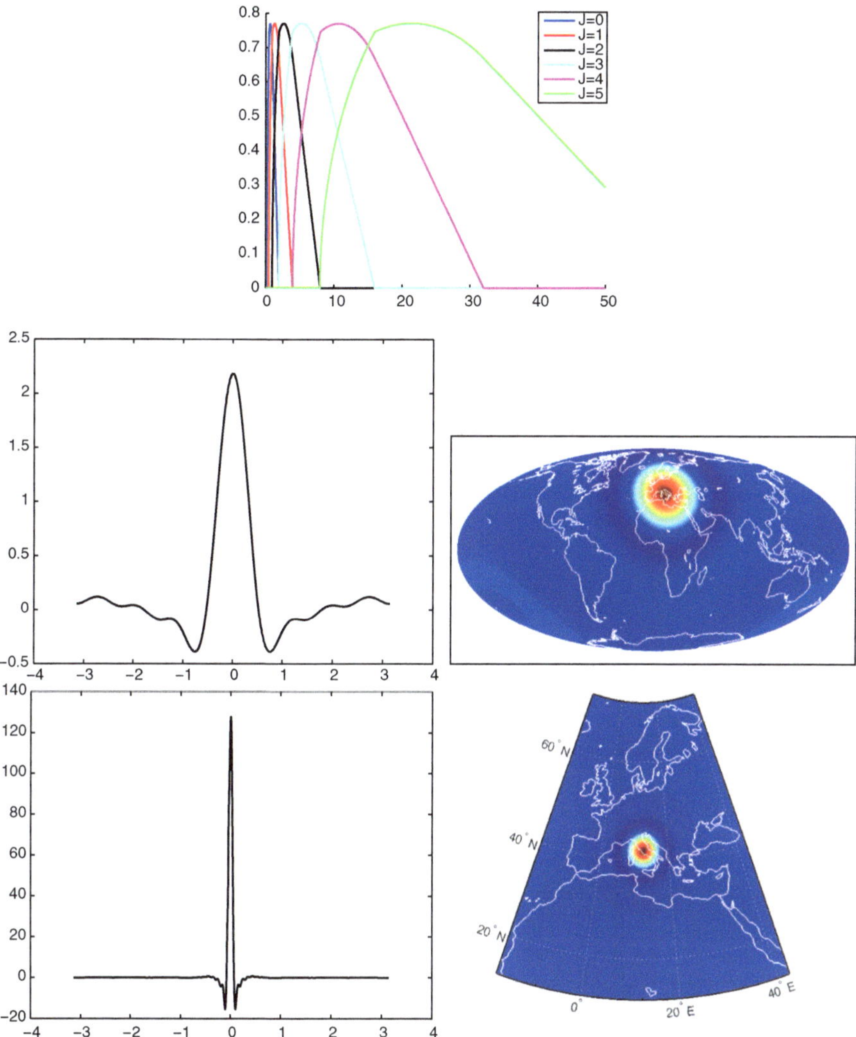

Fig. 7.14 Dilated generator of the modified Shannon P-wavelet with the chosen parameter $h = 0.25$ (*top*) as well as spatial plots of the wavelets Ψ_2 (*second row*) and Ψ_5 (*third row*) as 1D-function $[-\pi,\pi] \ni \vartheta \mapsto \Psi_J(\cos\vartheta)$ (*second and third row, left hand*) and zonal function $\Omega \ni \eta \mapsto \Psi_J(\xi\cdot\eta)$, where ξ is located in Rome (*second and third row, right hand*)

Figure 7.14 shows the (dilated) generator and the wavelet in the case $h = 0.25$.

(A1.3) **cp P-Wavelet**

Since

$$\Phi_J^{\wedge}(n) = \begin{cases} \left(1-2^{-J}n\right)^2\left(1+2^{1-J}n\right), & 0 \le n < 2^J \\ 0, & 2^J \le n \end{cases} \quad \forall J,n \in \mathbb{N}_0, \tag{7.90}$$

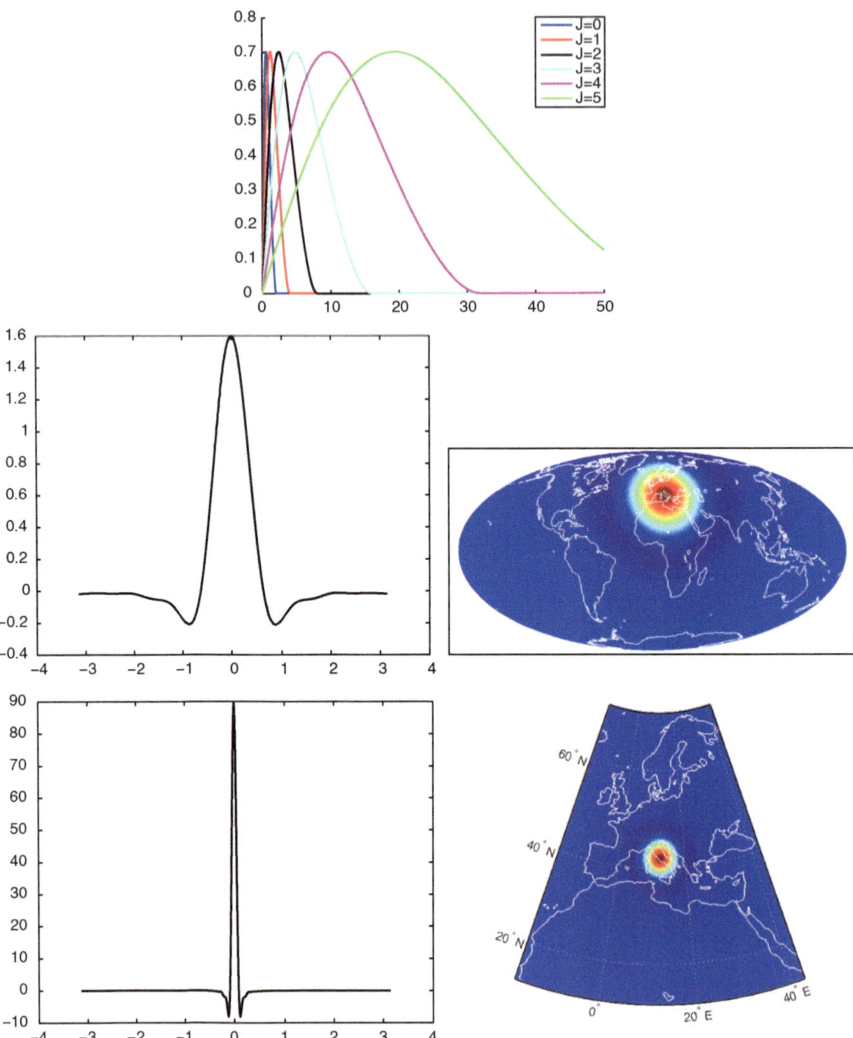

Fig. 7.15 Dilated generator of the cp P-wavelet (*top*) as well as spatial plots of the wavelets Ψ_2 (*second row*) and Ψ_5 (*third row*) as 1D-function $[-\pi,\pi] \ni \vartheta \mapsto \Psi_J(\cos\vartheta)$ (*second and third row, left hand*) and zonal function $\Omega \ni \eta \mapsto \Psi_J(\xi \cdot \eta)$, where ξ is located in Rome (*second and third row, right hand*)

we obtain

$$\Psi_J^\wedge(n) = \tilde{\Psi}_J^\wedge(n)$$

$$= \begin{cases} \left[\left(1-2^{-J-1}n\right)^4\left(1+2^{-J}n\right)^2 - \left(1-2^{-J}n\right)^4\left(1+2^{1-J}n\right)^2\right]^{1/2}, & 0 \leq n < 2^J \\ \left(1-2^{-J-1}n\right)^2\left(1+2^{-J}n\right), & 2^J \leq n < 2^{J+1} \\ 0, & 2^{J+1} \leq n \end{cases} \quad (7.91)$$

for all $J, n \in \mathbb{N}_0$; see Fig. 7.15.

(A1.4) **Blackman P-Wavelet**

The P-wavelet corresponding to the Blackman scaling function is given by the generator

$$\psi_0(x) = \tilde{\psi}_0(x) = \begin{cases} 0, & 0 \le x < \frac{1}{2}, \\ \sqrt{1 - \left[\frac{21}{50} - \frac{1}{2}\cos(2\pi x) + \frac{2}{25}\cos(4\pi x)\right]^2}, & \frac{1}{2} \le x < 1, \\ \frac{21}{50} - \frac{1}{2}\cos(\pi x) + \frac{2}{25}\cos(2\pi x), & 1 \le x < 2, \\ 0, & 2 \le x, \end{cases} \tag{7.92}$$

$x \in \mathbb{R}_0^+$. This generator and its wavelet are shown in Fig. 7.16.

(A2.1) **Rational P-Wavelet**

For a fixed parameter $s > 1$, we get

$$\psi_0(x) = \tilde{\psi}_0(x) = \sqrt{\left(1 + \frac{x}{2}\right)^{-2s} - (1+x)^{-2s}} \quad \forall x \in \mathbb{R}_0^+ \tag{7.93}$$

such that

$$\Psi_J^\wedge(n) = \tilde{\Psi}_J^\wedge(n) = \sqrt{\left(1 + 2^{-J-1}n\right)^{-2s} - \left(1 + 2^{-J}n\right)^{-2s}} \quad \forall n, J \in \mathbb{N}_0\,. \tag{7.94}$$

The generator and the wavelet are shown in Fig. 7.17 for $s = 2$.

(A2.2) **Modified Rational P-Wavelet**

For a fixed parameter $s > \frac{1}{2}$, we find

$$\psi_0(x) = \tilde{\psi}_0(x) = \sqrt{\left(1 + \frac{x^2}{4}\right)^{-2s} - (1+x^2)^{-2s}} \quad \forall x \in \mathbb{R}_0^+ \tag{7.95}$$

such that

$$\Psi_J^\wedge(n) = \tilde{\Psi}_J^\wedge(n) = \sqrt{\left(1 + 2^{-2J-2}n^2\right)^{-2s} - \left(1 + 2^{-2J}n^2\right)^{-2s}} \quad \forall n, J \in \mathbb{N}_0\,. \tag{7.96}$$

The corresponding illustration is Fig. 7.18, where $s = 1$.

(A2.3) **Abel–Poisson P-Wavelet**

A fixed parameter $R > 0$ is chosen. Then

$$\psi_0(x) = \tilde{\psi}_0(x) = \sqrt{\mathrm{e}^{-Rx} - \mathrm{e}^{-2Rx}} \quad \forall x \in \mathbb{R}_0^+. \tag{7.97}$$

Since this implies that

$$\Psi_J^\wedge(n) = \tilde{\Psi}_J^\wedge(n) = \sqrt{\mathrm{e}^{-R2^{-J}n} - \mathrm{e}^{-R2^{1-J}n}} \quad \forall n, J \in \mathbb{N}_0, \tag{7.98}$$

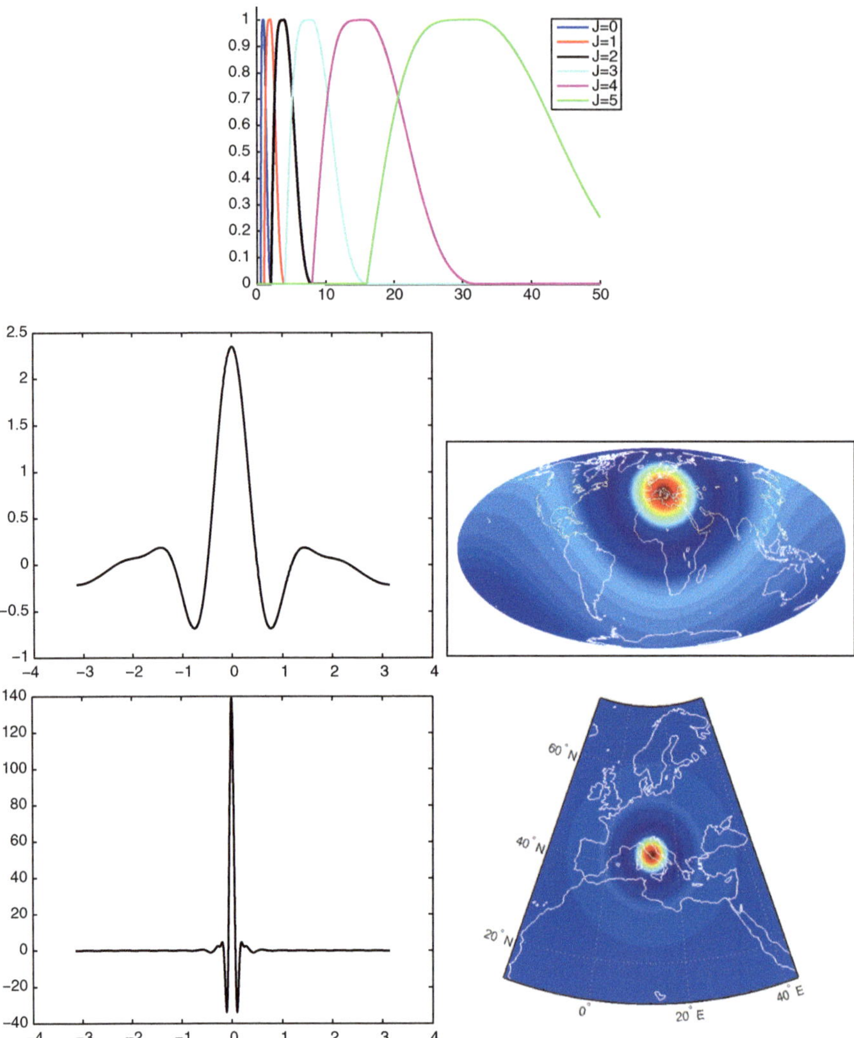

Fig. 7.16 Dilated generator of the Blackman P-wavelet (*top*) as well as spatial plots of the wavelets Ψ_2 (*second row*) and Ψ_5 (*third row*) as 1D-function $[-\pi,\pi] \ni \vartheta \mapsto \Psi_J(\cos\vartheta)$ (*second and third row, left hand*) and zonal function $\Omega \ni \eta \mapsto \Psi_J(\xi \cdot \eta)$, where ξ is located in Rome (*second and third row, right hand*)

we obtain

$$\Psi_J(t) = \tilde{\Psi}_J(t) = \sum_{n=0}^{\infty} \frac{2n+1}{4\pi} \sqrt{\mathrm{e}^{-R2^{-J}n} - \mathrm{e}^{-R2^{1-J}n}}\, P_n(t) \tag{7.99}$$

for all $t \in [-1,1]$; see Fig. 7.19. There is no closed representation for this series. However,

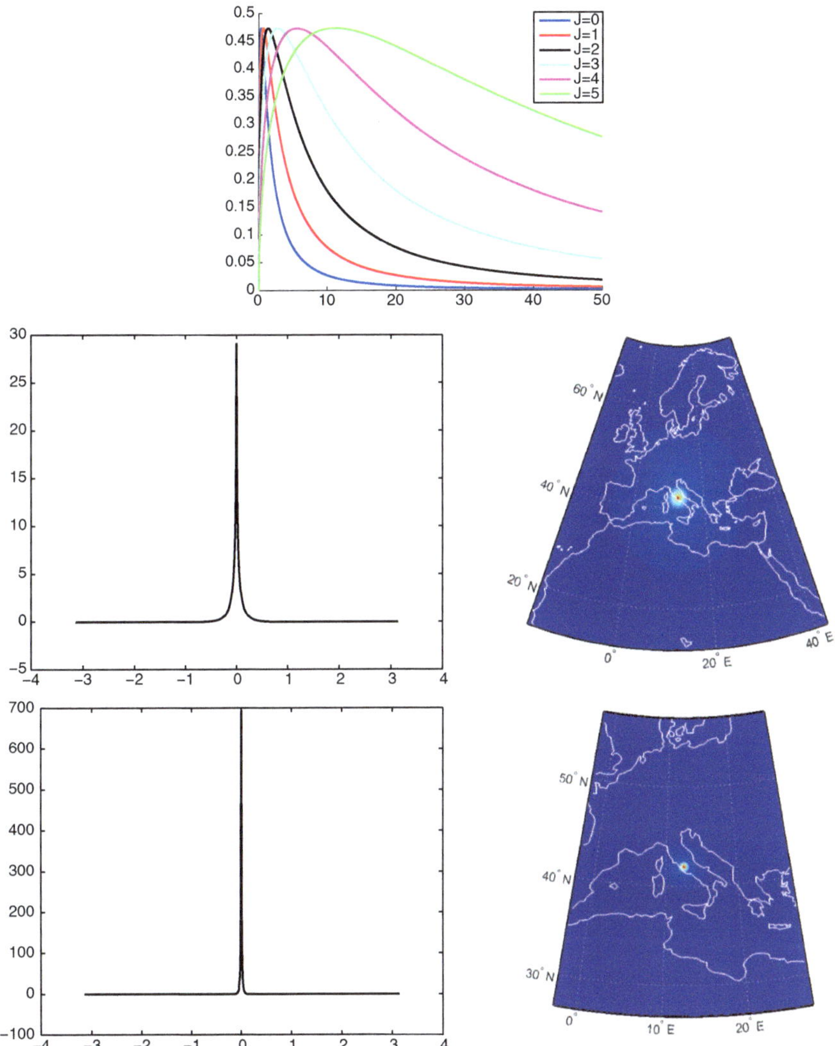

Fig. 7.17 Dilated generator of the rational P-wavelet with $s = 2$ (*top*) as well as spatial plots of the wavelets Ψ_2 (*second row*) and Ψ_5 (*third row*) as 1D-function $[-\pi, \pi] \ni \vartheta \mapsto \Psi_J(\cos\vartheta)$ (*second and third row, left hand*) and zonal function $\Omega \ni \eta \mapsto \Psi_J(\xi \cdot \eta)$, where ξ is located in Rome (*second and third row, right hand*). The series expansion of Ψ_J was truncated after degree 1,000

$$
\begin{aligned}
\left(\tilde{\Psi}_J * \Psi_J\right)(t) &= \sum_{n=0}^{\infty} \tilde{\Psi}_J^{\wedge}(n)\, \Psi_J^{\wedge}(n)\, \frac{2n+1}{4\pi} P_n(t) \\
&= \sum_{n=0}^{\infty} \left(\Phi_{J+1}^{\wedge}(n)\right)^2 \frac{2n+1}{4\pi} P_n(t) - \sum_{n=0}^{\infty} \left(\Phi_J^{\wedge}(n)\right)^2 \frac{2n+1}{4\pi} P_n(t)
\end{aligned}
$$

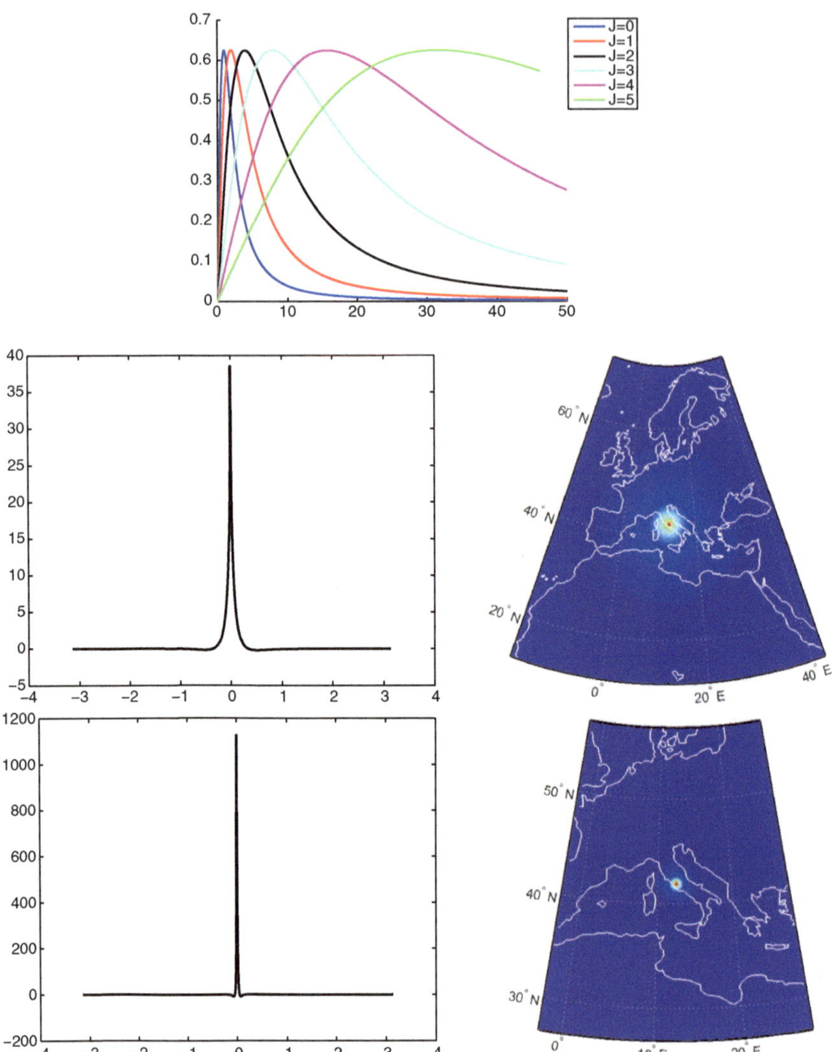

Fig. 7.18 Dilated generator of the modified rational P-wavelet with $s = 1$ (*top*) as well as spatial plots of the wavelets Ψ_2 (*second row*) and Ψ_5 (*third row*) as 1D-function $[-\pi,\pi] \ni \vartheta \mapsto \Psi_J(\cos\vartheta)$ (*second and third row, left hand*) and zonal function $\Omega \ni \eta \mapsto \Psi_J(\xi\cdot\eta)$, where ξ is located in Rome (*second and third row, right hand*). The series expansion of Ψ_J was truncated after degree 1,000

$$= \sum_{n=0}^{\infty}\left(\exp\left(-R2^{-J}\right)\right)^n \frac{2n+1}{4\pi}P_n(t) - \sum_{n=0}^{\infty}\left(\exp\left(-R2^{1-J}\right)\right)^n \frac{2n+1}{4\pi}P_n(t)$$

$$= \frac{1}{4\pi}\frac{1-\exp\left(-R2^{1-J}\right)}{\left(1+\exp(-R2^{1-J})-2\exp(-R2^{-J})t\right)^{3/2}}$$

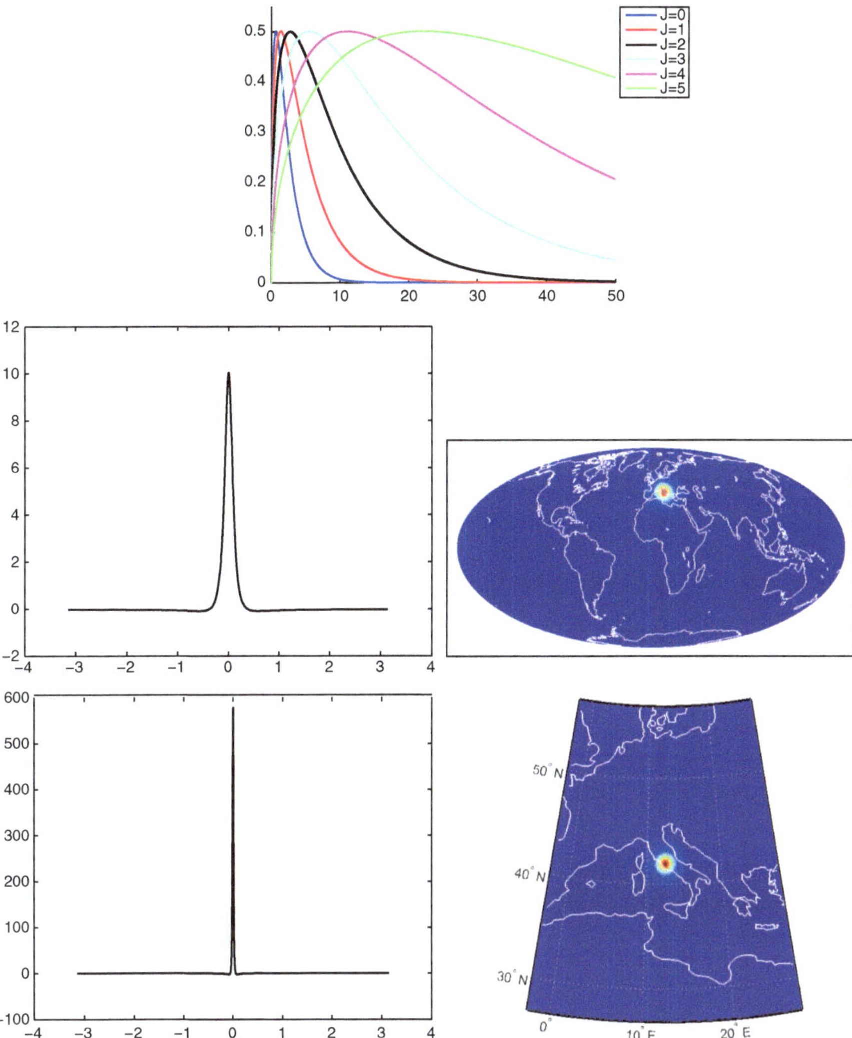

Fig. 7.19 Dilated generator of the Abel–Poisson P-wavelet with $R = 1$ (*top*) as well as spatial plots of the wavelets Ψ_2 (*second row*) and Ψ_5 (*third row*) as 1D-function $[-\pi, \pi] \ni \vartheta \mapsto \Psi_J(\cos\vartheta)$ (*second and third row, left hand*) and zonal function $\Omega \ni \eta \mapsto \Psi_J(\xi \cdot \eta)$, where ξ is located in Rome (*second and third row, right hand*). The series expansion of Ψ_J was truncated after degree 1,000

$$-\frac{1}{4\pi}\,\frac{1-\exp\left(-R2^{2-J}\right)}{\left(1+\exp(-R2^{2-J})-2\exp(-R2^{1-J})\,t\right)^{3/2}} \tag{7.100}$$

for all $t \in [-1, 1]$.

(A2.4) **Gauß–Weierstraß P-Wavelet**
In this case,

$$\psi_0(x) = \tilde{\psi}_0(x) = \sqrt{e^{-Rx(x/2+1)} - e^{-2Rx(x+1)}} \quad \forall x \in \mathbb{R}_0^+ \tag{7.101}$$

for a fixed parameter $R > 0$, such that

$$\Psi_J^\wedge(n) = \tilde{\Psi}_J^\wedge(n) = \sqrt{\exp\left(-R2^{-J}n\left(2^{-J-1}n+1\right)\right) - \exp\left(-R2^{1-J}n\left(2^{-J}n+1\right)\right)} \tag{7.102}$$

for all $n, J \in \mathbb{N}_0$; see Fig. 7.20.
(B) **Tykhonov–Philips P-Wavelet**
The definition of the P-wavelets in (7.79) yields here

$$\Psi_J^\wedge(n) = \tilde{\Psi}_J^\wedge(n) = \sqrt{\frac{1}{\left(1+\gamma_{J+1,n}^2\right)^2} - \frac{1}{\left(1+\gamma_{J,n}^2\right)^2}} \quad \forall n, J \in \mathbb{N}_0 . \tag{7.103}$$

The wavelets Ψ_2 and Ψ_5 are plotted in Fig. 7.21 for $\gamma_{J,n} := \frac{n}{J+1}$.

As we already observed, the scaling functions allow us to look at the investigated function F at different resolutions. The larger the scale J, the higher is the resolution of the obtained image. In other words, with increasing scale, more and more details of F become visible. It is, therefore, sometimes interesting to look for precisely those details which are added to the solution. This task can be solved by wavelets. This is based on the scale-step property, which is the fundamental property of wavelets.

Definition 7.29. Let $\{\Psi_J\}_{J\in\mathbb{N}_0\cup\{-1\}}$ and $\{\tilde{\Psi}_J\}_{J\in\mathbb{N}_0\cup\{-1\}}$ be a primal and a dual wavelet, respectively, corresponding to the same scaling function. Then the spaces

$$W_J := \left\{ \tilde{\Psi}_J * \Psi_J * F \,\middle|\, F \in \mathrm{L}^2(\Omega) \right\}, \quad J \in \mathbb{N}_0 \cup \{-1\}, \tag{7.104}$$

are the corresponding **detail spaces**.

Note that Theorem 7.8 in combination with Theorem 7.2 yields that $W_J \subset \mathrm{L}^2(\Omega)$ for all $J \in \mathbb{N}_0$.
A fundamental property of spherical wavelets (as well as most other wavelets) is the scale-step property (see also Theorem 3.35 on p. 74 for the case of Haar wavelets). We will prove this feature now.

Theorem 7.30 (Scale-Step Property). *Let the* $\mathrm{L}^2[-1,1]$*-families* $\{\Psi_J\}_{J\in\mathbb{N}_0\cup\{-1\}}$ *and* $\{\tilde{\Psi}_J\}_{J\in\mathbb{N}_0\cup\{-1\}}$ *be a primal and a dual wavelet corresponding to the scaling function* $\{\Phi_J\}_{J\in\mathbb{N}_0} \subset \mathrm{L}^2[-1,1]$*. If* $F \in \mathrm{L}^2(\Omega)$*, then*

$$\Phi_{J_2}^{(2)} * F = \Phi_{J_1}^{(2)} * F + \sum_{J=J_1}^{J_2-1} \tilde{\Psi}_J * \Psi_J * F \tag{7.105}$$

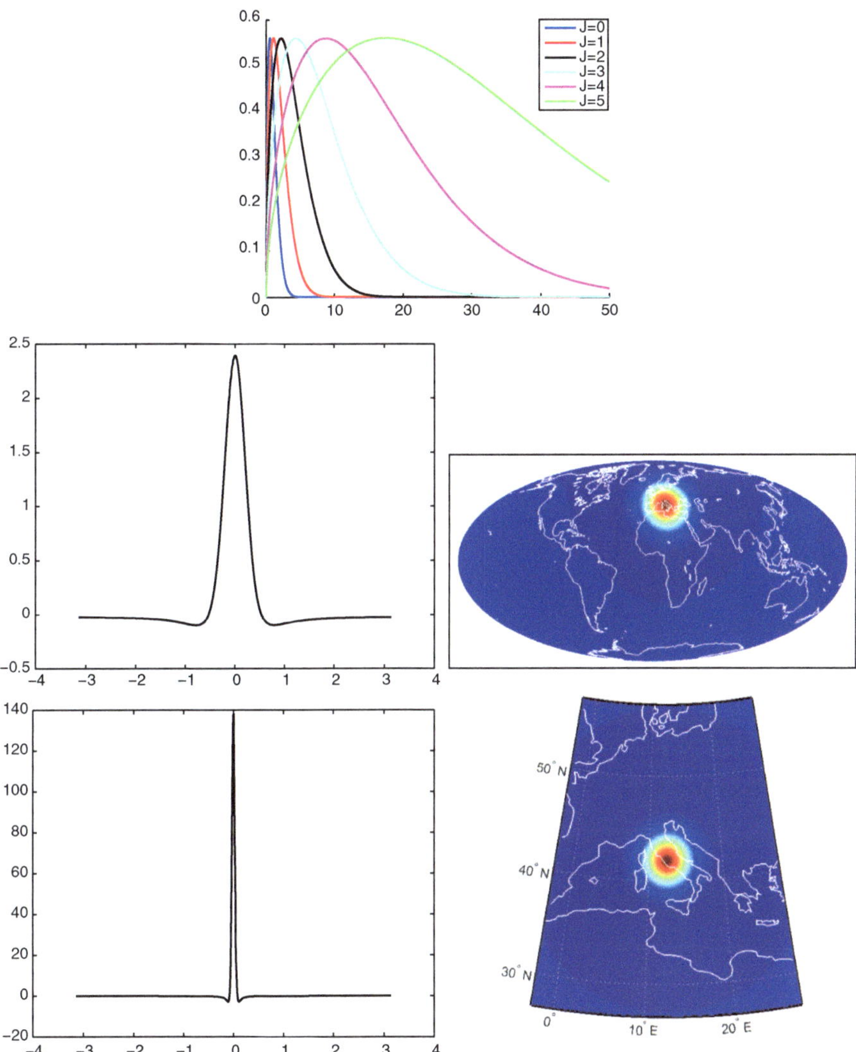

Fig. 7.20 Dilated generator of the Gauß–Weierstraß P-wavelet with $R = 1$ (*top*) as well as spatial plots of the wavelets Ψ_2 (*second row*) and Ψ_5 (*third row*) as 1D-function $[-\pi,\pi] \ni \vartheta \mapsto \Psi_J(\cos\vartheta)$ (*second and third row, left hand*) and zonal function $\Omega \ni \eta \mapsto \Psi_J(\xi\cdot\eta)$, where ξ is located in Rome (*second and third row, right hand*). The series expansion of Ψ_J was truncated after degree 1,000

for all $J_1, J_2 \in \mathbb{N}_0$ *with* $J_1 < J_2$ *and*

$$F = \Phi_{J_1}^{(2)} * F + \sum_{J=J_1}^{\infty} \tilde{\Psi}_J * \Psi_J * F \tag{7.106}$$

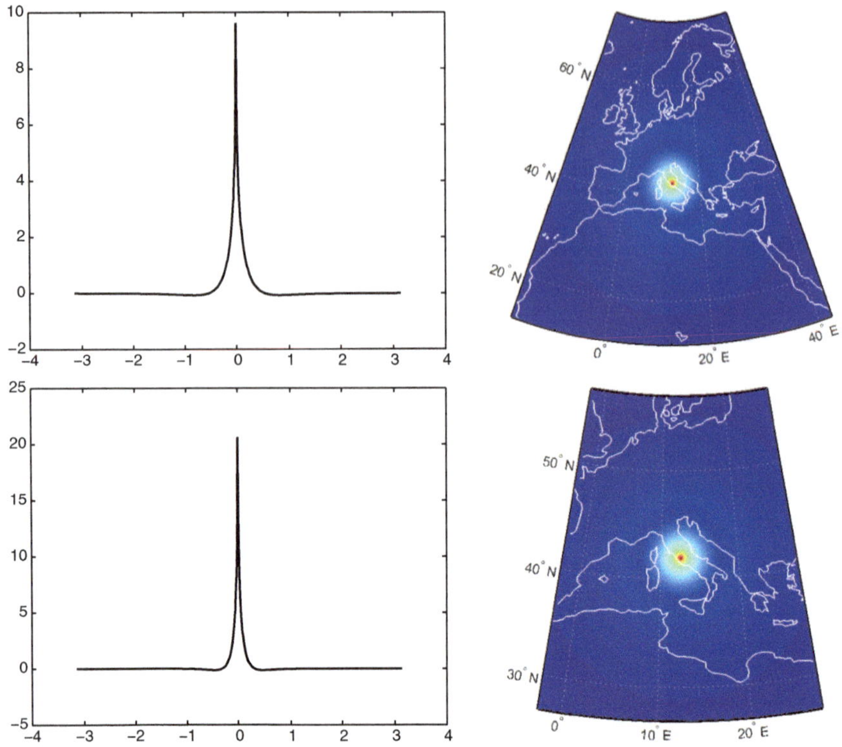

Fig. 7.21 Tykhonov–Philips P-wavelet Ψ_J with the chosen sequence $\gamma_{J,n} := \frac{n}{J+1}$: The *left-hand* plots show $[-\pi,\pi] \ni \vartheta \mapsto \Psi_J(\cos\vartheta)$, whereas the *right-hand* plots show the corresponding zonal functions $\Omega \ni \eta \mapsto \Psi_J(\xi \cdot \eta)$ with ξ located in Rome. The *first row* corresponds to $J = 2$ and the *second one* to $J = 5$. The series expansion was truncated after degree 1,000

in the sense of $\mathrm{L}^2(\Omega)$ *for all* $J_1 \in \mathbb{N}_0$. *Moreover,*

$$V_{J_2} = V_{J_1} + \sum_{J=J_1}^{J_2-1} W_J \tag{7.107}$$

for all $J_1, J_2 \in \mathbb{N}_0$ *with* $J_1 < J_2$.

Proof.

(1) Let us step one scale forward:
Let $F \in \mathrm{L}^2(\Omega)$ and $J \in \mathbb{N}_0$ be arbitrary. Due to (7.23) and (7.74), we obtain

$$\Phi_{J+1}^{(2)} * F = \sum_{n=0}^{\infty} \sum_{j=1}^{2n+1} \left(\Phi_{J+1}^{\wedge}(n)\right)^2 F^{\wedge}(n,j) Y_{n,j} \,,$$

$$\Phi_J^{(2)} * F = \sum_{n=0}^{\infty} \sum_{j=1}^{2n+1} \left(\Phi_J^{\wedge}(n)\right)^2 F^{\wedge}(n,j) Y_{n,j} \,, \tag{7.108}$$

and

$$\begin{aligned}\tilde{\Psi}_J * \Psi_J * F &= \sum_{n=0}^{\infty} \sum_{j=1}^{2n+1} \tilde{\Psi}_J^{\wedge}(n)\, \Psi_J^{\wedge}(n)\, F^{\wedge}(n,j)\, Y_{n,j} \\ &= \sum_{n=0}^{\infty} \sum_{j=1}^{2n+1} \left[\left(\Phi_{J+1}^{\wedge}(n)\right)^2 - \left(\Phi_J^{\wedge}(n)\right)^2 \right] F^{\wedge}(n,j)\, Y_{n,j} \\ &= \Phi_{J+1}^{(2)} * F - \Phi_J^{(2)} * F \end{aligned} \tag{7.109}$$

in the sense of $L^2(\Omega)$.

(2) Identity (7.105):
The first part of the theorem now follows from (7.109) by induction.

(3) Identity (7.106):
The second part is a combination of (7.105) and Theorem 7.21.

(4) Scale and detail spaces:
From (7.105) we already obtain that

$$V_{J_2} \subset V_{J_1} + \sum_{J=J_1}^{J_2-1} W_J\,. \tag{7.110}$$

However, the inclusion "$\supset$," as we will explain now for the case $J_2 = J_1 + 1$ (the rest is again an easy induction), is not trivial. An arbitrary element of V_{J_1} is represented by $\Phi_{J_1}^{(2)} * F_1$ with $F_1 \in L^2(\Omega)$, and an arbitrary element of W_{J_1} is represented by $\tilde{\Psi}_{J_1} * \Psi_{J_1} * F_2$ with $F_2 \in L^2(\Omega)$. The possibility that $F_1 \neq F_2$ is the reason why (7.105) does not suffice to show "$\supset$" in (7.107). We have to find, similarly to the proof of Theorem 7.22, a function $G \in L^2(\Omega)$ such that $\Phi_{J_1+1}^{(2)} * G = \Phi_{J_1}^{(2)} * F_1 + \tilde{\Psi}_{J_1} * \Psi_{J_1} * F_2$.

We will write J instead of J_1 to simplify the notation. Now let $G \in L^2(\Omega)$ be defined by

$$G^{\wedge}(n,j) := \begin{cases} \dfrac{\left(\Phi_J^{\wedge}(n)\right)^2 F_1^{\wedge}(n,j) + \left(\left(\Phi_{J+1}^{\wedge}(n)\right)^2 - \left(\Phi_J^{\wedge}(n)\right)^2\right) F_2^{\wedge}(n,j)}{\left(\Phi_{J+1}^{\wedge}(n)\right)^2}, & \text{if } \Phi_{J+1}^{\wedge}(n) \neq 0, \\ 0, & \text{else.} \end{cases} \tag{7.111}$$

First, we verify again that $G \in L^2(\Omega)$ is indeed valid. We observe that (S1) and (S3) (see Definition 7.11) imply

$$\left| \left(\frac{\Phi_J^{\wedge}(n)}{\Phi_{J+1}^{\wedge}(n)} \right)^2 F_1^{\wedge}(n,j) \right| \leq \left| F_1^{\wedge}(n,j) \right| \tag{7.112}$$

and

$$\left| \frac{\left(\Phi_{J+1}^{\wedge}(n)\right)^2 - \left(\Phi_J^{\wedge}(n)\right)^2}{\left(\Phi_{J+1}^{\wedge}(n)\right)^2} F_2^{\wedge}(n,j) \right| \leq \frac{\left(\Phi_{J+1}^{\wedge}(n)\right)^2}{\left(\Phi_{J+1}^{\wedge}(n)\right)^2} \left|F_2^{\wedge}(n,j)\right| = \left|F_2^{\wedge}(n,j)\right|. \tag{7.113}$$

Since $F_1, F_2 \in \mathrm{L}^2(\Omega)$ and $\mathrm{L}^2(\Omega)$ is a linear space, we can conclude that $G \in \mathrm{L}^2(\Omega)$. Finally, (7.23), (7.111), (7.72), (7.74), and again (7.23) yield (in the sense of $\mathrm{L}^2(\Omega)$)

$$\begin{aligned}
&\Phi_{J+1}^{(2)} * G \\
&\quad = \sum_{\substack{n=0 \\ \Phi_{J+1}^{\wedge}(n)\neq 0}}^{\infty} \sum_{j=1}^{2n+1} \left(\Phi_{J+1}^{\wedge}(n)\right)^2 G^{\wedge}(n,j)\, Y_{n,j} \\
&\quad = \sum_{\substack{n=0 \\ \Phi_{J+1}^{\wedge}(n)\neq 0}}^{\infty} \sum_{j=1}^{2n+1} \left[\left(\Phi_J^{\wedge}(n)\right)^2 F_1^{\wedge}(n,j) + \left(\left(\Phi_{J+1}^{\wedge}(n)\right)^2 - \left(\Phi_J^{\wedge}(n)\right)^2\right) F_2^{\wedge}(n,j)\right] Y_{n,j} \\
&\quad = \sum_{n=0}^{\infty} \sum_{j=1}^{2n+1} \left[\left(\Phi_J^{\wedge}(n)\right)^2 F_1^{\wedge}(n,j) + \left(\left(\Phi_{J+1}^{\wedge}(n)\right)^2 - \left(\Phi_J^{\wedge}(n)\right)^2\right) F_2^{\wedge}(n,j)\right] Y_{n,j} \\
&\quad = \sum_{n=0}^{\infty} \sum_{j=1}^{2n+1} \left[\left(\Phi_J^{\wedge}(n)\right)^2 F_1^{\wedge}(n,j) + \tilde{\Psi}_J^{\wedge}(n)\Psi_J^{\wedge}(n)\, F_2^{\wedge}(n,j)\right] Y_{n,j} \\
&\quad = \Phi_J^{(2)} * F_1 + \tilde{\Psi}_J * \Psi_J * F_2 .
\end{aligned} \tag{7.114}$$

Hence, (7.107) is proved. □

Note that, in analogy to the spherical Fourier transform (see Definition 5.27 on p. 123), one can define a spherical wavelet transform.

Definition 7.31. Let $\{\Psi_J\}_{J\in\mathbb{N}_0\cup\{-1\}} \subset \mathrm{L}^2[-1,1]$ be a primal wavelet. Then the mapping

$$\begin{aligned}
(\mathrm{SWT})_J : \mathrm{L}^2(\Omega) &\to \mathrm{L}^2(\Omega) \\
F &\mapsto \Psi_J * F
\end{aligned} \tag{7.115}$$

is called the corresponding **spherical wavelet transform** at scale $J \in \mathbb{N}_0 \cup \{-1\}$.

With the introduced notations, some of the results in Theorem 7.30 can be reformulated as follows: For all $F \in \mathrm{L}^2(\Omega)$ and all $J_1, J_2 \in \mathbb{N}_0$ with $J_1 < J_2$,

$$\Phi_{J_2}^{(2)} * F = \Phi_{J_1}^{(2)} * F + \sum_{J=J_1}^{J_2-1} \tilde{\Psi}_J * (\mathrm{SWT})_J(F), \tag{7.116}$$

$$\Phi_{J_2}^{(2)} * F = \sum_{J=-1}^{J_2-1} \tilde{\Psi}_J * (\mathrm{SWT})_J(F), \tag{7.117}$$

$$F = \Phi_{J_1}^{(2)} * F + \sum_{J=J_1}^{\infty} \tilde{\Psi}_J * (\mathrm{SWT})_J(F), \tag{7.118}$$

$$F = \sum_{J=-1}^{\infty} \tilde{\Psi}_J * (\mathrm{SWT})_J(F), \tag{7.119}$$

where the series converge with respect to $\mathrm{L}^2(\Omega)$.

Note that the last equation can, in particular, be interpreted as the wavelet analogue of the inverse Fourier transform

$$F = \sum_{n=0}^{\infty} \sum_{j=1}^{2n+1} F^{\wedge}(n,j)\, Y_{n,j} \tag{7.120}$$

(with respect to $\mathrm{L}^2(\Omega)$) and is, therefore, called the **inverse spherical wavelet transform**.

Remark 7.32. Often, wavelet theories require a so-called **frame** condition. This condition means that the wavelet transform and its inverse are continuous operations. Roughly speaking, if $(\mathrm{WT})(F)$ is the wavelet transform of a function F, then constants $C_1, C_2 > 0$ should exist such that

$$C_1 \|F\|_a \leq \|(\mathrm{WT})(F)\|_b \leq C_2 \|F\|_a \quad \forall F \tag{7.121}$$

with appropriately chosen norms $\|\cdot\|_a$ and $\|\cdot\|_b$ for the functions and their transforms, respectively. Since the Haar wavelets (see Sect. 3.4) yield an orthonormal basis, we even get in this case

$$\|F\|_{\mathrm{L}^2(\mathbb{R})} = \left(\sum_{j=-\infty}^{+\infty} \sum_{k=-\infty}^{+\infty} \langle F, H_{j,k} \rangle_{\mathrm{L}^2(\mathbb{R})}^2 \right)^{1/2}, \tag{7.122}$$

where the right-hand side is a norm for the Haar (wavelet) transform. Such a frame where $C_1 = C_2$ is called a **tight frame**.

Also for the spherical case, wavelets which satisfy a corresponding frame condition were constructed. Since this condition guarantees a stable transform and a stable inverse transform, particular examples of these wavelets are also called S-wavelets with "S" for "stability." For further details, see [77]. A generalized approach which is not only valid on the sphere can be found in [125].

It is beyond the scope of this textbook to completely cover the topic of frames on the sphere—in particular, also because this would require lengthy preparations including certain other theoretical concepts such as results on Marcinkiewicz–Zygmund inequalities (a technique useful for the construction of methods of numerical integration; see also the end of Sect. 7.4). Moreover, there do exist not only **wavelet frames** but also **polynomial frames**, though both concepts cannot be completely distinguished. For this reason, we refer to the excellent survey article

[124], where a general concept on polynomial frames (i.e., frames corresponding to bandlimited kernels) is presented. This paper also covers and cites a variety of works on spherical frames. Examples of investigations of frames on the sphere include [122, 123, 141], where frames on the sphere with particular properties of localization are constructed via Marcinkiewicz–Zygmund inequalities; [148], where a different approach (based on a discrete spherical Fourier transform) for localized spherical trial functions constituting a frame is developed; [22], where two concepts for the construction of frames based on the (group-theoretic) spherical wavelets from [10,11] are presented; and [38,39], where a general frame concept is developed and applied to the sphere. Frame experts hopefully excuse this brief treatment of the frames, which already stops here.

Remember Fig. 7.11 on p. 209. We observed that the scaling functions of different scales are able to represent a given function at different resolutions. With increasing scale, more and more details are added. These details can be extracted as $\tilde{\Psi}_J * \Psi_J * F$; see Fig. 7.22.
The numerical implementation of the spherical wavelet analysis consists of two essential parts:

(a) The calculation of the involved kernel: If the Abel–Poisson kernel is not chosen, the computation of a finite expansion in Legendre polynomials is required, for which the Clenshaw algorithm is a well-established method (see Theorem 3.17 on p. 47). For the iterated kernels $\Phi_J^{(2)}$ or the convolved kernels $\tilde{\Psi}_J * \Psi_J$, Theorem 7.8 is used.
(b) The implementation of the spherical convolution: This involves a numerical integration method on Ω. We will, therefore, study this topic in the next section. Note that, in practice, one has a point grid $\{\xi_{r,s}\}_{\substack{r=0,\ldots,N\\ s=0,\ldots,M}}$ for plotting the result. Hence, every pair (r,s) requires the calculation of the integral

$$(K * F)(\xi_{r,s}) = \int_\Omega K(\xi_{r,s} \cdot \eta)\, F(\eta)\, \mathrm{d}\omega(\eta), \tag{7.123}$$

where $K = \Phi_J^{(2)}$ or $K = \tilde{\Psi}_J * \Psi_J$.

As the plots of the scaling functions and the wavelets show, the advantage of this method is the localization of the involved kernels. If, for example, F is noisy in a small subregion $R \subset \Omega$, then the integrals

$$\int_\Omega K(\xi \cdot \eta)\, F(\eta)\, \mathrm{d}\omega(\eta) \tag{7.124}$$

will only be influenced by the noise if ξ is in R or a small neighborhood of R, since $K(\xi \cdot \eta) \approx 0$, if η is far away from ξ.

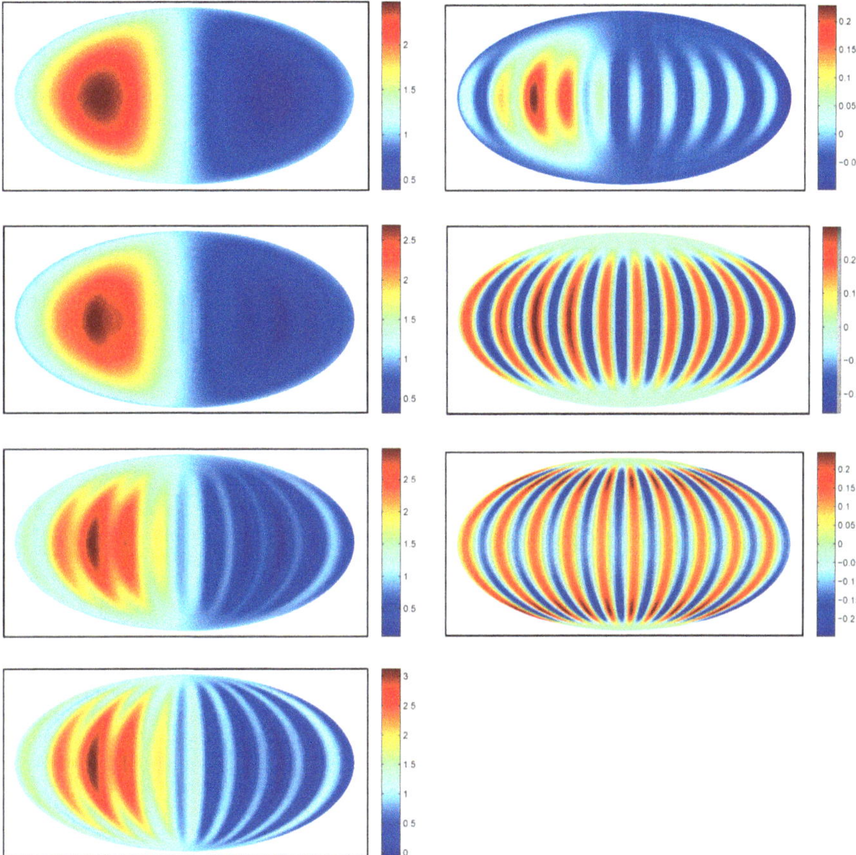

Fig. 7.22 The *left-hand* column shows $\Phi_J^{(2)} * F$, as in Fig. 7.11 (see p. 209), from $J = 3$ (*top*) to $J = 6$ (*bottom*). The *right-hand* column shows $\tilde{\Psi}_J * \Psi_J * F$ for the corresponding scales. Note that the scale-step property (Theorem 7.30) yields $\Phi_{J+1}^{(2)} * F = \Phi_J^{(2)} * F + \tilde{\Psi}_J * \Psi_J * F$

Furthermore, the localization of K has the advantage that the numerical calculation of $(K * F)(\xi)$ can be accelerated by merely integrating over a small neighborhood of ξ instead of the whole sphere, where the error of the inaccurate integration remains small. A strategy for the reduction of the domain of the integration is described in [133].

Moreover, the multiresolution analysis allows us to extract features of a chosen spatial extension from the signal. Spherical wavelets are, for example, a common tool for the denoising of the monthly gravity models obtained from the satellite mission GRACE (see [205] for details on the mission and [52, 54, 56, 57] for a wavelet-based denoising of GRACE gravity models).

7.4 Numerical Integration on the Sphere (Very Briefly)

This section gives a small insight into the numerical integration based on regular or irregular grids on the sphere. This topic alone could fill a whole book. For this reason, we can only deliver some selected ideas here and refer to a few examples of further publications at the end of this section.

A common quadrature method on the sphere uses an equiangular grid (see Example 5.35 on p. 135) and is due to J.R. Driscoll and R.M. Healy (see [46]).

Theorem 7.33 (Driscoll–Healy Method). *Let* $F \in \mathrm{Harm}_{0\ldots m}(\Omega)$ *for a given odd number* $m \in \mathbb{N}$. *Moreover, let the points* $\eta_{p,q} = \eta(\vartheta_p, \varphi_q) \in \Omega$ *be given by the polar coordinates*

$$\vartheta_p := \frac{\pi p}{m+1}, \quad p = 0, \ldots, m \quad \textit{(latitude)} \tag{7.125}$$

$$\varphi_q := \frac{2\pi q}{m+1}, \quad q = 0, \ldots, m \quad \textit{(longitude)} \tag{7.126}$$

and the weights $a_0, \ldots, a_m \in \mathbb{R}$ *be given by the formula*

$$a_p := \frac{4}{m+1} \sin\left(\frac{p\pi}{m+1}\right) \sum_{s=0}^{\frac{m+1}{2}-1} \frac{1}{2s+1} \sin\left((2s+1)\frac{p\pi}{m+1}\right), \quad p = 0, \ldots, m \, . \tag{7.127}$$

Then

$$\int_\Omega F(\eta)\, \mathrm{d}\omega(\eta) = \frac{2\pi}{m+1} \sum_{p=0}^{m} a_p \sum_{q=0}^{m} F(\eta_{p,q}) \, . \tag{7.128}$$

Note that the degree for the polynomial exactness has to be odd.

Not every application, however, allows a free choice of the quadrature grid. For this reason, we will study here how general grids can be treated (see also [66, Sect. 3.3 and Chap. 7]).

A quadrature method is often given in the following form:

$$\int_\Omega F(\xi)\, \mathrm{d}\omega(\xi) \approx \sum_{l=1}^{N} w_l F(\eta_l) \, , \tag{7.129}$$

where $\{\eta_l\}_{l=1,\ldots,N} \subset \Omega$ is a point grid (the quadrature grid) and $w_1, \ldots, w_N \in \mathbb{R}$ are weights. If we require a polynomial accuracy of the method up to degree $d \in \mathbb{N}_0$, then

$$\int_\Omega \sum_{n=0}^{d} \sum_{j=1}^{2n+1} a_{n,j} Y_{n,j}(\xi)\, \mathrm{d}\omega(\xi) = \sum_{l=1}^{N} w_l \sum_{n=0}^{d} \sum_{j=1}^{2n+1} a_{n,j} Y_{n,j}(\eta_l) \tag{7.130}$$

must be valid for every choice of Fourier coefficients $a_{0,1}, a_{1,1}, a_{1,2}, \ldots, a_{d,2d+1} \in \mathbb{R}$. As a consequence, (choose $a_{n,j} = \delta_{nm}\delta_{jk}$), we obtain

$$\int_\Omega Y_{m,k}(\xi)\,\mathrm{d}\omega(\xi) = \sum_{l=1}^{N} w_l\,Y_{m,k}(\eta_l)\ \forall m = 0,\ldots,d\ \forall k = 1,\ldots,2m+1\,. \tag{7.131}$$

Since $\{Y_{n,j}\}_{n\in\mathbb{N}_0,\,j=1,\ldots,2n+1}$ is an $\mathrm{L}^2(\Omega)$-orthonormal system and $Y_{0,1} \equiv (4\pi)^{-1/2}$, we have

$$\int_\Omega Y_{m,k}(\xi)\cdot 1\,\mathrm{d}\omega(\xi) = \delta_{m0}\delta_{k1}\sqrt{4\pi}\,. \tag{7.132}$$

Hence, (7.131) is equivalent to

$$\sum_{l=1}^{N} w_l\,Y_{m,k}(\eta_l) = \delta_{m0}\delta_{k1}\sqrt{4\pi} \quad \forall m = 0,\ldots,d \quad \forall k = 1,\ldots,2m+1\,. \tag{7.133}$$

This is a system of linear equations with the matrix

$$\left(Y_{m,k}(\eta_l)\right)_{\substack{m=0,\ldots,d;\,k=1,\ldots,2m+1\\ l=1,\ldots,N}}\,. \tag{7.134}$$

It appears to be reasonable to have a quadratic matrix such that the size of the point grid should be $N = (d+1)^2$ (see Corollary 5.7 on p. 101). From our considerations regarding the Extended Haar's theorem (Theorem 5.34 on p. 133), we already know that not every point grid produces here a regular matrix. If, however, the matrix is regular, then (7.133) shows us how to determine the weights of the quadrature method (provided that the matrix is also well conditioned such that a stable inversion can be performed numerically).

Let us summarize what we obtained so far.

Definition 7.34. A point set $X_N := \{\eta_1,\ldots,\eta_N\} \subset \Omega$ with $N = (d+1)^2$, $d \in \mathbb{N}_0$, is called a **fundamental system** relative to $\mathrm{Harm}_{0\ldots d}(\Omega)$ if the matrix

$$\mathrm{matr}_{X_N}(Y_{0,1},\ldots,Y_{d,2d+1}) = \left(Y_{n,j}(\eta_l)\right)_{\substack{n=0,\ldots,d;\,j=1,\ldots,2n+1\\ l=1,\ldots,N}} \tag{7.135}$$

is regular.

Theorem 7.35. *If $X_N \subset \Omega$ (with $N = (d+1)^2$, $d \in \mathbb{N}_0$) is a fundamental system relative to* $\mathrm{Harm}_{0\ldots d}(\Omega)$ *and $w = (w_1,\ldots,w_N)^{\mathrm{T}} \in \mathbb{R}^N$ is the solution of the system*

$$\sum_{l=1}^{N} w_l\,Y_{m,k}(\eta_l) = \delta_{m0}\delta_{k1}\sqrt{4\pi} \quad \forall m = 0,\ldots,d \quad \forall k = 1,\ldots,2m+1, \tag{7.136}$$

then

$$\int_\Omega F(\xi)\,\mathrm{d}\omega(\xi) = \sum_{l=1}^{N} w_l\,F(\eta_l) \tag{7.137}$$

for all $F \in \mathrm{Harm}_{0\ldots d}(\Omega)$. *Moreover, (7.136) is equivalent to*

$$\left(\sum_{n=0}^{d} \frac{2n+1}{4\pi} P_n(\eta_i \cdot \eta_k) \right)_{i,k=1,\ldots,N} w = \begin{pmatrix} 1 \\ \vdots \\ 1 \end{pmatrix}. \tag{7.138}$$

Proof. It only remains to show that (7.136) and (7.138) are equivalent. Note that (7.136) can be represented as

$$\mathrm{matr}_{X_N}\left(Y_{0,1},\ldots,Y_{d,2d+1}\right) w = \left(\sqrt{4\pi}, 0, \ldots, 0\right)^{\mathrm{T}}. \tag{7.139}$$

If we multiply this equation by the (invertible) transposed matrix from the left-hand side, we get

$$\begin{aligned}
&\left(\mathrm{matr}_{X_N}\left(Y_{0,1},\ldots,Y_{d,2d+1}\right)\right)^{\mathrm{T}} \left(\mathrm{matr}_{X_N}\left(Y_{0,1},\ldots,Y_{d,2d+1}\right)\right) w \\
&= \begin{pmatrix} Y_{0,1}(\eta_1) & \ldots\ldots\ldots & Y_{d,2d+1}(\eta_1) \\ \vdots & & \vdots \\ Y_{0,1}(\eta_i) & \ldots\ldots\ldots & Y_{d,2d+1}(\eta_i) \\ \vdots & & \vdots \\ Y_{0,1}(\eta_N) & \ldots\ldots\ldots & Y_{d,2d+1}(\eta_N) \end{pmatrix} \\
&\quad \times \begin{pmatrix} Y_{0,1}(\eta_1) & \ldots & Y_{0,1}(\eta_k) & \ldots & Y_{0,1}(\eta_N) \\ \vdots & & \vdots & & \vdots \\ \vdots & & \vdots & & \vdots \\ \vdots & & \vdots & & \vdots \\ Y_{d,2d+1}(\eta_1) & \ldots & Y_{d,2d+1}(\eta_k) & \ldots & Y_{d,2d+1}(\eta_N) \end{pmatrix} w \\
&= \left(\sum_{n=0}^{d} \sum_{j=1}^{2n+1} Y_{n,j}(\eta_i)\, Y_{n,j}(\eta_k) \right)_{i,k=1,\ldots,N} w \\
&= \left(\sum_{n=0}^{d} \frac{2n+1}{4\pi} P_n(\eta_i \cdot \eta_k) \right)_{i,k=1,\ldots,N} w
\end{aligned} \tag{7.140}$$

due to the addition theorem (Theorem 5.11 on p. 103) as well as

$$\begin{aligned}
&\left(\mathrm{matr}_{X_N}\left(Y_{0,1},\ldots,Y_{d,2d+1}\right)\right)^{\mathrm{T}} \left(\sqrt{4\pi}, 0, \ldots, 0\right)^{\mathrm{T}} \\
&= \left(Y_{0,1}(\eta_1)\sqrt{4\pi}, \ldots, Y_{0,1}(\eta_N)\sqrt{4\pi} \right)^{\mathrm{T}} \\
&= (1, \ldots, 1)^{\mathrm{T}}.
\end{aligned} \tag{7.141}$$

□

The Extended Haar's theorem brought bad news. However, there is also good news: Not every point system fails.

Theorem 7.36. *For every $d \in \mathbb{N}_0$, there exists a fundamental system $X_{(d+1)^2}$ relative to* $\mathrm{Harm}_{0\ldots d}(\Omega)$.

Proof. This theorem can be proved by induction. For this purpose, we change the enumeration (not the order) of $Y_{0,1}, Y_{1,1}, \ldots, Y_{1,3}, Y_{2,1}, \ldots, Y_{d,2d+1}$ to $Y^{(1)}, \ldots, Y^{(N)}$, where $N = (d+1)^2$. Obviously, $Y^{(1)} = Y_{0,1} = (4\pi)^{-1/2}$ allows every single point in Ω to be a fundamental system in $\mathrm{Harm}_0(\Omega)$. Now let us assume that we found n pairwise distinct[2] points (with $n < N$) $\eta_1, \ldots, \eta_n$ such that the matrix $\mathrm{matr}_{\{\eta_1,\ldots,\eta_n\}}(Y^{(1)}, \ldots, Y^{(n)})$ is regular. We continue with an indirect proof and assume that every $\xi \in \Omega$ causes a *singular* matrix

$$A_{n+1} := \mathrm{matr}_{\{\eta_1,\ldots,\eta_n,\xi\}}\left(Y^{(1)}, \ldots, Y^{(n+1)}\right). \tag{7.142}$$

If we can show that this implies a contradiction, then we are done.
The following conclusions are now important: Due to the assumption of the induction, the first n rows of A_{n+1} are linearly independent. However, the last row must, consequently, depend linearly on these n rows. In other words, there exist coefficients $a_1, \ldots, a_n$ such that

$$\begin{aligned}&\left(Y^{(n+1)}(\eta_1), \ldots, Y^{(n+1)}(\eta_n), Y^{(n+1)}(\xi)\right)\\ &\quad = \sum_{k=1}^{n} a_k \left(Y^{(k)}(\eta_1), \ldots, Y^{(k)}(\eta_n), Y^{(k)}(\xi)\right).\end{aligned} \tag{7.143}$$

Using once again the assumption of the induction, we know that the submatrix consisting of the upper left n rows and n columns of A_{n+1} is regular. We conclude that all rows of the submatrix, that is,

$$\left(Y^{(k)}(\eta_1), \ldots, Y^{(k)}(\eta_n)\right), \quad k = 1, \ldots, n, \tag{7.144}$$

are linearly independent. Hence, they represent a basis of $\mathbb{R}^n$. Thus, there can only be one set of coefficients $a_1, \ldots, a_n$ such that

$$\left(Y^{(n+1)}(\eta_1), \ldots, Y^{(n+1)}(\eta_n)\right) = \sum_{k=1}^{n} a_k \left(Y^{(k)}(\eta_1), \ldots, Y^{(k)}(\eta_n)\right). \tag{7.145}$$

As a consequence, the coefficients in (7.143) are independent of ξ. Hence,

$$Y^{(n+1)}(\xi) = \sum_{k=1}^{n} a_k Y^{(k)}(\xi) \textbf{ for all } \xi \in \Omega. \tag{7.146}$$

[2]Note that equal points produce identical corresponding columns of the matrix such that we, anyway, have to search for pairwise distinct points.

This is the desired contradiction since the functions $Y^{(1)},\dots,Y^{(n+1)}$ are $\mathrm{L}^2(\Omega)$-orthogonal and, therefore, linearly independent. □

Remember that we briefly discussed numerical integration in the context of splines (see Remark 3.27 on p. 61). A simple, but nevertheless sometimes (where "sometimes" needs to be further concretized) useful, way to choose a functional

$$\mathscr{G}_N F := \sum_{j=1}^{N} a_j F(\eta_j) \tag{7.147}$$

to approximate the functional

$$\mathscr{F} F := \int_{\Omega} F(\eta)\, \mathrm{d}\omega(\eta) \tag{7.148}$$

is to set $a_j := \frac{4\pi}{N}$ for all j. This is, at least, exact if F is constant. In [33, Theorem 3.1], it is shown that

$$\left| \frac{1}{4\pi} \int_{\Omega} F(\eta)\, \mathrm{d}\omega(\eta) - \sum_{j=1}^{N} \frac{1}{N} F(\eta_j) \right| \leq \frac{1}{N} \left(\sum_{i,j=1}^{N} \sum_{n=1}^{\infty} \frac{2n+1}{4\pi A_n^2} P_n(\eta_i \cdot \eta_j) \right)^{1/2} \|F\|_{\mathscr{H}} \tag{7.149}$$

for all $F \in \mathscr{H}$ if the sequence (A_n) of the Sobolev space $\mathscr{H} = \mathscr{H}((A_n);\Omega)$ satisfies

$$\lim_{n\to\infty} \frac{|A_n|}{\left(n+\frac{1}{2}\right)^s} = \mathrm{const} \neq 0 \tag{7.150}$$

for a fixed $s > 1$ and $A_n \neq 0$ for all $n \in \mathbb{N}_0$. Note that (7.150) implies the summability of (A_n) (see Definition 6.13 on p. 153). The term

$$D(X_N; (A_n)) := \frac{1}{N} \left(\sum_{i,j=1}^{N} \sum_{n=1}^{\infty} \frac{2n+1}{4\pi A_n^2} P_n(\eta_i \cdot \eta_j) \right)^{1/2}, \tag{7.151}$$

where $X_N = \{\eta_1,\dots,\eta_N\} \subset \Omega$, consequently, quantifies the accuracy of the numerical integration and is called the **generalized discrepancy**. Appropriate numerical integration methods of the kind (7.147) with equal weights a_j, therefore, use sequences of point sets $(X_N)_N$ such that

$$\lim_{N\to\infty} D(X_N; (A_n)) = 0\,. \tag{7.152}$$

One can imagine that point sets X_N which have strong agglomerations which are concentrated to certain areas are not good for this purpose, since all points are equally weighted in the quadrature formula. For this reason, this topic always has to be regarded in the context of **equidistributed point sets**. For tables of generalized discrepancies of selected point systems, see [33].

Let us allow general coefficients a_j again. Following [179], one can derive an interesting property for the worst-case error:

$$\mathrm{e}(\mathscr{G}_N;\mathscr{H}) := \sup\left\{\left|(\mathscr{F}-\mathscr{G}_N)(F)\right| : F\in\mathscr{H}, \|F\|_{\mathscr{H}}\le 1\right\}, \tag{7.153}$$

where we still assume that (7.150) is valid, and require $A_0 = 1$. Since $\mathscr{F}$ and $\mathscr{G}_N$ are continuous on $\mathscr{H}$ (see also Corollary 6.20 on p. 160), Theorem 6.4 yields

$$(\mathscr{F}-\mathscr{G}_N)F = \left\langle F, (\mathscr{F}-\mathscr{G}_N)_\eta\, K_{\mathscr{H}}(\eta\cdot)\right\rangle_{\mathscr{H}} \tag{7.154}$$

for all $F\in\mathscr{H}$, where $K_{\mathscr{H}}$ is the reproducing kernel of $\mathscr{H}$. We set now

$$\begin{aligned} G(\xi) &:= (\mathscr{F}-\mathscr{G}_N)_\eta\, K_{\mathscr{H}}(\xi\cdot\eta) \\ &= \int_\Omega K_{\mathscr{H}}(\xi\cdot\eta)\,\mathrm{d}\omega(\eta) - \sum_{j=1}^N a_j K_{\mathscr{H}}(\xi\cdot\eta_j), \quad \xi\in\Omega, \end{aligned} \tag{7.155}$$

as an abbreviation. As a consequence of the previous considerations, we get (using again Theorem 6.4)

$$\begin{aligned} \mathrm{e}(\mathscr{G}_N;\mathscr{H}) &= \sup\left\{\left|\langle F,G\rangle_{\mathscr{H}}\right| : F\in\mathscr{H}, \|F\|_{\mathscr{H}}\le 1\right\} \\ &= \|G\|_{\mathscr{H}} \\ &= \left\langle (\mathscr{F}-\mathscr{G}_N)_\xi\, K_{\mathscr{H}}(\xi\cdot), (\mathscr{F}-\mathscr{G}_N)_\eta\, K_{\mathscr{H}}(\eta\cdot)\right\rangle_{\mathscr{H}}^{1/2} \\ &= \left[(\mathscr{F}-\mathscr{G}_N)_\xi\,(\mathscr{F}-\mathscr{G}_N)_\eta\, K_{\mathscr{H}}(\xi\cdot\eta)\right]^{1/2} \\ &= \left[\int_\Omega\int_\Omega K_{\mathscr{H}}(\xi\cdot\eta)\,\mathrm{d}\omega(\eta)\,\mathrm{d}\omega(\xi)\right. \\ &\qquad - 2\sum_{j=1}^N a_j \int_\Omega K_{\mathscr{H}}(\eta_j\cdot\xi)\,\mathrm{d}\omega(\xi) \\ &\qquad \left. + \sum_{j,k=1}^N a_j a_k K_{\mathscr{H}}(\eta_j\cdot\eta_k)\right]^{1/2}. \end{aligned} \tag{7.156}$$

Note that the reproducing kernel satisfies (see Theorem 6.18 on p. 159)

$$\begin{aligned} \int_\Omega K_{\mathscr{H}}(\xi\cdot\eta)\,\mathrm{d}\omega(\eta) &= \int_\Omega K_{\mathscr{H}}(\xi\cdot\eta)\cdot 1\,\mathrm{d}\omega(\eta) \\ &= \frac{1}{4\pi A_0^2}\int_\Omega \underbrace{P_0(\xi\cdot\eta)}_{=1}\cdot 1\,\mathrm{d}\omega(\eta) \\ &= 1 \end{aligned} \tag{7.157}$$

for all $\xi \in \Omega$ (remember that we required $A_0 = 1$ above). Furthermore, the quadrature formula should, at least, be exact for constant functions, that is, $\sum_{j=1}^{N} a_j = 4\pi$. Hence, we get

$$\mathrm{e}(\mathscr{G}_N; \mathscr{H}) = \left[-4\pi + \sum_{j,k=1}^{N} a_j a_k K_{\mathscr{H}}(\eta_j \cdot \eta_k) \right]^{1/2} \tag{7.158}$$

as a formula for the worst-case error.[3] From [33], we know that the choice

$$A_n := \begin{cases} 1, & \text{if } n = 0 \\ \sqrt{(2n+1)n(n+1)}, & \text{if } n > 0 \end{cases} \tag{7.160}$$

yields a closed representation for the reproducing kernel

$$K_{\mathscr{H}}(\xi \cdot \eta) = \frac{1}{2\pi} \left[1 - \ln\left(1 + \sqrt{\frac{1 - \xi \cdot \eta}{2}} \right) \right]; \quad \xi, \eta \in \Omega. \tag{7.161}$$

Moreover, in [179] it is shown for this particular choice of (A_n) that any quadrature rule $\mathscr{G}_{(n+1)^2}$ of the form (7.147) with $a_j > 0$ for all $j = 1, \ldots, (n+1)^2$ and $\mathscr{G}_{(n+1)^2} F = \int_\Omega F(\eta)\, \mathrm{d}\omega(\eta)$ for all $F \in \mathrm{Harm}_{0\ldots n}(\Omega)$ satisfies

$$\mathrm{e}\left(\mathscr{G}_{(n+1)^2}; \mathscr{H} \right) \le \sqrt{\frac{4\pi}{n+1}}\,. \tag{7.162}$$

There exist numerous other results on quadrature rules on the sphere Ω. For further details, the survey article [92] is recommended. Further examples of elaborated integration methods can be found, for example, in [109, 149]. Note that the so-called Marcinkiewicz–Zygmund inequalities are popular tools in the context of numerical integration on the sphere. They yield relations between the integral to be approximated and the quadrature formula. These inequalities resemble the concept of a norm equivalence from functional analysis. For example, if $\|\cdot\|$ is an integral-

---

[3] Note that the particular case $a_j = \frac{4\pi}{N}\ \forall j$ (where again $A_0 = 1$) corresponds to

$$\begin{aligned} \mathrm{e}(\mathscr{G}_N; \mathscr{H}) &= \left[-4\pi + \sum_{j,k=1}^{N} \left(\frac{4\pi}{N} \right)^2 \sum_{n=0}^{\infty} \frac{2n+1}{4\pi A_n^2} P_n(\eta_j \cdot \eta_k) \right]^{1/2} \\ &= \left[\sum_{j,k=1}^{N} \left(\frac{4\pi}{N} \right)^2 \sum_{n=1}^{\infty} \frac{2n+1}{4\pi A_n^2} P_n(\eta_j \cdot \eta_k) \right]^{1/2} \\ &= 4\pi D(X_N; (A_n))\,. \end{aligned} \tag{7.159}$$

based norm (such as an L^p-norm) and $Q(f)$ is a quadrature formula for f based on discrete samples, then one would try to prove that there exist constants $C_1, C_2 > 0$ such that

$$C_1 \|f\| \leq Q(f) \leq C_2 \|f\| \tag{7.163}$$

for a certain class of functions f. This guarantees a stability of the numerical integration. The association with the norm equivalence also makes clear, why Marcinkiewicz–Zygmund inequalities play an important role in the construction of particular frames on the sphere (see also Remark 7.32).

Moreover, there also exist publications which develop numerical integration methods for subsets of the sphere such as spherical caps or spherical triangles. Examples are [15, 120, 121].

7.5 Questions for Understanding

- What is a spherical convolution?
- To which function space does the result of a convolution belong?
- What is the Funk–Hecke formula? How is it connected to spherical convolutions?
- Which other kind of a convolution is needed for the spherical wavelet analysis presented here?
- What is an iterated kernel?
- What can you say about the Fourier expansion of a spherical convolution? How is this result proved? Why is this property important for the spherical wavelet analysis?
- What is a spherical scaling function?
- We saw that there is a particular way to construct scaling functions. Can you say more about this?
- We distinguish bandlimited and non-bandlimited scaling functions. What does this mean?
- If you convolve a bandlimited scaling function with an arbitrary function in $L^2(\Omega)$, what kind of a function do you always get? Why?
- Which examples of scaling functions do you know? What are their advantages and disadvantages?
- Why is the Abel–Poisson scaling function so special?
- What are the fundamental properties of a scaling function? How are they proved? Which of the conditions (S1), (S2), and (S3) for a scaling function can be linked to which of the fundamental properties?
- What is a spherical wavelet?
- Are there similar ways to construct wavelets as they exist for the scaling functions? If yes, how is it done?
- There is a link between scaling functions and wavelets. What is it? Is a wavelet uniquely given by a scaling function? Is a scaling function uniquely given by a wavelet?

- What is the fundamental property of a wavelet? How is it proved? If you interchange the primal and the dual wavelet in this property, does it remain valid?
- What is the spherical wavelet transform? What is the motivation for its introduction?
- You learned a basic quadrature method on the sphere. What kind of a point grid is used by this method?
- If you want or need to use a different point grid, how can you construct a quadrature method for your point grid? Can you always construct such a method (no matter, what your point grid looks like)?

Chapter 8
Spherical Slepian Functions

8.1 Spherical Slepian Functions

We have seen that localized basis functions provide us with several advantages in comparison to spherical harmonics. On the other hand, spherical harmonics are ideal to represent global phenomena or structures with, at least, a very large spatial extension. The localized basis functions discussed so far are isotropic, that is, they are associated to zonal functions. For instance, a spline basis function $\Omega \ni \xi \mapsto K_{\mathscr{H}}(\eta_k \cdot \xi)$ depends on the distance $|\eta_k - \xi|$ only since $|\eta_k - \xi|^2 = |\eta_k|^2 + |\xi|^2 - 2\eta_k \cdot \xi = 2(1 - \eta_k \cdot \xi)$. For some applications, however, non-isotropic trial functions are more useful. This occurs, for example, if phenomena with a preferred direction in space or structures with sophisticated geometries (like oceans or single continents) are investigated. For this purpose, the concept of Slepian functions was developed by F.J. Simons et al. in [34, 172–174, 198, 199]. The basic principles are summarized here. For further details, please consult the listed references. The functions are named after D. Slepian, who introduced such a concept for the Euclidean setting together with H.J. Landau and H.O. Pollak (see [107, 177, 178]).

The idea behind spherical Slepian functions is as follows: for a given subset $R \subset \Omega$, we are looking for a function F which shows the best possible concentration on R. This is quantified by comparing $\|F\|_{\mathrm{L}^2(R)}$ with $\|F\|_{\mathrm{L}^2(\Omega)}$. A function F which is optimally localized at R maximizes the ratio of $\|F\|_{\mathrm{L}^2(R)}$ and $\|F\|_{\mathrm{L}^2(\Omega)}$.

Definition 8.1. Let $R \subset \Omega$ be a measurable subset and $F \in \mathrm{L}^2(\Omega) \setminus \{0\}$ be a given function. Then

$$\lambda_R(F) := \frac{\|F\|^2_{\mathrm{L}^2(R)}}{\|F\|^2_{\mathrm{L}^2(\Omega)}} \tag{8.1}$$

is called the **energy ratio** of F with respect to R.

Obviously,

$$0 \le \lambda_R(F) \le 1. \tag{8.2}$$

V. Michel, *Lectures on Constructive Approximation*, Applied and Numerical Harmonic Analysis, DOI 10.1007/978-0-8176-8403-7_8,

If we restrict our attention to a bandlimited function F, then the ansatz

$$F = \sum_{n=0}^{N} \sum_{j=1}^{2n+1} F^{\wedge}(n,j) Y_{n,j} \tag{8.3}$$

leads us to

$$\|F\|^2_{\mathrm{L}^2(\Omega)} = \sum_{n=0}^{N} \sum_{j=1}^{2n+1} \left(F^{\wedge}(n,j)\right)^2 \tag{8.4}$$

and

$$\|F\|^2_{\mathrm{L}^2(R)} = \sum_{n,m=0}^{N} \sum_{j=1}^{2n+1} \sum_{k=1}^{2m+1} F^{\wedge}(n,j) \left\langle Y_{n,j}, Y_{m,k} \right\rangle_{\mathrm{L}^2(R)} F^{\wedge}(m,k). \tag{8.5}$$

As a consequence, we end up with a finite-dimensional optimization problem, where we are looking for a vector $a = (a_{n,j})_{n=0,\ldots,N;\, j=1,\ldots,2n+1} \in \mathbb{R}^{(N+1)^2}$, which maximizes

$$\frac{a^{\mathrm{T}} D a}{a^{\mathrm{T}} a}, \tag{8.6}$$

where the matrix D consists of the entries

$$d_{(n,j),(m,k)} := \left\langle Y_{n,j}, Y_{m,k} \right\rangle_{\mathrm{L}^2(R)}. \tag{8.7}$$

F is then represented by $F = \sum_{n=0}^{N} \sum_{j=1}^{2n+1} a_{n,j} Y_{n,j}$, that is, $a = (\mathrm{SFT})(F)$ (see also Definition 5.27 on p. 123).

Since $\lambda_R(rF) = \lambda_R(F)$ for all $F \in \mathrm{L}^2(\Omega) \setminus \{0\}$ and all $r \in \mathbb{R} \setminus \{0\}$, it suffices to look for functions $F \in \mathrm{L}^2(\Omega)$ with $\|F\|_{\mathrm{L}^2(\Omega)} = 1$ such that $\lambda_R(F) = \|F\|^2_{\mathrm{L}^2(R)}$ is maximal. For this constrained optimization problem, the Lagrangian function is $L(a,\mu) := a^{\mathrm{T}} D a + \mu(a^{\mathrm{T}} a - 1)$. A necessary condition for a maximum is $\frac{\partial L}{\partial a}(a,\mu) = 0$, that is, we obtain

$$2Da + 2\mu a = 0, \tag{8.8}$$

where we used the symmetry of D. Hence, we have to find eigenvectors of D. If $Da = \lambda a$, then $a^{\mathrm{T}} D a = \lambda a^{\mathrm{T}} a$. Thus, the energy ratio $\lambda_R(F)$ is the corresponding eigenvalue λ. As a consequence, we have to solve

$$Da = \lambda_R(F) a. \tag{8.9}$$

Thus, we are looking for all eigenvectors of D, where we know that the largest eigenvalue yields the best possible localization to R. Moreover, note that D is symmetric (due to the symmetry of the inner product) such that we can find $(N+1)^2$ orthonormal eigenvectors of D. Since

$$\langle F, G \rangle_{\mathrm{L}^2(\Omega)} = \sum_{n=0}^{\infty} \sum_{j=1}^{2n+1} F^{\wedge}(n,j) G^{\wedge}(n,j) \tag{8.10}$$

for all $F,G \in \mathrm{L}^2(\Omega)$, the functions associated to such orthonormal eigenvectors are also orthonormal in the $\mathrm{L}^2(\Omega)$-sense. Note that the eigenvalues may be degenerated. In this case, there exists more than one linearly independent solution of the maximization problem.

Let us summarize what we obtained so far.

Theorem 8.2. *The maximizers $F \in \mathrm{Harm}_{0\ldots N}(\Omega) \setminus \{0\}$ of $\lambda_R(F)$ for a given measurable set $R \subset \Omega$ are associated to these coefficient vectors*

$$a = (F^\wedge(n,j))_{n=0,\ldots,N;\, j=1,\ldots,2n+1} \in \mathbb{R}^{(N+1)^2} \tag{8.11}$$

which are eigenvectors of the matrix $D \in \mathbb{R}^{(N+1)^2 \times (N+1)^2}$ [defined in (8.7)] corresponding to the largest eigenvalue of D. This largest eigenvalue is the maximal energy ratio $\lambda_R(F)$ on $\mathrm{Harm}_{0\ldots N}(\Omega)$.

Corollary 8.3. *The maximizers $F \in \mathrm{Harm}_{0\ldots N}(\Omega) \setminus \{0\}$ of $\lambda_R(F)$ are the solutions of the Fredholm integral equation of the second kind*

$$\int_R \mathscr{D}(\xi \cdot \eta)\, F(\eta)\, \mathrm{d}\omega(\eta) = \lambda F(\xi) \quad \forall \xi \in \Omega, \tag{8.12}$$

where λ is the maximal energy ratio and

$$\mathscr{D}(\xi \cdot \eta) := \sum_{n=0}^{N} \frac{2n+1}{4\pi} P_n(\xi \cdot \eta), \quad \xi, \eta \in \Omega. \tag{8.13}$$

Note that the function $\mathscr{D}$ is a Shannon scaling function of scale $J = \log_2(N+1)$.

Proof. By inserting the Fourier expansion of F and using the addition theorem (Theorem 5.11 on p. 103), we get

$$\begin{aligned}
&\int_R \mathscr{D}(\xi \cdot \eta)\, F(\eta)\, \mathrm{d}\omega(\eta) \\
&= \sum_{n,m=0}^{N} \sum_{k=1}^{2m+1} \frac{2n+1}{4\pi} F^\wedge(m,k) \int_R P_n(\xi \cdot \eta)\, Y_{m,k}(\eta)\, \mathrm{d}\omega(\eta) \\
&= \sum_{n,m=0}^{N} \sum_{j=1}^{2n+1} \sum_{k=1}^{2m+1} F^\wedge(m,k) \underbrace{\int_R Y_{n,j}(\eta)\, Y_{m,k}(\eta)\, \mathrm{d}\omega(\eta)}_{=d_{(n,j),(m,k)}} Y_{n,j}(\xi).
\end{aligned} \tag{8.14}$$

Hence, the vector consisting of the Fourier coefficients $F^\wedge(n,j)$ is an eigenvector of the matrix D corresponding to the eigenvalue $\lambda_R(F)$ if and only if

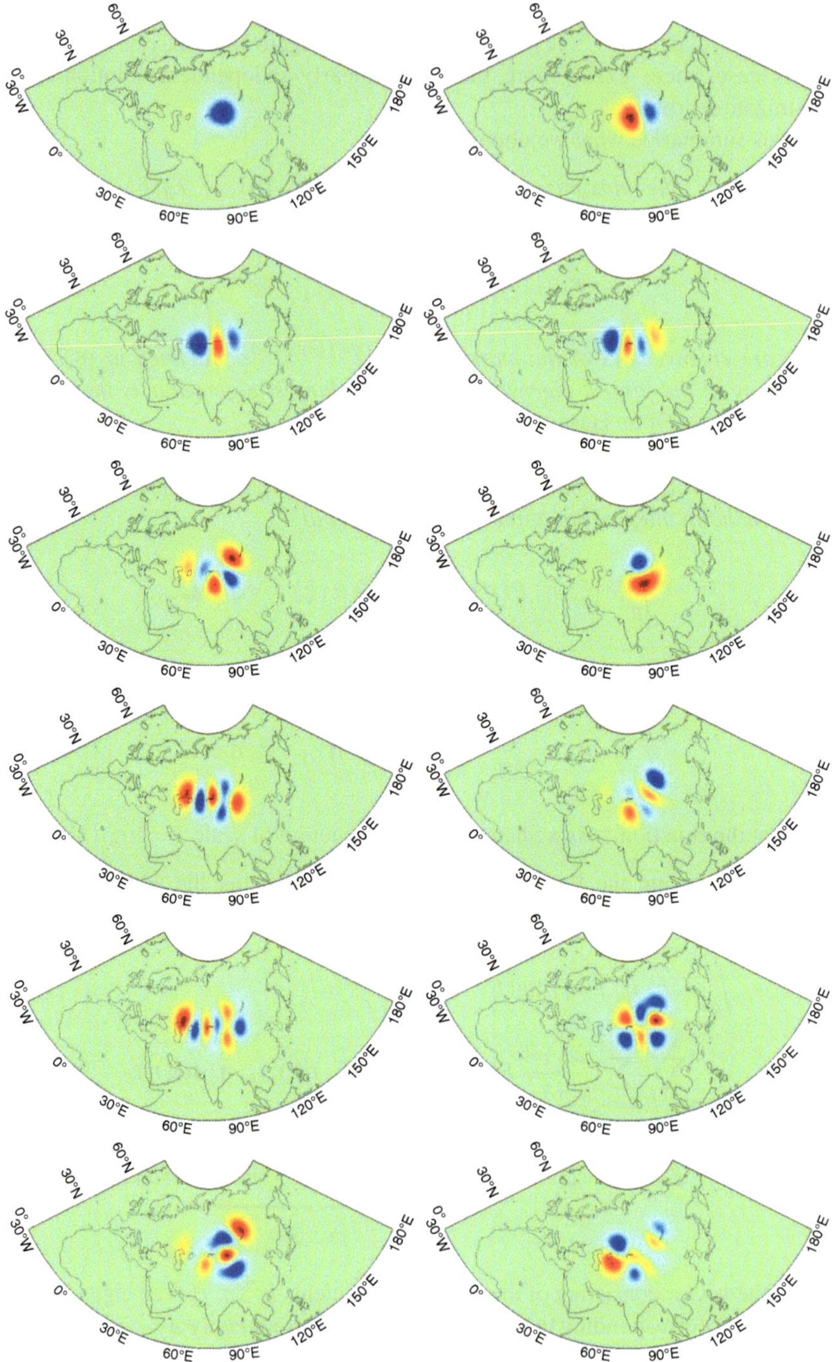

Fig. 8.1 Examples of optimally localized spatial Slepian functions for Eurasia in $\mathrm{Harm}_{0\ldots 50}(\Omega)$

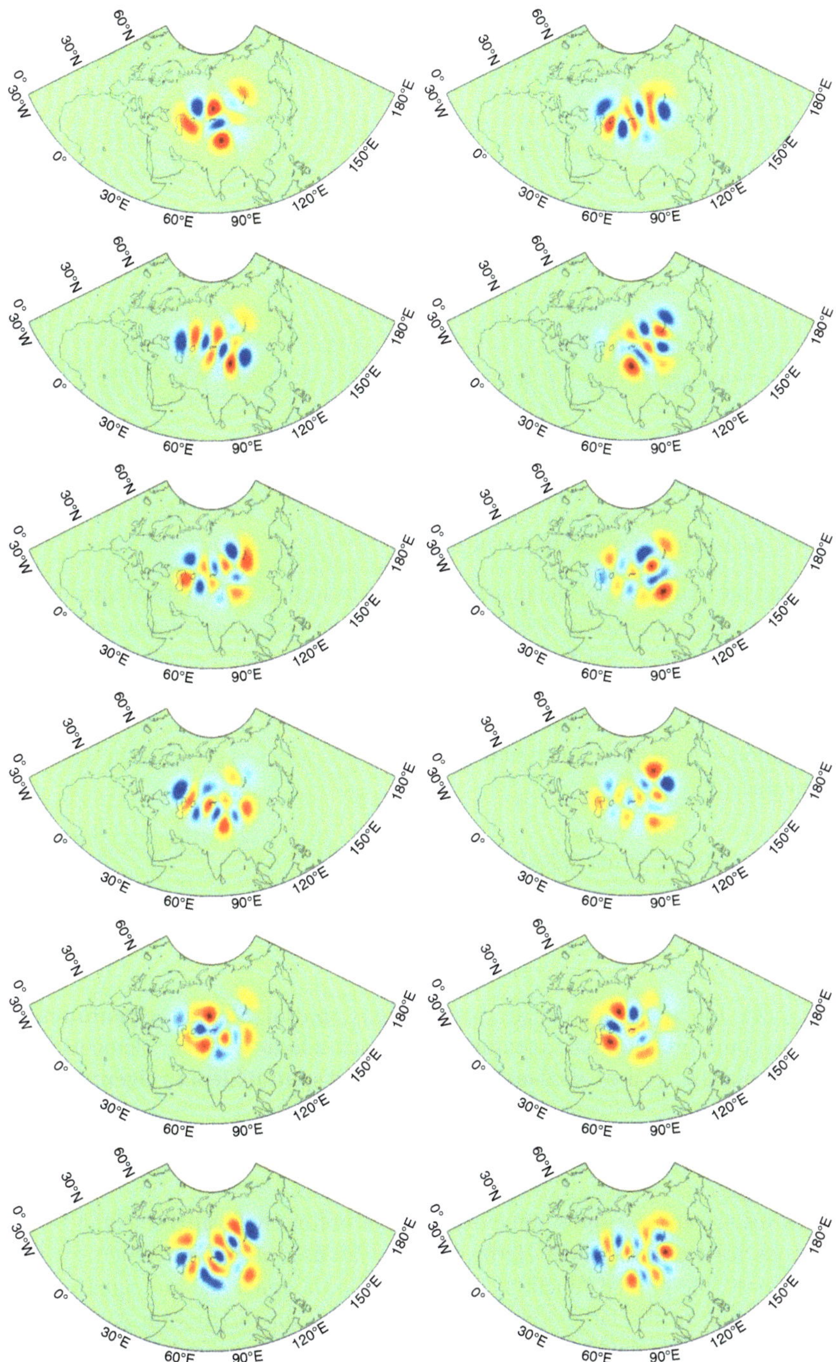

Fig. 8.2 Examples of optimally localized spatial Slepian functions for Eurasia in $\mathrm{Harm}_{0\dots50}(\Omega)$

$$\int_R \mathscr{D}(\xi\cdot\eta)\,F(\eta)\,\mathrm{d}\omega(\eta) = \sum_{n=0}^{N}\sum_{j=1}^{2n+1} \lambda_R(F)\,F^\wedge(n,j)\,Y_{n,j}(\xi)$$
$$= \lambda_R(F)F(\xi)\ \forall\xi\in\Omega. \tag{8.15}$$

□

Definition 8.4. The subspace of $\mathrm{Harm}_{0\ldots N}(\Omega)$ which is the eigenspace of (8.12) corresponding to the largest eigenvalue is denoted by $\mathrm{Slepian}(R,N)$, that is,

$$\mathrm{Slepian}(R,N) \tag{8.16}$$
$$:= \left\{ F\in\mathrm{Harm}_{0\ldots N}(\Omega)\ \middle|\ \int_R \mathscr{D}(\xi\cdot\eta)\,F(\eta)\,\mathrm{d}\omega(\eta) = \lambda_{R,\max}F(\xi)\quad \forall\xi\in\Omega \right\},$$

where

$$\lambda_{R,\max} := \max\{\lambda_R(F)\,|\,F\in\mathrm{Harm}_{0\ldots N}(\Omega)\setminus\{0\}\} \tag{8.17}$$

and $R\subset\Omega$ is a given measurable subset.

Note that the symmetry of the matrix D allows us to find orthonormal eigenvectors which represent a basis of $\mathbb{R}^{(N+1)^2}$, which is the space into which the Fourier coefficients $(\mathrm{SFT})(F)$ of functions $F\in\mathrm{Harm}_{0\ldots N}(\Omega)$ can be embedded. As we also observed above, these eigenvectors then correspond to an orthonormal basis of $\mathrm{Harm}_{0\ldots N}(\Omega)$. One advantage of this new basis is the quantification of the localization of each basis function with respect to R, which is given by the associated eigenvalues.

Definition 8.5. If $g_1,\ldots,g_{(N+1)^2}\in\mathrm{Harm}_{0\ldots N}(\Omega)$ are $\mathrm{L}^2(\Omega)$-orthonormal functions and eigenfunctions of the operator

$$F\mapsto\int_R \mathscr{D}(\eta\cdot)\,F(\eta)\,\mathrm{d}\omega(\eta), \tag{8.18}$$

then these functions are called (bandlimited) **spatial Slepian eigenfunctions**. Moreover, the orthonormal eigenfunctions which additionally are elements of the space $\mathrm{Slepian}\,(R,N)$ are called **optimally localized spatial Slepian eigenfunctions**.

Note that, since the eigenvalues equal the energy ratio, eigenfunctions corresponding to, for example, the second-largest or third-largest eigenvalue still show a strong localization.

Figures 8.1 and 8.2 show some examples of optimally localized spatial Slepian functions for Eurasia, where $\mathrm{Harm}_{0\ldots 50}(\Omega)$ was used as the reference space. In other words, the plotted functions are pairwise orthogonal and are eigenfunctions corresponding to the largest eigenvalue $\lambda_{\mathrm{Eurasia,\,max}}$ in (8.16) with $N=50$. The functions were calculated with F.J. Simons' software at www.frederik.net.

8.2 Questions for Understanding

- What is the energy ratio?
- Why is $0 \leq \lambda_R(F) \leq 1$?
- Why do we want to maximize $\lambda_R(F)$?
- Why does it suffice to solve the problem for functions F with $\|F\|_{\mathrm{L}^2(\Omega)} = 1$?
- The maximization problem is equivalent to two eigenvalue problems which are, on their part, equivalent to each other. Which eigenvalue problems are these? Explain the equivalences!
- Why is it important that the matrix D is symmetric? What are the consequences?
- Why do orthonormal eigenvectors to the matrix D correspond to orthonormal functions in $\mathrm{L}^2(\Omega)$? Are they also orthonormal in $\mathrm{L}^2(R)$?
- For which applications do you consider spherical Slepian functions to be useful?

Part III
Approximation on the 3D Ball

Chapter 9
Orthonormal Bases

Within Part III, we consider functions on the closed ball

$$\mathscr{B} := \left\{ x \in \mathbb{R}^3 \,\middle|\, |x| \leq \beta \right\} \tag{9.1}$$

with the radius $\beta > 0$. Typical applications, where such functions have to be approximated, are tomographic problems in geophysics or medical imaging. Since the Earth's interior and the human brain do—roughly speaking—consist of layers with concentric spheres as boundaries, a tensor product ansatz for the Cartesian coordinates x, y, and z (see Sect. 3.5) is not useful here. Instead, a separation $x = r\xi$, $r = |x| \in \mathbb{R}_0^+$, $\xi \in \Omega$, is more appropriate for practical purposes.

We will first see how orthonormal bases for $\mathrm{L}^2(\mathscr{B})$ can be constructed. Then we will study in the following Chaps. 10 and 11 how the concepts of spherical splines and spherical wavelets (see Chaps. 6 and 7) can be extended to the ball.

For a survey article on constructive approximation on the 3D ball, see [132].

9.1 Construction and Basic Properties

For finding an orthonormal system, we use the separation ansatz

$$G(x) = F(|x|)\, Y\left(\frac{x}{|x|}\right), \quad x \in \mathscr{B} \setminus \{0\}, \tag{9.2}$$

for the functions. Note that already here, one problem becomes visible: The functions might not be defined in the origin. For $x = 0$, the direction $\frac{x}{|x|}$ is not defined. If $F(0) \neq 0$, then the limit $\lim_{x \to 0} G(x)$ does not exist. However, in the sense of $\mathrm{L}^2(\mathscr{B})$, this is not a problem.

The separation ansatz fits to the corresponding separation of integrals on the ball (see Theorem 4.13 on p. 92). Assume that we have two functions $G, \tilde{G} \in \mathrm{L}^2(\mathscr{B})$ with

V. Michel, *Lectures on Constructive Approximation*, Applied and Numerical Harmonic Analysis, DOI 10.1007/978-0-8176-8403-7_9,

$$G(x) = F(|x|)\, Y\left(\frac{x}{|x|}\right), \quad \tilde{G}(x) = \tilde{F}(|x|)\, \tilde{Y}\left(\frac{x}{|x|}\right), \quad x \in \mathscr{B}. \tag{9.3}$$

Then (with $r := |x|$ and $\xi := \frac{x}{|x|}$)

$$\begin{aligned} \langle G, \tilde{G} \rangle_{\mathrm{L}^2(\mathscr{B})} &= \int_0^\beta r^2 \int_\Omega F(r) Y(\xi)\, \tilde{F}(r) \tilde{Y}(\xi)\, \mathrm{d}\omega(\xi)\, \mathrm{d}r \\ &= \int_0^\beta r^2 F(r) \tilde{F}(r)\, \mathrm{d}r \cdot \int_\Omega Y(\xi)\, \tilde{Y}(\xi)\, \mathrm{d}\omega(\xi). \end{aligned} \tag{9.4}$$

For this reason, we choose $\{Y_{n,j}\}_{n\in\mathbb{N}_0;\, j=1,\ldots,2n+1}$ for the angular part. For F, we have to choose functions which are orthogonal with respect to $\langle \cdot, \cdot \rangle_{\mathrm{L}^2_w[0,\beta]}$ with $w(r) := r^2$, $r \in [0, \beta]$. Moreover, for obtaining a *complete* orthonormal system in $\mathrm{L}^2(\mathscr{B})$, we have to take into account the following fact: We do not know yet which index range we need for the radial part. Let

$$G_{\alpha,n,j}(x) := F_\alpha(|x|)\, Y_{n,j}\left(\frac{x}{|x|}\right), \quad x \in \mathscr{B} \setminus \{0\}, \tag{9.5}$$

for $n \in \mathbb{N}_0$ and $j \in \{1, \ldots, 2n+1\}$ and a yet not further specified index α. Assume that $f \in \mathrm{L}^2(\mathscr{B})$ with

$$\langle f, G_{\alpha,n,j} \rangle_{\mathrm{L}^2(\mathscr{B})} = 0 \quad \text{for all } \alpha, n, j. \tag{9.6}$$

Then

$$\begin{aligned} 0 &= \int_{\mathscr{B}} f(x)\, G_{\alpha,n,j}(x)\, \mathrm{d}x \\ &= \int_\Omega Y_{n,j}(\xi) \int_0^\beta r^2\, F_\alpha(r)\, f(r\xi)\, \mathrm{d}r\, \mathrm{d}\omega(\xi). \end{aligned} \tag{9.7}$$

Since $\{Y_{n,j}\}_{n\in\mathbb{N}_0;\, j=1,\ldots,2n+1}$ is complete in $\mathrm{L}^2(\Omega)$, this implies that

$$\int_0^\beta r^2\, F_\alpha(r)\, f(r\xi)\, \mathrm{d}r = 0 \tag{9.8}$$

for almost every $\xi \in \Omega$. For the final conclusion "$f = 0$ almost everywhere in $\mathscr{B}$," we need, consequently, a complete system in $\mathrm{L}^2_w[0, \beta]$.

Two approaches for the construction of such systems are common.

I. Following Ballani, Dufour, Engels, and Grafarend (see [12, 47]), we set

$$F^{\mathrm{I}}_\alpha(r) = r^n\, \tilde{F}^{\mathrm{I}}_{m,n}\left(r^2\right), \quad r \in [0, \beta];\ m, n \in \mathbb{N}_0; \tag{9.9}$$

such that

$$G^{\mathrm{I}}_{m,n,j}(r\xi) = r^n\, \tilde{F}^{\mathrm{I}}_{m,n}\left(r^2\right)\, Y_{n,j}(\xi) \tag{9.10}$$

for $r \in [0,\beta]$, $\xi \in \Omega$; $m,n \in \mathbb{N}_0$; $j \in \{1,\ldots,2n+1\}$. This has the advantage that we get $r^n Y_{n,j}(\xi)$ in the basis functions, which is a homogeneous harmonic polynomial on $\mathscr{B}$. Therefore, these basis functions are defined in $r=0$. Moreover, we also get algebraic polynomials in x_1,x_2,x_3. The disadvantage is a new coupling of the radial part and the angular part: For each degree n of the angular part, a whole sequence of basis functions $(\tilde{F}^{\mathrm{I}}_{m,n})_m$ has to be computed. This is numerically expensive.

II. Following Tscherning (see [186]), we really decouple the radial and the angular part and take radial parts

$$F^{\mathrm{II}}_m(r) \quad \text{for } m \in \mathbb{N}_0. \tag{9.11}$$

The advantage is the numerical efficiency. The disadvantages are the indefinite values at $x=0$ and the non-polynomial structure (note that $r=\sqrt{x_1^2+x_2^2+x_3^2}$).

For system I, we substitute $r=\beta\sqrt{\frac{t+1}{2}}$, and for system II, we set $r=\frac{\beta}{2}(t+1)$. The orthonormality requirement is then equivalent to the conditions

$$\begin{aligned}
\delta_{m_1,m_2} &= \int_0^\beta r^{2n+2}\,\tilde{F}^{\mathrm{I}}_{m_1,n}\left(r^2\right)\tilde{F}^{\mathrm{I}}_{m_2,n}\left(r^2\right)\mathrm{d}r\\
&= \int_{-1}^1 \left[\frac{\beta^2}{2}(t+1)\right]^{n+1}\tilde{F}^{\mathrm{I}}_{m_1,n}\left(\frac{\beta^2}{2}(t+1)\right)\tilde{F}^{\mathrm{I}}_{m_2,n}\left(\frac{\beta^2}{2}(t+1)\right)\frac{\beta}{\sqrt{8(t+1)}}\,\mathrm{d}t\\
&= \frac{\beta^{2n+3}}{2^{n+5/2}}\int_{-1}^1 (t+1)^{n+1/2}\,\tilde{F}^{\mathrm{I}}_{m_1,n}\left(\frac{\beta^2}{2}(t+1)\right)\tilde{F}^{\mathrm{I}}_{m_2,n}\left(\frac{\beta^2}{2}(t+1)\right)\mathrm{d}t
\end{aligned} \tag{9.12}$$

for all $m_1,m_2,n \in \mathbb{N}_0$ and

$$\begin{aligned}
\delta_{m_1,m_2} &= \int_0^\beta r^2\,F^{\mathrm{II}}_{m_1}(r)\,F^{\mathrm{II}}_{m_2}(r)\,\mathrm{d}r\\
&= \int_{-1}^1 \left[\frac{\beta}{2}(t+1)\right]^2 F^{\mathrm{II}}_{m_1}\left(\frac{\beta}{2}(t+1)\right)F^{\mathrm{II}}_{m_2}\left(\frac{\beta}{2}(t+1)\right)\frac{\beta}{2}\,\mathrm{d}t\\
&= \frac{\beta^3}{8}\int_{-1}^1 (t+1)^2\,F^{\mathrm{II}}_{m_1}\left(\frac{\beta}{2}(t+1)\right)F^{\mathrm{II}}_{m_2}\left(\frac{\beta}{2}(t+1)\right)\mathrm{d}t,
\end{aligned} \tag{9.13}$$

for all $m_1,m_2 \in \mathbb{N}_0$, respectively. Hence, the following choices provide us with complete orthogonal systems due to the considerations in Sect. 3.1 (see, in particular, Definition 3.10 on p. 41):

$$\begin{aligned}
\tilde{F}^{\mathrm{I}}_{m,n}\left(r^2\right) &= a_{m,n}\,P^{(0,n+1/2)}_m\left(2\frac{r^2}{\beta^2}-1\right), \quad r\in[0,\beta], \quad m,n\in\mathbb{N}_0\,,\\
F^{\mathrm{II}}_m(r) &= b_m\,P^{(0,2)}_m\left(2\frac{r}{\beta}-1\right), \quad r\in[0,\beta], \quad m\in\mathbb{N}_0\,.
\end{aligned} \tag{9.14}$$

The constants $a_{m,n}$ and b_m are determined (up to the sign) by the requirement that the norm is 1:

$$1 = \frac{\beta^{2n+3}}{2^{n+5/2}} a_{m,n}^2 \int_{-1}^{1} (t+1)^{n+1/2} \left(P_m^{(0,n+1/2)}(t)\right)^2 \, \mathrm{d}t,$$

$$1 = \frac{\beta^3}{8} b_m^2 \int_{-1}^{1} (t+1)^2 \left(P_m^{(0,2)}(t)\right)^2 \, \mathrm{d}t \, . \tag{9.15}$$

Due to Theorem 3.14, we have (note that the degree of the polynomial is here m and not n):

$$1 = \frac{\beta^{2n+3}}{2^{n+5/2}} a_{m,n}^2 \frac{2^{n+3/2}}{2m+n+\frac{3}{2}} \frac{\Gamma(m+1)\,\Gamma\left(m+n+\frac{3}{2}\right)}{m!\,\Gamma\left(m+n+\frac{3}{2}\right)} \Leftrightarrow a_{m,n}^2 = \frac{4m+2n+3}{\beta^{2n+3}}, \tag{9.16}$$

and

$$1 = \frac{\beta^3}{8} b_m^2 \frac{2^3}{2m+3} \frac{\Gamma(m+1)\,\Gamma(m+3)}{m!\,\Gamma(m+3)} \Leftrightarrow b_m^2 = \frac{2m+3}{\beta^3}. \tag{9.17}$$

We obtain the following result (see also [12, 47, 69, 127, 132, 186]):

Theorem 9.1. *The following systems are complete orthonormal systems in* $\mathrm{L}^2(\mathscr{B})$*:*

$$G_{m,n,j}^{\mathrm{I}}(r\xi) := \sqrt{\frac{4m+2n+3}{\beta^3}}\, P_m^{(0,n+1/2)}\left(2\frac{r^2}{\beta^2}-1\right)\left(\frac{r}{\beta}\right)^n Y_{n,j}(\xi)\,;$$

$$r \in [0,\beta],\ \xi \in \Omega,\ m,n \in \mathbb{N}_0,\ j \in \{1,\ldots,2n+1\}\,;$$

$$G_{m,n,j}^{\mathrm{II}}(r\xi) := \sqrt{\frac{2m+3}{\beta^3}}\, P_m^{(0,2)}\left(2\frac{r}{\beta}-1\right) Y_{n,j}(\xi)\,;$$

$$r \in\ ]0,\beta],\ \xi \in \Omega,\ m,n \in \mathbb{N}_0,\ j \in \{1,\ldots,2n+1\}\,. \tag{9.18}$$

In the following, $G_{m,n,j}$ without “I” or “II” means that both systems may be used.

The following property is important for the inverse gravimetric problem, which is concerned with the determination of the mass density distribution inside the Earth out of the gravitational potential outside the Earth. It is known that only the harmonic part of the density can be determined uniquely (see, e.g., the survey [134]).

Theorem 9.2. *The function* $G_{m,n,j}^{\mathrm{I}}$ *is harmonic if and only if* $m=0$.

Proof. The requirement

$$\Delta_x \left(F(|x|)\, Y_{n,j}\left(\frac{x}{|x|}\right)\right) = 0 \tag{9.19}$$

yields (together with Theorem 4.5 and Lemma 5.8, see pp. 87 and 101, respectively) that

$$\left(F''(r) + \frac{2}{r}F'(r) - \frac{n(n+1)}{r^2}F(r)\right) Y_{n,j}(\xi) = 0 \tag{9.20}$$

for all $r \in]0,\beta]$, $\xi \in \Omega$. The resulting ordinary differential equation[1] is

$$r^2 F''(r) + 2rF'(r) - n(n+1)F(r) = 0, \quad r \in [0,\beta]. \tag{9.21}$$

A fundamental system for this equation is $\{r^n, r^{-n-1}\}$. Since we need a solution which exists at $r=0$, the general solution has the form

$$F(r) = cr^n, \quad c \in \mathbb{R} \text{ constant.} \tag{9.22}$$

Hence, $m = 0$. □

Let us summarize the advantages and disadvantages of systems I and II. In general, we cannot say that a particular system is always better.

System I consists of algebraic polynomials and, therefore, in particular, of $\mathrm{C}^{(\infty)}$-functions and is defined in $x=0$. Moreover, if the inverse gravimetric problem is to be solved, then this system allows an easy distinction of the null-space and its orthogonal complement. On the other hand, the functions are more expensive to compute, because every degree n of the angular part requires a whole new function system for the radial part.

System II has a complete decoupling of the radial and the angular part. This saves CPU time and RAM use. On the other hand, this system is neither defined in nor continuously extendable to $x=0$, if $n>0$. In order to have at least a value at $x=0$ with which we can deal, we make the following arrangement.

Definition 9.3. For every $m \in \mathbb{N}_0$, $n \in \mathbb{N}$, and $j \in \{1,\ldots,2n+1\}$, we set

$$G^{\mathrm{II}}_{m,n,j}(0) := \sqrt{\frac{3}{4\pi\beta^3}} \quad \text{and} \quad F^{\mathrm{II}}_{m,n}(0) := \sqrt{\frac{3}{\beta^3}}. \tag{9.23}$$

Figures 9.1–9.6 show $G^{\mathrm{I}}_{m,n,j}$ and $G^{\mathrm{II}}_{m,n,j}$ for $m \le 2$ and $n \le 2$ on the plane through the origin with the normal vector $(1,1,1)^{\mathrm{T}}$. Note the discontinuity of $G^{\mathrm{II}}_{m,n,j}$ at $x=0$ for $n>0$.

We now derive a few further properties of the orthonormal bases.

Theorem 9.4. *Each $G^{\mathrm{I}}_{m,n,j}$; $m,n \in \mathbb{N}_0$, $j=1,\ldots,2n+1$; is an algebraic polynomial in x_1,x_2,x_3, where*

$$\deg G^{\mathrm{I}}_{m,n,j} = 2m+n. \tag{9.24}$$

Proof. We have already mentioned that $r^n Y_{n,j}(\xi)$ with $x = r\xi$ is a homogeneous harmonic polynomial of degree n in x_1,x_2,x_3 (see also the proof of Lemma 5.8 on p. 101). Furthermore, $P_m^{(0,n+1/2)}(2\frac{r^2}{\beta^2}-1)$ is a polynomial of degree m in $r^2 = x_1^2+x_2^2+x_3^2$. Thus, the total degree is $2m+n$. □

[1]Note that the function F has to be continuous on $[0,\beta]$. This continuity argument allows us to extend the resulting equation to $r=0$ again, though this point was lost in the previous decomposition of the Laplace operator.

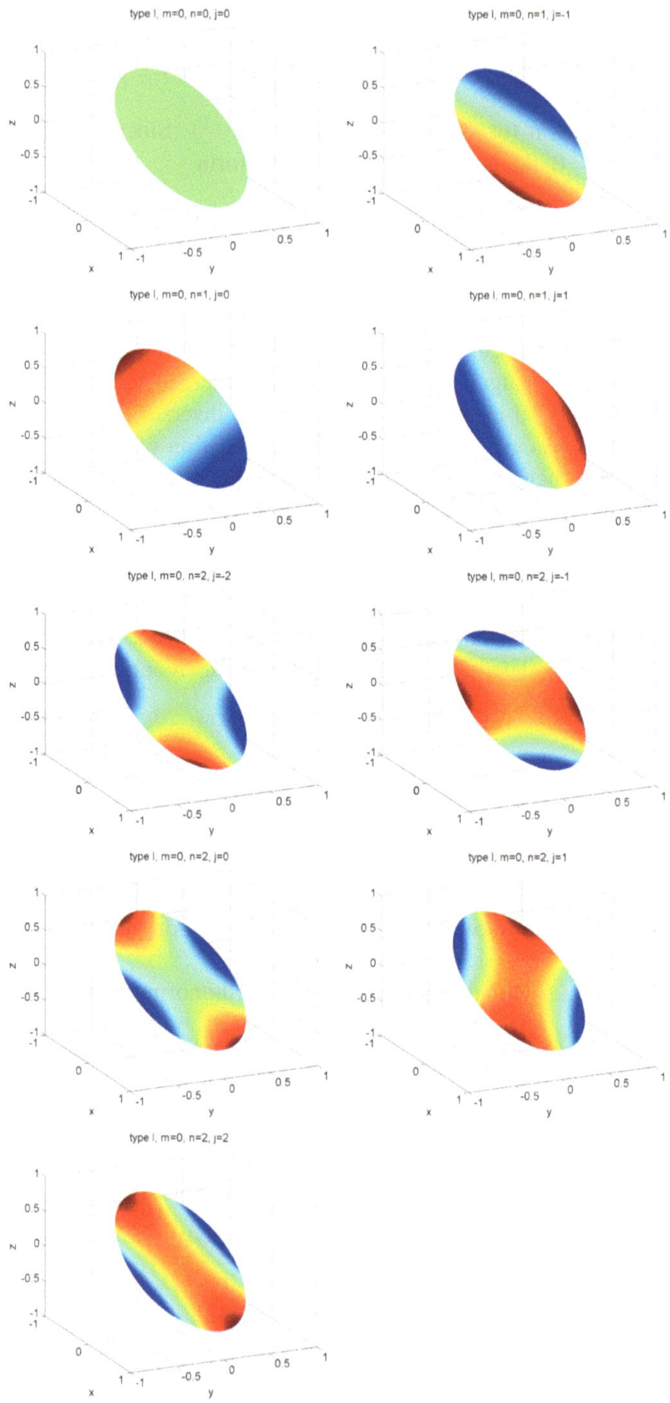

Fig. 9.1 Plots of $G^{\mathrm{I}}_{m,n,j}$ at the plane through the origin with normal vector $(1,1,1)^{\mathrm{T}}$ for $m=0$ and $n \leq 2$. For the particular parameters, see the headline of each plot. The maximum is always *red* and the minimum is *blue*. Here, $Y_{n,j}$ is a fully normalized spherical harmonic (see Sect. 5.2)

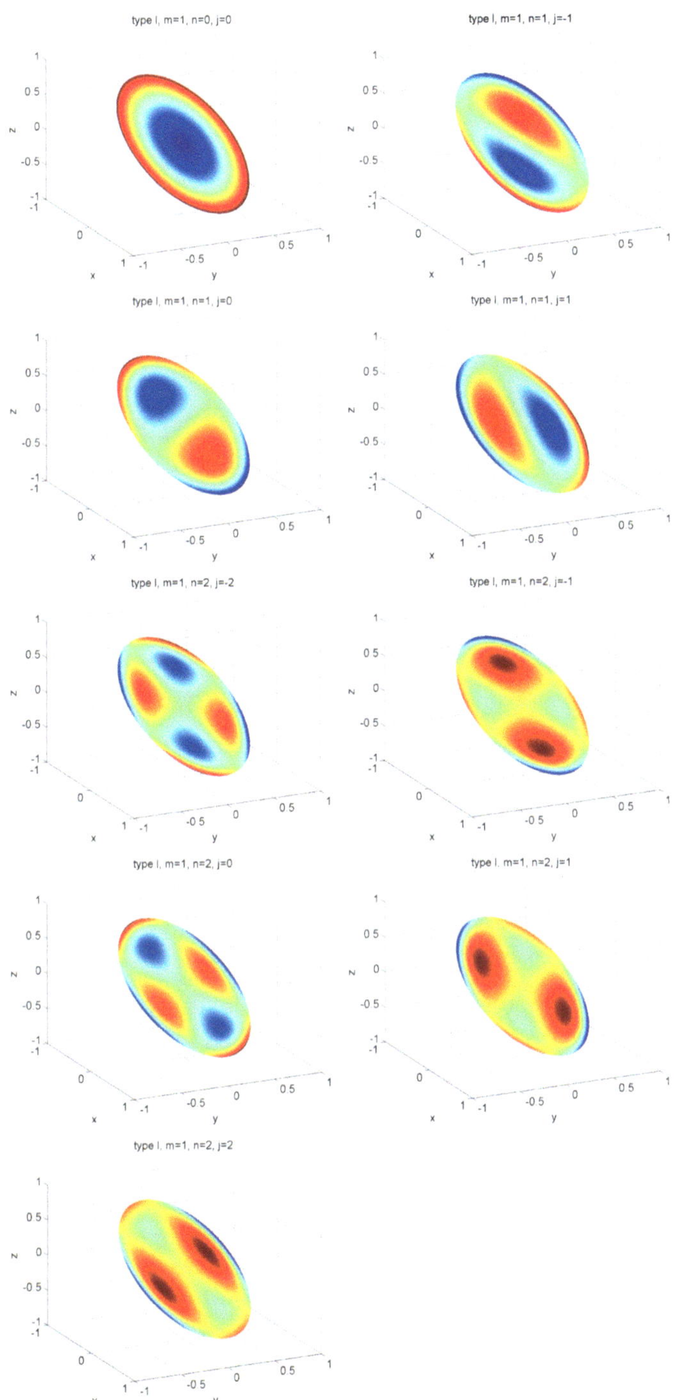

Fig. 9.2 Plots of $G^{\mathrm{I}}_{m,n,j}$ at the plane through the origin with normal vector $(1,1,1)^{\mathrm{T}}$ for $m=1$ and $n \leq 2$. For the particular parameters, see the headline of each plot. The maximum is always *red* and the minimum is *blue*. Here, $Y_{n,j}$ is a fully normalized spherical harmonic (see Sect. 5.2)

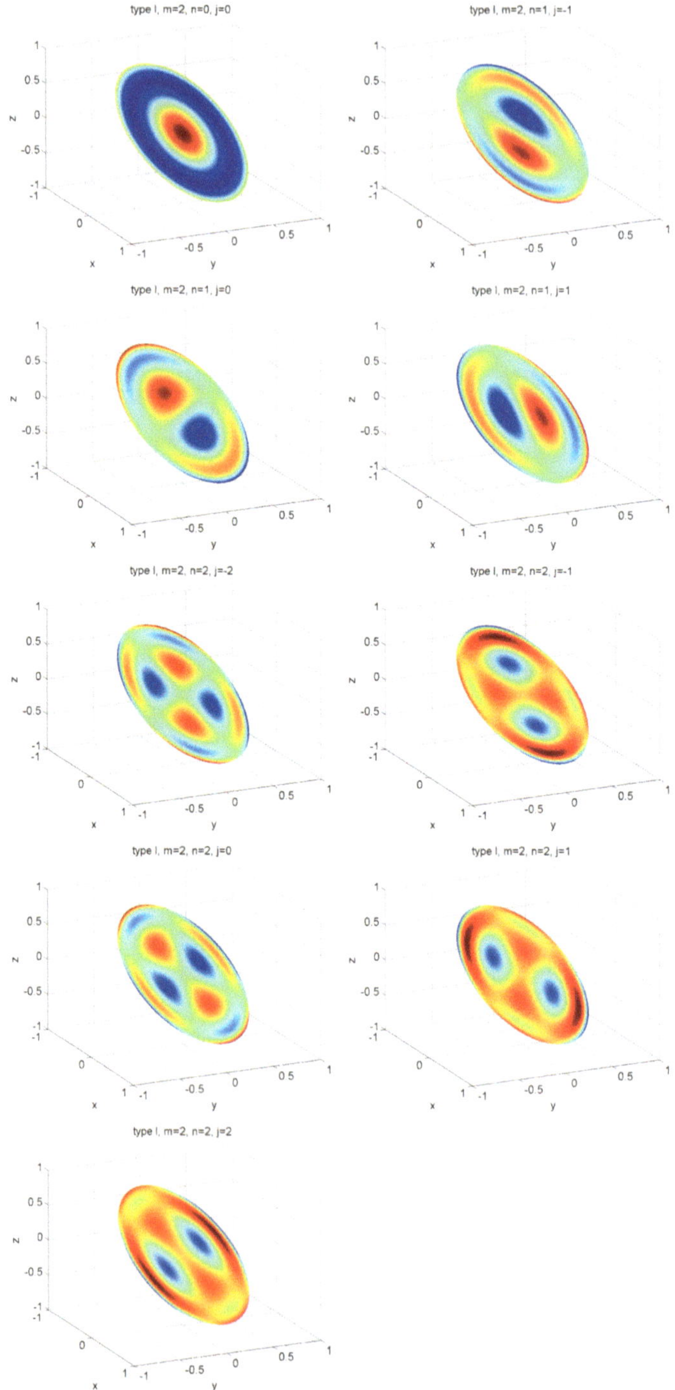

Fig. 9.3 Plots of $G^{\mathrm{I}}_{m,n,j}$ at the plane through the origin with normal vector $(1,1,1)^{\mathrm{T}}$ for $m=2$ and $n \leq 2$. For the particular parameters, see the headline of each plot. The maximum is always *red* and the minimum is *blue*. Here, $Y_{n,j}$ is a fully normalized spherical harmonic (see Sect. 5.2)

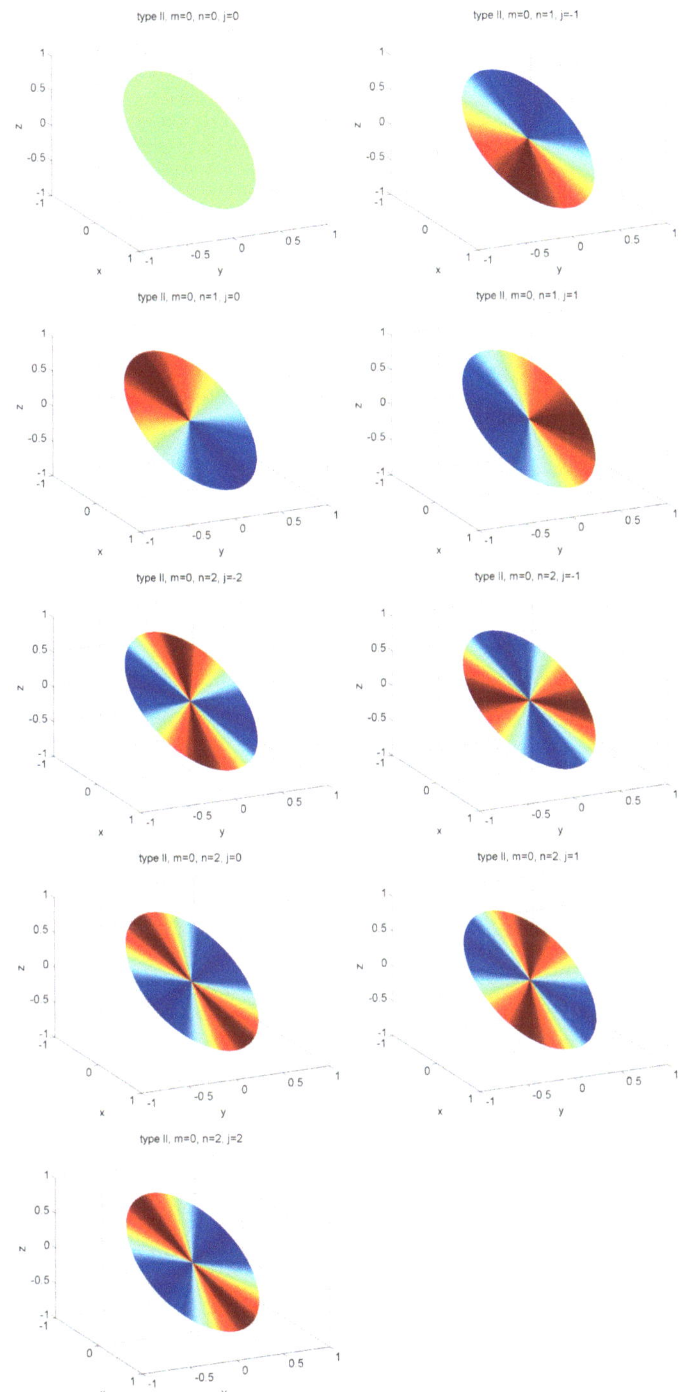

Fig. 9.4 Plots of $G^{\mathrm{II}}_{m,n,j}$ at the plane through the origin with normal vector $(1,1,1)^{\mathrm{T}}$ for $m=0$ and $n \leq 2$. For the particular parameters, see the headline of each plot. The maximum is always *red* and the minimum is *blue*. Here, $Y_{n,j}$ is a fully normalized spherical harmonic (see Sect. 5.2)

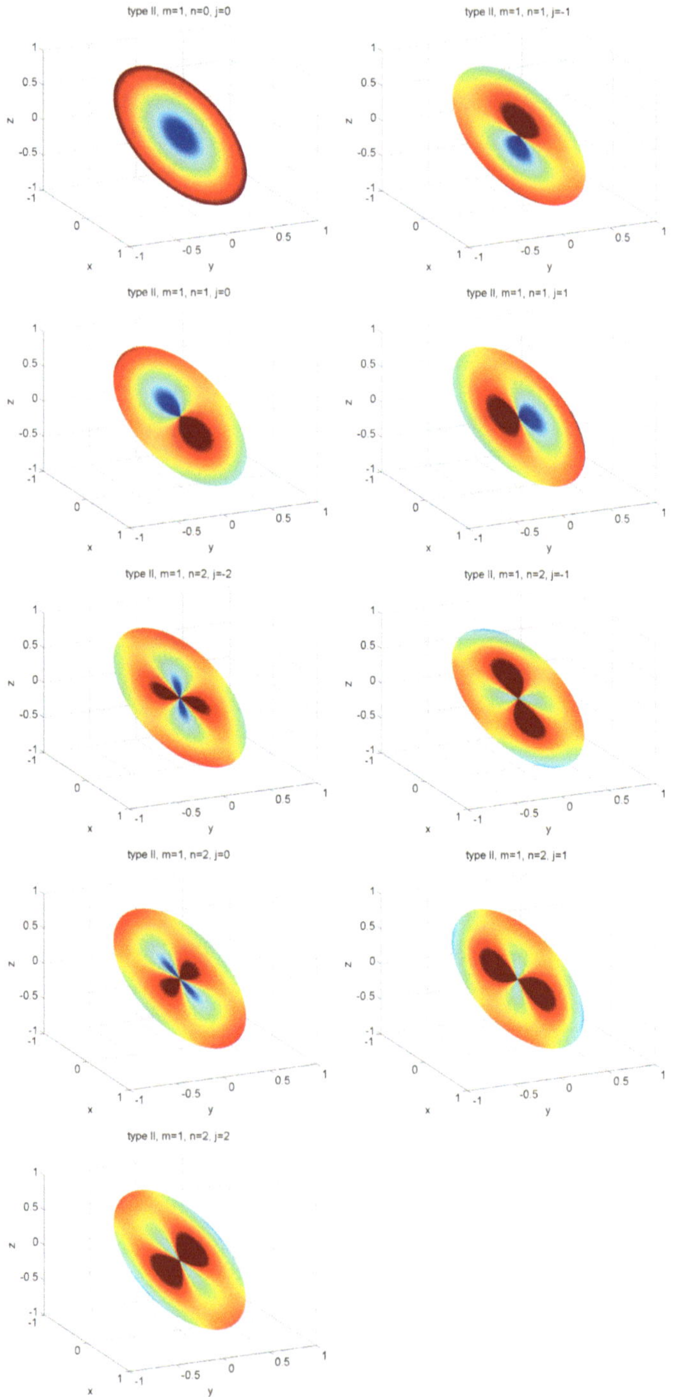

Fig. 9.5 Plots of $G^{\mathrm{II}}_{m,n,j}$ at the plane through the origin with normal vector $(1,1,1)^{\mathrm{T}}$ for $m=1$ and $n \leq 2$. For the particular parameters, see the headline of each plot. The maximum is always *red* and the minimum is *blue*. Here, $Y_{n,j}$ is a fully normalized spherical harmonic (see Sect. 5.2)

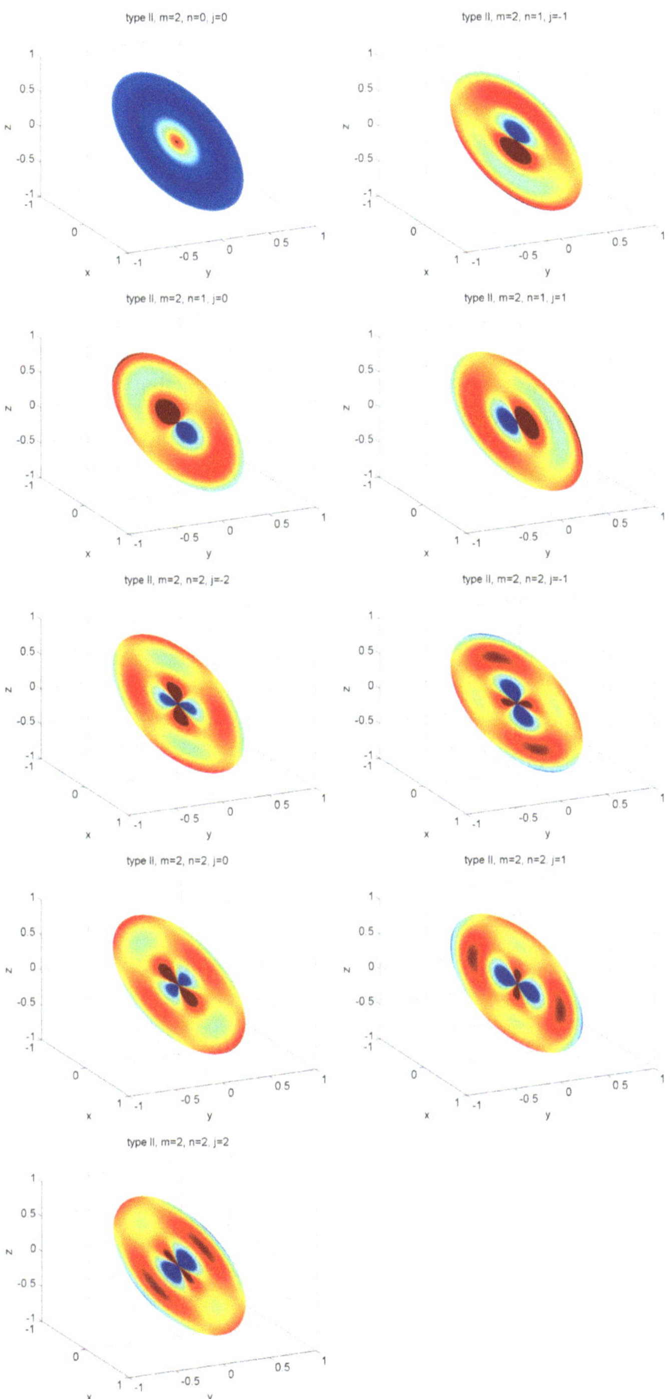

Fig. 9.6 Plots of $G^{\mathrm{II}}_{m,n,j}$ at the plane through the origin with normal vector $(1,1,1)^{\mathrm{T}}$ for $m=2$ and $n\leq 2$. For the particular parameters, see the headline of each plot. The maximum is always *red* and the minimum is *blue*. Here, $Y_{n,j}$ is a fully normalized spherical harmonic (see Sect. 5.2)

Theorem 9.5. *The maximum for each basis function on the ball satisfies*

$$\left\|G^{\mathrm{I}}_{m,n,j}\right\|_{\mathrm{C}(\mathscr{B})} \leq \sqrt{\frac{(4m+2n+3)(2n+1)}{4\pi\beta^3}\,\frac{\Gamma\left(m+n+\frac{3}{2}\right)}{m!\,\Gamma\left(n+\frac{3}{2}\right)}} \tag{9.25}$$

and

$$\left\|G^{\mathrm{II}}_{m,n,j}\right\|_{\mathrm{C}(\mathscr{B})} \leq \sqrt{\frac{(2m+3)(2n+1)}{4\pi\beta^3}\,\frac{(m+2)(m+1)}{2}}, \tag{9.26}$$

respectively, for $m,n \in \mathbb{N}_0$*;* $j = 1,\ldots,2n+1$.

Proof. From Theorem 5.17 on p. 111, we already know that $\|Y_{n,j}\|_{\mathrm{C}(\Omega)} \leq \sqrt{\frac{2n+1}{4\pi}}$. Moreover, $\frac{r}{\beta} \leq 1$. Furthermore, Theorem 3.18 on p. 49 yields

$$\left\|P_m^{(0,n+1/2)}\right\|_{\mathrm{C}[-1,1]} = \binom{m+n+\frac{1}{2}}{m} = \frac{\Gamma\left(m+n+\frac{3}{2}\right)}{m!\,\Gamma\left(n+\frac{3}{2}\right)}, \tag{9.27}$$

$$\left\|P_m^{(0,2)}\right\|_{\mathrm{C}[-1,1]} = \binom{m+2}{m} = \frac{(m+2)(m+1)}{2}. \tag{9.28}$$

□

Note that, in the case of type I, the absolute maximum of the Jacobi polynomial and the maximum of $\frac{r}{\beta}$ occur at the opposite ends of the interval $[0,\beta]$ (see Theorem 3.18 and the lines below the theorem). Since we estimate both factors from above by their corresponding maximum, the estimate for $G^{\mathrm{I}}_{m,n,j}$ can be expected to be larger than the true maximum.

9.2 Eigenfunctions of Differential Operators

As in the 1D case (Theorem 3.22 on p. 53) and the spherical case (Lemma 5.8 on p. 101), we can derive differential operators for which the basis functions are eigenfunctions (see [5]).

Lemma 9.6. *For every* $m,n \in \mathbb{N}_0$*, the function* $F^{\mathrm{I}}_{m,n}$ *is an eigenfunction of the differential operator*

$$D_r^{\mathrm{I},n} := \left(\beta^2 - r^2\right)\frac{\mathrm{d}^2}{\mathrm{d}r^2} + 2\left(\frac{\beta^2}{r} - 2r\right)\frac{\mathrm{d}}{\mathrm{d}r} - n(n+1)\frac{\beta^2}{r^2} \tag{9.29}$$

corresponding to the eigenvalue $-(n(n+3)+4m(m+n+\frac{3}{2}))$*, and each* F^{II}_m *is an eigenfunction of the differential operator*

$$D_r^{\mathrm{II}} := r(\beta - r)\,\frac{\mathrm{d}^2}{\mathrm{d}r^2} + (3\beta - 4r)\,\frac{\mathrm{d}}{\mathrm{d}r} \tag{9.30}$$

corresponding to the eigenvalue $-m(m+3)$.

Proof.

(1) Derivatives of $F^{\mathrm{I}}_{m,n}$:
We abbreviate the normalizing factor of $F^{\mathrm{I}}_{m,n}$ by $a_{m,n} := \sqrt{\frac{4m+2n+3}{\beta^3}}$, that is, we have

$$F^{\mathrm{I}}_{m,n}(r) = a_{m,n}\, P_m^{(0,n+1/2)}\left(2\frac{r^2}{\beta^2} - 1\right)\left(\frac{r}{\beta}\right)^n . \tag{9.31}$$

Consequently, the first and second derivatives are

$$\begin{aligned}\frac{\mathrm{d}}{\mathrm{d}r}\, F^{\mathrm{I}}_{m,n}(r) &= a_{m,n}\, P_m^{(0,n+1/2)\prime}\left(2\frac{r^2}{\beta^2} - 1\right) 4\,\frac{r^{n+1}}{\beta^{n+2}}\\ &\quad + a_{m,n}\, P_m^{(0,n+1/2)}\left(2\frac{r^2}{\beta^2} - 1\right) n\,\frac{r^{n-1}}{\beta^n}\end{aligned} \tag{9.32}$$

and

$$\begin{aligned}\frac{\mathrm{d}^2}{\mathrm{d}r^2}\, F^{\mathrm{I}}_{m,n}(r) &= a_{m,n}\, P_m^{(0,n+1/2)\prime\prime}\left(2\frac{r^2}{\beta^2} - 1\right) 16\,\frac{r^{n+2}}{\beta^{n+4}}\\ &\quad + a_{m,n}\, P_m^{(0,n+1/2)\prime}\left(2\frac{r^2}{\beta^2} - 1\right) 4(n+1)\,\frac{r^n}{\beta^{n+2}}\\ &\quad + a_{m,n}\, P_m^{(0,n+1/2)\prime}\left(2\frac{r^2}{\beta^2} - 1\right) 4n\,\frac{r^n}{\beta^{n+2}}\\ &\quad + a_{m,n}\, P_m^{(0,n+1/2)}\left(2\frac{r^2}{\beta^2} - 1\right) n(n-1)\,\frac{r^{n-2}}{\beta^n} .\end{aligned} \tag{9.33}$$

(2) The differential equation of the Jacobi polynomials:
From Theorem 3.22, we know that

$$\begin{aligned}&\left(1 - s^2\right) P_m^{(0,n+1/2)\prime\prime}(s) + \left[n + \frac{1}{2} - \left(n + \frac{5}{2}\right) s\right] P_m^{(0,n+1/2)\prime}(s)\\ &+ m\left(m + n + \frac{3}{2}\right) P_m^{(0,n+1/2)}(s) = 0 .\end{aligned} \tag{9.34}$$

Here, we have $s = 2\frac{r^2}{\beta^2} - 1$ such that

$$1 - s^2 = 4\frac{r^2}{\beta^2} - 4\frac{r^4}{\beta^4}. \tag{9.35}$$

(3) The second-order term in $D^{\mathrm{I},n}$:
If we use (9.33) and insert (9.35) in (9.34), then the second-order term in $D^{\mathrm{I},n}$ yields

$$\begin{aligned}
&a_{m,n}^{-1}\left(\beta^2 - r^2\right)\frac{\mathrm{d}^2}{\mathrm{d}r^2}F_{m,n}^{\mathrm{I}}(r)\\
&= 4\frac{r^n}{\beta^n}\left(4\frac{r^2}{\beta^2} - 4\frac{r^4}{\beta^4}\right)P_m^{(0,n+1/2)''}\left(2\frac{r^2}{\beta^2} - 1\right)\\
&+4\frac{r^n}{\beta^n}(2n+1)\left(1 - \frac{r^2}{\beta^2}\right)P_m^{(0,n+1/2)'}\left(2\frac{r^2}{\beta^2} - 1\right)\\
&+n(n-1)\frac{r^n}{\beta^n}\left(\frac{\beta^2}{r^2} - 1\right)P_m^{(0,n+1/2)}\left(2\frac{r^2}{\beta^2} - 1\right)\\
&= 4\frac{r^n}{\beta^n}\left[-n - \frac{1}{2} + \left(n + \frac{5}{2}\right)\left(2\frac{r^2}{\beta^2} - 1\right)\right.\\
&\left.+ (2n+1)\left(1 - \frac{r^2}{\beta^2}\right)\right]P_m^{(0,n+1/2)'}\left(2\frac{r^2}{\beta^2} - 1\right)\\
&+\frac{r^n}{\beta^n}\left[-4m\left(m + n + \frac{3}{2}\right) + n(n-1)\left(\frac{\beta^2}{r^2} - 1\right)\right]P_m^{(0,n+1/2)}\left(2\frac{r^2}{\beta^2} - 1\right)\\
&= 4\frac{r^n}{\beta^n}\left(-2 + 4\frac{r^2}{\beta^2}\right)P_m^{(0,n+1/2)'}\left(2\frac{r^2}{\beta^2} - 1\right)\\
&+\frac{r^n}{\beta^n}\left[n(n-1)\frac{\beta^2}{r^2} - \left(4m\left(m + n + \frac{3}{2}\right) + n(n-1)\right)\right]\\
&\times P_m^{(0,n+1/2)}\left(2\frac{r^2}{\beta^2} - 1\right).
\end{aligned} \tag{9.36}$$

(4) The first-order term in $D^{\mathrm{I},n}$:
Using (9.32), we obtain

$$\begin{aligned}
2a_{m,n}^{-1}\left(\frac{\beta^2}{r} - 2r\right)\frac{\mathrm{d}}{\mathrm{d}r}F_{m,n}^{\mathrm{I}}(r) &= 4\frac{r^n}{\beta^n}\left(2 - 4\frac{r^2}{\beta^2}\right)P_m^{(0,n+1/2)'}\left(2\frac{r^2}{\beta^2} - 1\right)\\
&+4\frac{r^n}{\beta^n}n\left(\frac{\beta^2}{2r^2} - 1\right)P_m^{(0,n+1/2)}\left(2\frac{r^2}{\beta^2} - 1\right).
\end{aligned} \tag{9.37}$$

(5) The eigenvalue:
We sum up (9.36) and (9.37) and obtain

$$\begin{aligned} a_{m,n}^{-1} D_r^{\mathrm{I},n} F_{m,n}^{\mathrm{I}}(r) &= \frac{r^n}{\beta^n}\left[\frac{\beta^2}{r^2}\left(n\,(n-1)+2n-n\,(n+1)\right)\right. \\ &\quad \left. -\left(4m\left(m+n+\frac{3}{2}\right)+n\,(n-1)+4n\right)\right] P_m^{(0,n+1/2)}\left(2\frac{r^2}{\beta^2}-1\right) \\ &= -\left[4m\left(m+n+\frac{3}{2}\right)+n\,(n+3)\right]\frac{r^n}{\beta^n} P_m^{(0,n+1/2)}\left(2\frac{r^2}{\beta^2}-1\right). \end{aligned} \tag{9.38}$$

(6) Type II:
The result for D_r^{II} can be derived as follows: We have

$$F_m^{\mathrm{II}}(r) = \sqrt{\frac{2m+3}{\beta^3}}\, P_m^{(0,2)}\left(2\frac{r}{\beta}-1\right) \tag{9.39}$$

and Theorem 3.22 yields

$$\left(1-t^2\right) P_m^{(0,2)\,''}(t) + (2-4t)\, P_m^{(0,2)\,'}(t) + m\,(m+3)\, P_m^{(0,2)}(t) = 0, \tag{9.40}$$

where $t = 2\frac{r}{\beta}-1$ and

$$\begin{aligned} 1-t^2 &= 4\frac{r}{\beta}-4\frac{r^2}{\beta^2}, \\ 2-4t &= 6-8\frac{r}{\beta}. \end{aligned} \tag{9.41}$$

The derivative terms of D^{II} can now be calculated as $\left(b_m := \sqrt{\frac{2m+3}{\beta^3}}\right)$

$$\begin{aligned} & b_m^{-1} r\,(\beta-r)\,\frac{\mathrm{d}^2}{\mathrm{d}r^2} F_m^{\mathrm{II}}(r) \\ &= 4\left(\frac{r}{\beta}-\frac{r^2}{\beta^2}\right)\frac{\beta^2}{4}\frac{\mathrm{d}^2}{\mathrm{d}r^2} P_m^{(0,2)}\left(2\frac{r}{\beta}-1\right) \\ &= 4\left(\frac{r}{\beta}-\frac{r^2}{\beta^2}\right) P_m^{(0,2)\,''}\left(2\frac{r}{\beta}-1\right) \\ &= \left(8\frac{r}{\beta}-6\right) P_m^{(0,2)\,'}\left(2\frac{r}{\beta}-1\right) - m\,(m+3)\, P_m^{(0,2)}\left(2\frac{r}{\beta}-1\right) \end{aligned} \tag{9.42}$$

and

$$b_m^{-1}\,(3\beta-4r)\,\frac{\mathrm{d}}{\mathrm{d}r}\,F_m^{\mathrm{II}}(r) = (3\beta-4r)\,\frac{2}{\beta}\,P_m^{(0,2)\prime}\left(2\frac{r}{\beta}-1\right)$$
$$= \left(6-8\frac{r}{\beta}\right)\,P_m^{(0,2)\prime}\left(2\frac{r}{\beta}-1\right). \tag{9.43}$$

Hence,

$$D_r^{\mathrm{II}}\,F_m^{\mathrm{II}}(r) = -m\,(m+3)\,F_m^{\mathrm{II}}(r). \tag{9.44}$$

□

Theorem 9.7. *Let $m,n\in\mathbb{N}_0$ and $j\in\{1,\ldots,2n+1\}$. Then each $G^{\mathrm{I}}_{m,n,j}$ is an eigenfunction of the operator $\Delta^{\mathrm{I},n}$ given by*

$$\Delta_x^{\mathrm{I},n}\,F(x) := D_{|x|}^{\mathrm{I},n}\,\Delta^*_{x/|x|}\,F(x), \quad F\in \mathrm{C}^{(4)}(\mathscr{B}), \tag{9.45}$$

corresponding to the eigenvalue $(n(n+3)+4m(m+n+\frac{3}{2}))n(n+1)$. Moreover, each $G^{\mathrm{II}}_{m,n,j}$ is an eigenfunction of the operator Δ^{II} given by

$$\Delta_x^{\mathrm{II}}\,F(x) := D_{|x|}^{\mathrm{II}}\,\Delta^*_{x/|x|}\,F(x), \quad F\in \mathrm{C}^{(4)}(\mathscr{B}), \tag{9.46}$$

corresponding to the eigenvalue $m(m+3)n(n+1)$.

Theorem 9.7 is an immediate consequence of Lemmata 5.8 and 9.6.

Remark 9.8. An abstract theoretical treatise on orthogonal polynomials of several variables, including the case of polynomials on a d-dimensional unit ball, can be found in [48]. The functions $P^n_{j;\beta}(W^B_{1/2};\cdot)$ in the notation of [48] correspond to the basis of type I in our notations. Further properties, including a differential equation in Cartesian coordinates, of the onb of type I can, therefore, be found in this reference.

9.3 Questions for Understanding

- Which orthonormal basis systems for $\mathrm{L}^2(\mathscr{B})$ do you know? How are they constructed? What are their advantages and disadvantages?
- What is the inverse gravimetric problem?
- Why does $\{G^{\mathrm{I}}_{m,n,j}\}_{m,n\in\mathbb{N}_0;\,j=1,\ldots,2n+1}$ play a particular role in the inverse gravimetric problem?
- Besides the orthogonality, the $\mathrm{L}^2(\mathscr{B})$-bases have another analogous feature in comparison to their 1D and spherical counterparts. What is this feature?

Chapter 10
Splines

As we already mentioned in Sect. 6.1, the data of unknown functions $F \in \mathrm{L}^2(\mathscr{B})$ are usually only indirectly given as data of tomographic problems. For example, the inverse gravimetric problem consists of finding the mass density $\rho \in \mathrm{L}^2(\mathscr{B})$ for a given gravitational potential V on $\mathbb{R}^3 \setminus \mathscr{B}$ such that

$$\int_{\mathscr{B}} \frac{\rho(x)}{|x-y|}\,\mathrm{d}x = V(y) \quad \forall y \in \mathbb{R}^3 \setminus \mathscr{B}. \tag{10.1}$$

Data could be, for instance, values or derivatives of the potential at a point grid, that is,

$$\begin{aligned}
\mathscr{F}^k\rho &:= \int_{\mathscr{B}} \frac{\rho(x)}{|x-y_k|}\,\mathrm{d}x\,; \quad k = 1,\ldots,N_1; \\
\mathscr{F}^k\rho &:= \frac{y_k}{|y_k|}\cdot\left(\nabla_y \int_{\mathscr{B}} \frac{\rho(x)}{|x-y|}\,\mathrm{d}x\right)\Bigg|_{y=y_k}; \quad k = N_1+1,\ldots,N_2; \\
\mathscr{F}^k\rho &:= \left(\frac{y_k}{|y_k|}\right)^{\mathrm{T}}\left(\nabla_y \otimes \nabla_y \int_{\mathscr{B}} \frac{\rho(x)}{|x-y|}\,\mathrm{d}x\right)\Bigg|_{y=y_k} \frac{y_k}{|y_k|}; \quad k = N_2+1,\ldots,N_3.
\end{aligned} \tag{10.2}$$

Note that $\mathscr{F}^{N_1+1}\rho,\ldots,\mathscr{F}^{N_2}\rho$ are values of the first radial derivative of the potential V and $\mathscr{F}^{N_2+1}\rho,\ldots,\mathscr{F}^{N_3}\rho$ are values of the second radial derivative of V. Such data can be derived from satellite missions such as CHAMP, GRACE, and GOCE (see, e.g., [53,71]). For further details on the inverse gravimetric problem, see the survey [134].

In the following, we will get to know a spline method developed in [8, 9, 17–19,53,132,135] based on the principles of the spherical spline method by W. Freeden (see Sect. 6).

V. Michel, *Lectures on Constructive Approximation*, Applied and Numerical Harmonic Analysis, DOI 10.1007/978-0-8176-8403-7_10,

10.1 Sobolev Spaces

In the spherical case, we took all basis functions to construct the Sobolev spaces (though we would not have needed to). In the case of the ball, practical applications such as the inverse gravimetric problem with a significant null-space require the possibility to avoid certain basis functions.[1]

Definition 10.1. Let $X \in \{\mathrm{I},\mathrm{II}\}$ and $(A_{m,n})_{m,n\in\mathbb{N}_0}$ be a given real sequence. Then the space $\mathscr{E}((A_{m,n}),X,\mathscr{B})$ consists of all $F \in \mathrm{C}^{(\infty)}(\mathscr{B})$ such that

$$\left\langle F, G^X_{m,n,j}\right\rangle_{\mathrm{L}^2(\mathscr{B})} = 0 \quad \text{for all } (m,n,j) \text{ with } A_{m,n} = 0 \tag{10.3}$$

and

$$\sum_{m,n=0}^{\infty} \sum_{j=1}^{2n+1} A^2_{m,n} \left\langle F, G^X_{m,n,j}\right\rangle^2_{\mathrm{L}^2(\mathscr{B})} < +\infty. \tag{10.4}$$

The space is equipped with the inner product

$$\langle F_1, F_2\rangle_{\mathscr{H}} := \sum_{m,n=0}^{\infty} \sum_{j=1}^{2n+1} A^2_{m,n} \left\langle F_1, G^X_{m,n,j}\right\rangle_{\mathrm{L}^2(\mathscr{B})} \left\langle F_2, G^X_{m,n,j}\right\rangle_{\mathrm{L}^2(\mathscr{B})}; \tag{10.5}$$

$F_1, F_2 \in \mathscr{E}((A_{m,n}),X,\mathscr{B})$.
$\mathscr{H}((A_{m,n}),X,\mathscr{B})$ denotes the completion of $\mathscr{E}((A_{m,n}),X,\mathscr{B})$ with respect to $\langle\cdot,\cdot\rangle_{\mathscr{H}}$ and is called the **Sobolev space** on $\mathscr{B}$ with respect to $(A_{m,n})_{m,n\in\mathbb{N}_0}$. If no confusion is likely to arise, we will simply write $\mathscr{H}$ instead of $\mathscr{H}((A_{m,n}),X,\mathscr{B})$. The induced norm is denoted by $\|\cdot\|_{\mathscr{H}}$.

Note that (10.4) and (10.5) are analogues of (6.27) and (6.28) on p. 149. Obviously, (10.5) is an inner product due to the Cauchy–Schwarz inequality:

$$\begin{aligned}
&\left| \sum_{m,n=0}^{\infty} \sum_{j=1}^{2n+1} A^2_{m,n} \left\langle F_1, G^X_{m,n,j}\right\rangle_{\mathrm{L}^2(\mathscr{B})} \left\langle F_2, G^X_{m,n,j}\right\rangle_{\mathrm{L}^2(\mathscr{B})} \right| \\
&\leq \left(\sum_{m,n=0}^{\infty} \sum_{j=1}^{2n+1} A^2_{m,n} \left\langle F_1, G^X_{m,n,j}\right\rangle^2_{\mathrm{L}^2(\mathscr{B})} \right)^{1/2} \left(\sum_{m,n=0}^{\infty} \sum_{j=1}^{2n+1} A^2_{m,n} \left\langle F_2, G^X_{m,n,j}\right\rangle^2_{\mathrm{L}^2(\mathscr{B})} \right)^{1/2} \\
&< +\infty,
\end{aligned} \tag{10.6}$$

$F_1, F_2 \in \mathscr{H}((A_{m,n}),X,\mathscr{B})$.

---

[1] In the case of the inverse gravimetric problem, a harmonicity constraint is one way to get a unique solution (see [134]). According to Theorem 9.2, this means that only $G^{\mathrm{I}}_{0,n,j}$; $n \in \mathbb{N}_0$, $j = 1,\ldots,2n+1$; is used.

Assumption 10.2. *Depending on the choice of the orthonormal basis of type I or type II, we assume that a real sequence $(A_{m,n})_{m,n\in\mathbb{N}_0}$ is given which satisfies the* ***summability condition*** *of type I, namely,*

$$\sum_{\substack{m,n=0\\A_{m,n}\neq 0}}^{\infty} A_{m,n}^{-2}\, n\,(2m+n)\, \frac{\left(n+m+\frac{1}{2}\right)^{2m}}{(m!)^2} < +\infty, \tag{10.7}$$

and of type II, namely

$$\sum_{\substack{m,n=0\\A_{m,n}\neq 0}}^{\infty} A_{m,n}^{-2}\, n m^5 < +\infty, \tag{10.8}$$

respectively. We also briefly say that $(A_{m,n})_{m,n\in\mathbb{N}_0}$ is ***I-summable*** *and* ***II-summable****, respectively.*

Now let $H\in\mathscr{H}=\mathscr{H}((A_{m,n}),X,\mathscr{B})$, $x\in\mathscr{B}$, and $N\in\mathbb{N}_0$. Then we get, using the Cauchy–Schwarz inequality, the addition theorem for spherical harmonics (Theorem 5.11 on p. 103), Definition 3.10 (see p. 41), Theorem 9.1, and Theorem 3.18 (see p. 49),

$$\left|\sum_{\substack{m+n\geq N\\A_{m,n}\neq 0}}\sum_{j=1}^{2n+1} \left\langle H, G_{m,n,j}^X\right\rangle_{\mathrm{L}^2(\mathscr{B})} G_{m,n,j}^X(x)\right|$$

$$\leq \left(\sum_{\substack{m+n\geq N\\A_{m,n}\neq 0}}\sum_{j=1}^{2n+1} A_{m,n}^2 \left\langle H, G_{m,n,j}^X\right\rangle_{\mathrm{L}^2(\mathscr{B})}^2\right)^{1/2} \left(\sum_{\substack{m+n\geq N\\A_{m,n}\neq 0}}\sum_{j=1}^{2n+1} A_{m,n}^{-2} \left(G_{m,n,j}^X(x)\right)^2\right)^{1/2}$$

$$\leq \|H\|_{\mathscr{H}}\left(\sum_{\substack{m+n\geq N\\A_{m,n}\neq 0}} A_{m,n}^{-2}\left(F_{m,n}^X(|x|)\right)^2 \frac{2n+1}{4\pi} \underbrace{P_n\left(\frac{x}{|x|}\cdot\frac{x}{|x|}\right)}_{=1}\right)^{1/2}$$

$$\leq \begin{cases} \|H\|_{\mathscr{H}} \sum_{\substack{m+n\geq N\\A_{m,n}\neq 0}} A_{m,n}^{-2}\,\frac{4m+2n+3}{\beta^3}\binom{m+n+\frac{1}{2}}{m}^2\frac{2n+1}{4\pi}, & \text{if } X=\mathrm{I},\\ \|H\|_{\mathscr{H}} \sum_{\substack{m+n\geq N\\A_{m,n}\neq 0}} A_{m,n}^{-2}\,\frac{2m+3}{\beta^3}\binom{m+2}{m}^2\frac{2n+1}{4\pi}, & \text{if } X=\mathrm{II}. \end{cases} \tag{10.9}$$

Since

$$\binom{m+n+\frac{1}{2}}{m} = \frac{\Gamma\left(m+n+\frac{3}{2}\right)}{m!\,\Gamma\left(n+\frac{3}{2}\right)} = \frac{\Gamma\left(n+\frac{3}{2}\right)}{m!\,\Gamma\left(n+\frac{3}{2}\right)}\prod_{k=3/2}^{m+1/2}(n+k) \leq \frac{\left(m+n+\frac{1}{2}\right)^m}{m!} \tag{10.10}$$

and

$$\binom{m+2}{m} = \frac{(m+2)!}{m!2} = \frac{(m+1)(m+2)}{2}, \tag{10.11}$$

we see that the summability conditions (10.7) and (10.8) guarantee that, in each case, the series behind $\|H\|_{\mathscr{H}}$ in (10.9) is finite and tends to 0, if $N \to \infty$. This yields two important results. First, we get an analogue of the Sobolev Lemma.

Lemma 10.3 (Sobolev Lemma). *Every $F \in \mathscr{H}$ has a uniformly convergent Fourier series on $\mathscr{B}$ and is continuous on $\mathscr{B} \setminus \{0\}$. In the case of type I, F is also continuous on $\mathscr{B}$.*

Second, for every arbitrary but fixed $x \in \mathscr{B}$, the evaluation functional

$$L_x : \mathscr{H} \to \mathbb{R}$$

$$F \mapsto F(x) \tag{10.12}$$

is bounded and, therefore, continuous. Hence, Aronszajn's theorem (Theorem 6.1 on p. 145) implies the existence of a reproducing kernel. Moreover, our considerations corresponding to (10.9) show that

$$\sum_{\substack{m,n=0\\A_{m,n}\neq 0}}^{\infty} \sum_{j=1}^{2n+1} \left(A_{m,n}^{-1} G_{m,n,j}^{X}(x)\right)^2 < +\infty \tag{10.13}$$

for all $x \in \mathscr{B}$. Furthermore,

$$\begin{aligned}
&\left\langle A_{\mu,\nu}^{-1} G_{\mu,\nu,\iota}^{X}, A_{r,p}^{-1} G_{r,p,q}^{X} \right\rangle_{\mathscr{H}} \\
&= \sum_{\substack{m,n=0\\A_{m,n}\neq 0}}^{\infty} \sum_{j=1}^{2n+1} A_{m,n}^{2} \left\langle A_{\mu,\nu}^{-1} G_{\mu,\nu,\iota}^{X}, G_{m,n,j}^{X} \right\rangle_{\mathrm{L}^2(\mathscr{B})} \left\langle A_{r,p}^{-1} G_{r,p,q}^{X}, G_{m,n,j}^{X} \right\rangle_{\mathrm{L}^2(\mathscr{B})} \\
&= \delta_{\mu r} \delta_{\nu p} \delta_{\iota q}.
\end{aligned} \tag{10.14}$$

In addition,

$$\left\langle F, A_{m,n}^{-1} G_{m,n,j}^{X} \right\rangle_{\mathscr{H}} = A_{m,n} \left\langle F, G_{m,n,j}^{X} \right\rangle_{\mathrm{L}^2(\mathscr{B})},$$

such that (10.5) is the Parseval identity for $(A_{m,n}^{-1} G_{m,n,j}^{X})_{m,n \in \mathbb{N}_0, A_{m,n}\neq 0, j=1,\ldots,2n+1}$ in $\mathscr{H}$ (i.e., this system is complete in $\mathscr{H}$). As a consequence, we can use Theorem 6.2 (see p. 147) to derive an explicit formula for the reproducing kernel.

Theorem 10.4. *Every $\mathscr{H}$ has a unique reproducing kernel, which is given by*

$$\begin{aligned}
K_{\mathscr{H}}(x,y) &= \sum_{\substack{m,n=0\\A_{m,n}\neq 0}}^{\infty} \sum_{j=1}^{2n+1} A_{m,n}^{-2} G_{m,n,j}^{X}(x) G_{m,n,j}^{X}(y) \\
&= \sum_{\substack{m,n=0\\A_{m,n}\neq 0}}^{\infty} A_{m,n}^{-2} F_{m,n}^{X}(|x|) F_{m,n}^{X}(|y|) \frac{2n+1}{4\pi} P_n\left(\frac{x}{|x|} \cdot \frac{y}{|y|}\right),
\end{aligned} \tag{10.15}$$

$x, y \in \mathscr{B}$, where the latter formula is only valid for $x, y \in \mathscr{B} \setminus \{0\}$, if $X = \mathrm{II}$.

Let us discuss which sequences $(A_{m,n})$ satisfy the summability conditions. Of course, the conditions become trivial, if only a finite number of components is non-vanishing. So we need not discuss such examples.

In practice, one often uses sequences of the form $A_{m,n} = B_m C_n$. In the case of type II, this separation ansatz not only decouples the summability condition to

$$\sum_{\substack{m=0\\B_m\neq 0}}^{\infty} B_m^{-2} m^5 < +\infty \quad \text{and} \quad \sum_{\substack{n=0\\C_n\neq 0}}^{\infty} C_n^{-2} n < +\infty \tag{10.16}$$

but also the reproducing kernel:

$$\begin{aligned} K_{\mathscr{H}}(x,y) = {} & \left(\sum_{m=0}^{\infty} B_m^{-2} \frac{2m+3}{\beta^3} P_m^{(0,2)}\left(2\frac{|x|}{\beta}-1\right) P_m^{(0,2)}\left(2\frac{|y|}{\beta}-1\right)\right) \\ & \times \left(\sum_{n=0}^{\infty} C_n^{-2} \frac{2n+1}{4\pi} P_n\left(\frac{x}{|x|}\cdot\frac{y}{|y|}\right)\right); \quad x,y \in \mathscr{B}\setminus\{0\}. \end{aligned} \tag{10.17}$$

Nevertheless, the separation $A_{m,n} = B_m C_n$ is also popular for kernels with basis functions of type I, though the kernels themselves cannot be separated. There is still the advantage that the radial localization can be controlled independently from the angular localization.

It appears to be relatively easy to find sequences (B_m) and (C_n) satisfying (10.16). Therefore, we will further discuss the summability condition for type I, that is, (10.7). We adapt here the investigations in [68, Example 6.4] though there are some differences in the details.

Example 10.5. First, we use Stirling's formula (see, e.g., [208, p. 600]) to estimate $m!$: For every $m \in \mathbb{N}$, there exists $\vartheta(m) \in]0,1[$ such that

$$m! = \sqrt{2\pi m}\left(\frac{m}{e}\right)^m \mathrm{e}^{\vartheta(m)/(12m)}. \tag{10.18}$$

The coefficients of $A_{m,n}^{-2}$ in (10.7) can now be treated as follows (for $m > 0$):

$$\begin{aligned} n(2m+n)\frac{\left(n+m+\frac{1}{2}\right)^{2m}}{(m!)^2} &\leq n(2m+n)\frac{(n+2m)^{2m}}{(m!)^2} \\ &= \frac{n(2m+n)}{2\pi m}\mathrm{e}^{2m-\vartheta(m)/(6m)}\left(\frac{n+2m}{m}\right)^{2m} \\ &= \frac{n}{2\pi}\mathrm{e}^{2m-\vartheta(m)/(6m)}\left(\frac{n}{m}+2\right)^{2m+1}. \end{aligned} \tag{10.19}$$

We now use the ansatz

$$A_{m,n} = \mathrm{e}^{b_m + c_n} \tag{10.20}$$

and further replace the summability condition of type I by stronger (i.e., sufficient) conditions:

$$\begin{aligned}
&\sum_{m,n=0}^{\infty} A_{m,n}^{-2} n(2m+n) \frac{\left(n+m+\frac{1}{2}\right)^{2m}}{(m!)^2} < +\infty \\
&\Leftarrow \sum_{m,n=1}^{\infty} \mathrm{e}^{-2b_m - 2c_n} n \mathrm{e}^{2m-\vartheta(m)/(6m)} \left(\frac{n}{m}+2\right)^{2m+1} < +\infty \\
&\Leftrightarrow \sum_{n=1}^{\infty} n \mathrm{e}^{-2c_n} \sum_{m=1}^{\infty} \mathrm{e}^{-2b_m + 2m - \vartheta(m)/(6m)} 2^{2m+1} \left(\frac{n}{2m}+1\right)^{2m+1} < +\infty.
\end{aligned} \tag{10.21}$$

In order to continue, we investigate the term

$$Z(x) := \left(1+\frac{n}{x}\right)^x, \quad x > 0. \tag{10.22}$$

By differentiating Z, we get

$$\begin{aligned}
Z'(x) &= \left(1+\frac{n}{x}\right)^x \left(\ln\left(1+\frac{n}{x}\right) + \frac{x}{1+\frac{n}{x}} \cdot \left(-\frac{n}{x^2}\right)\right) \\
&= \left(1+\frac{n}{x}\right)^x \left(\ln\left(1+\frac{n}{x}\right) - \frac{n}{x+n}\right).
\end{aligned} \tag{10.23}$$

For $\frac{n}{x} \geq 2$, we now get

$$\begin{aligned}
Z'(x) &= \left(1+\frac{n}{x}\right)^x \left(\ln\left(1+\frac{n}{x}\right) - \frac{n}{x+n}\right) \\
&\geq \left(1+\frac{n}{x}\right)^x (\ln 3 - 1) > 0.
\end{aligned} \tag{10.24}$$

For $\frac{n}{x} < 1$, we obtain by using the Taylor expansion of $\ln(1+\cdot)$ and the Leibniz criterion for alternating series

$$\begin{aligned}
Z'(x) &= \left(1+\frac{n}{x}\right)^x \left(\frac{n}{x} - \sum_{\nu=2}^{\infty} (-1)^{\nu} \frac{\left(\frac{n}{x}\right)^{\nu}}{\nu} - \frac{n}{x+n}\right) \\
&\geq \left(1+\frac{n}{x}\right)^x \left(\frac{n}{x} - \frac{1n^2}{2x^2} - \frac{n}{x+n}\right) \\
&= \left(1+\frac{n}{x}\right)^x \frac{2nx^2 + 2n^2x - n^2x - n^3 - 2nx^2}{2x^2(x+n)} \\
&= \left(1+\frac{n}{x}\right)^x \left(\frac{n^2}{x^2} - \frac{n^3}{x^3}\right) \frac{x}{2(x+n)} > 0.
\end{aligned} \tag{10.25}$$

For $2 > \frac{n}{x} \geq 1$, we get

$$\begin{aligned} Z'(x) &= \left(1+\frac{n}{x}\right)^x \left(\ln\left(1+\frac{n}{x}\right) - \frac{\frac{n}{x}}{1+\frac{n}{x}}\right) \\ &= \left(1+\frac{n}{x}\right)^x \left(\ln\left(1+\frac{n}{x}\right) - 1 + \frac{1}{1+\frac{n}{x}}\right) \\ &> \left(1+\frac{n}{x}\right)^x \left(\ln 2 - 1 + \frac{1}{3}\right) > 0. \end{aligned} \tag{10.26}$$

Consequently, the sequence $(\frac{n}{2m}+1)^{2m}$ is monotonically increasing with a well-known limit, namely, e^n. We continue replacing the summability condition by sufficient conditions, where we use $0 \leq \frac{n}{2m}+1 \leq 2n$ (for $n,m \geq 1$) and $0 < \vartheta(m) < 1$,

$$\begin{aligned} &\sum_{n=1}^{\infty} n\mathrm{e}^{-2c_n} \sum_{m=1}^{\infty} \mathrm{e}^{-2b_m+2m-\vartheta(m)/(6m)} 2^{2m+1} \left(\frac{n}{2m}+1\right)^{2m+1} < +\infty \\ &\Leftarrow \sum_{n=1}^{\infty} n\mathrm{e}^{-2c_n} \sum_{m=1}^{\infty} \mathrm{e}^{-2b_m+2m-\vartheta(m)/(6m)+2m\ln 2} n\mathrm{e}^n < +\infty \\ &\Leftarrow \sum_{n=1}^{\infty} n^2 \mathrm{e}^{n-2c_n} < +\infty \quad \text{and} \\ &\sum_{m=1}^{\infty} \mathrm{e}^{2(1+\ln 2)m-2b_m} < +\infty. \end{aligned} \tag{10.27}$$

Obviously, $c_n = \alpha n$ with $1-2\alpha < 0$, that is, $\alpha > \frac{1}{2}$ satisfies the condition for the angular part. Hence, we may use coefficients for the angular part of the reproducing kernel which we know from the Abel–Poisson kernel:

$$C_n^{-2} = \mathrm{e}^{-2c_n} = \mathrm{e}^{-2\alpha n} = \left(\mathrm{e}^{-2\alpha}\right)^n. \tag{10.28}$$

For the radial part, it suffices to achieve that

$$(1+\ln 2)\, m - b_m \leq -\ln m + c \tag{10.29}$$

for sufficiently large m, where $c \in \mathbb{R}$ is a constant, since, in this case,

$$\exp\left[2\,(1+\ln 2)\, m - 2b_m\right] \leq \exp\left[-2\ln m + 2c\right] = \frac{1}{m^2}\mathrm{e}^{2c}. \tag{10.30}$$

This is, for example, valid, if $b_m = \tilde{\alpha} m$ with $\tilde{\alpha} > 1+\ln 2$. Finally, we can conclude that the summability condition of type I is, for example, satisfied if

$$A_{m,n} = h_1^{-m} h_2^{-n} \tag{10.31}$$

and $h_1 < \mathrm{e}^{-1-\ln 2} = \frac{1}{2e} \approx 0.18$ and $h_2 < \mathrm{e}^{-1/2} \approx 0.61$. However, note that we replaced the summability condition by sufficient conditions, where the summability condition itself included non-sharp estimates for the maximum norm of the Jacobi polynomials. Hence, larger values of h_1 and h_2 are probably also possible. Moreover, note that numerical implementations require anyway the truncation of the series.

Note that we saw that the orthonormal basis functions are eigenfunctions of a differential operator. Based on this knowledge, we can define particular Sobolev spaces, where the spaces $\mathscr{H}_s(\Omega)$ (see Definition 6.9 on p. 151) are spherical counterparts. This was first published in [5].

Definition 10.6. For any $s \in \mathbb{R}_0^+$, we define the spaces

$$\mathscr{H}_s^{\mathrm{I}}(\mathscr{B}) := \mathscr{H}\left(\left(\left(n+2m+\frac{3}{2}\right)^s\left(n+\frac{1}{2}\right)^s\right), \mathrm{I}, \mathscr{B}\right) \tag{10.32}$$

and

$$\mathscr{H}_s^{\mathrm{II}}(\mathscr{B}) := \mathscr{H}\left(\left(\left(m+\frac{3}{2}\right)^s\left(n+\frac{1}{2}\right)^s\right), \mathrm{II}, \mathscr{B}\right). \tag{10.33}$$

Obviously, $\mathscr{H}_{s_1}^X(\mathscr{B}) \subset \mathscr{H}_{s_2}^X(\mathscr{B})$ for $s_1 \geq s_2$ and $X \in \{\mathrm{I}, \mathrm{II}\}$. Furthermore, $\mathscr{H}_0^{\mathrm{I}}(\mathscr{B}) = \mathscr{H}_0^{\mathrm{II}}(\mathscr{B}) = \mathrm{L}^2(\mathscr{B})$. Note that the corresponding sequences are not necessarily summable.

Definition 10.7. Let $s, t \in \mathbb{R}_0^+$ with $s \geq 2t$. Then we formally define the operators

$$\left({}^{**}\Delta^{\mathrm{I}}\right)^t : \mathscr{H}_s^{\mathrm{I}}(\mathscr{B}) \to \mathscr{H}_{s-2t}^{\mathrm{I}}(\mathscr{B}) \tag{10.34}$$

and

$$\left({}^{**}\Delta^{\mathrm{II}}\right)^t : \mathscr{H}_s^{\mathrm{II}}(\mathscr{B}) \to \mathscr{H}_{s-2t}^{\mathrm{II}}(\mathscr{B}) \tag{10.35}$$

by

$$\begin{aligned}\left({}^{**}\Delta^{\mathrm{I}}\right)^t F_1 &= \sum_{m,n=0}^{\infty}\sum_{j=1}^{2n+1}\left(\left(n+\frac{1}{2}\right)\left(n+2m+\frac{3}{2}\right)\right)^{2t}\left\langle F_1, G_{m,n,j}^{\mathrm{I}}\right\rangle_{\mathrm{L}^2(\mathscr{B})} G_{m,n,j}^{\mathrm{I}},\\ \left({}^{**}\Delta^{\mathrm{II}}\right)^t F_2 &= \sum_{m,n=0}^{\infty}\sum_{j=1}^{2n+1}\left(\left(n+\frac{1}{2}\right)\left(m+\frac{3}{2}\right)\right)^{2t}\left\langle F_2, G_{m,n,j}^{\mathrm{II}}\right\rangle_{\mathrm{L}^2(\mathscr{B})} G_{m,n,j}^{\mathrm{II}}\end{aligned} \tag{10.36}$$

for $F_1 \in \mathscr{H}_s^{\mathrm{I}}(\mathscr{B})$ and $F_2 \in \mathscr{H}_s^{\mathrm{II}}(\mathscr{B})$.

Note that we get, due to Lemmata 9.6 and 5.8 (see p. 101), for $m, n \in \mathbb{N}_0$; $j \in \{1, \ldots, 2n+1\}, x \in \mathscr{B} \setminus \{0\}$,

$$
\begin{aligned}
&\left(-D^{\mathrm{I},n}_{|x|}+\frac{9}{4}\right)\left(-\Delta^{*}_{x/|x|}+\frac{1}{4}\right)\left(F^{\mathrm{I}}_{m,n}(|x|)\,Y_{n,j}\left(\frac{x}{|x|}\right)\right)\\
&\quad=\left(4m\left(m+n+\frac{3}{2}\right)+n(n+3)+\frac{9}{4}\right)\left(n(n+1)+\frac{1}{4}\right)\left(F^{\mathrm{I}}_{m,n}(|x|)\,Y_{n,j}\left(\frac{x}{|x|}\right)\right)\\
&\quad=\left(n+2m+\frac{3}{2}\right)^{2}\left(n+\frac{1}{2}\right)^{2}\left(F^{\mathrm{I}}_{m,n}(|x|)\,Y_{n,j}\left(\frac{x}{|x|}\right)\right)\\
&\quad=\left({}^{**}\Delta^{\mathrm{I}}_{x}\right)^{1}G^{\mathrm{I}}_{m,n,j}(x)
\end{aligned}
\tag{10.37}
$$

and

$$
\begin{aligned}
&\left(-D^{\mathrm{II},n}_{|x|}+\frac{9}{4}\right)\left(-\Delta^{*}_{x/|x|}+\frac{1}{4}\right)\left(F^{\mathrm{II}}_{m}(|x|)\,Y_{n,j}\left(\frac{x}{|x|}\right)\right)\\
&\quad=\left(m(m+3)+\frac{9}{4}\right)\left(n(n+1)+\frac{1}{4}\right)\left(F^{\mathrm{II}}_{m}(|x|)\,Y_{n,j}\left(\frac{x}{|x|}\right)\right)\\
&\quad=\left(m+\frac{3}{2}\right)^{2}\left(n+\frac{1}{2}\right)^{2}\left(F^{\mathrm{II}}_{m}(|x|)\,Y_{n,j}\left(\frac{x}{|x|}\right)\right)\\
&\quad=\left({}^{**}\Delta^{\mathrm{II}}_{x}\right)^{1}G^{\mathrm{II}}_{m,n,j}(x).
\end{aligned}
\tag{10.38}
$$

Obviously, ${}^{**}\Delta^{\mathrm{I}}$ and ${}^{**}\Delta^{\mathrm{II}}$ are injective operators. Note that Definition 10.7 has already said that ${}^{**}\Delta^{X}$ maps into $\mathscr{H}^{X}_{s-2t}(\mathscr{B})$. This can be verified as follows for type I:

$$
\begin{aligned}
&\left\|\left({}^{**}\Delta^{\mathrm{I}}\right)^{t}F_1\right\|^{2}_{\mathscr{H}^{\mathrm{I}}_{s-2t}(\mathscr{B})}\\
&\quad=\sum_{m,n=0}^{\infty}\sum_{j=1}^{2n+1}\left(n+\frac{1}{2}\right)^{2s-4t}\left(n+2m+\frac{3}{2}\right)^{2s-4t}\left\langle\left({}^{**}\Delta^{\mathrm{I}}\right)^{t}F_1,G^{\mathrm{I}}_{m,n,j}\right\rangle^{2}_{\mathrm{L}^2(\mathscr{B})}\\
&\quad=\sum_{m,n=0}^{\infty}\sum_{j=1}^{2n+1}\left(n+\frac{1}{2}\right)^{2s}\left(n+2m+\frac{3}{2}\right)^{2s}\left\langle F_1,G^{\mathrm{I}}_{m,n,j}\right\rangle^{2}_{\mathrm{L}^2(\mathscr{B})}\\
&\quad=\|F_1\|^{2}_{\mathscr{H}^{\mathrm{I}}_{s}(\mathscr{B})}<+\infty.
\end{aligned}
\tag{10.39}
$$

Analogous considerations can be made for type II. Note that (10.39) also proves the following theorem.

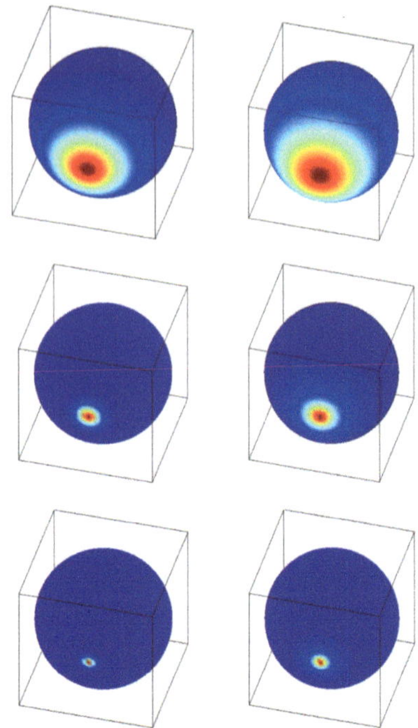

Fig. 10.1 Spline basis functions $K_{\mathscr{H}}(x_k,\cdot)$ for $x_k=(0,-\beta,0)^{\mathrm{T}}$ with $h=0.5$ (*top*), $h=0.8$ (*middle*), and $h=0.9$ (*bottom*) as well as $X=\mathrm{I}$ (*left hand*) and $X=\mathrm{II}$ (*right hand*); the functions are plotted on the surface $\partial\mathscr{B}$ of the ball $\mathscr{B}$ (from [54])

Theorem 10.8. *Let $s,t\in\mathbb{R}_0^+$ with $s\geq 2t$. If $F_1\in\mathscr{H}_s^{\mathrm{I}}(\mathscr{B})$ and $F_2\in\mathscr{H}_s^{\mathrm{II}}(\mathscr{B})$, then*

$$\left\|\left({}^{**}\Delta^{\mathrm{I}}\right)^t F_1\right\|_{\mathscr{H}^{\mathrm{I}}_{s-2t}(\mathscr{B})}=\|F_1\|_{\mathscr{H}_s^{\mathrm{I}}(\mathscr{B})},$$
$$\left\|\left({}^{**}\Delta^{\mathrm{II}}\right)^t F_2\right\|_{\mathscr{H}^{\mathrm{II}}_{s-2t}(\mathscr{B})}=\|F_2\|_{\mathscr{H}_s^{\mathrm{II}}(\mathscr{B})},\tag{10.40}$$

that is, $({}^{**}\Delta^{\mathrm{I}})^t$ and $({}^{**}\Delta^{\mathrm{II}})^t$ are isometric isomorphisms. Moreover, the particular case $t=\frac{s}{2}$ yields

$$\left\|\left({}^{**}\Delta^{\mathrm{I}}\right)^{s/2} F_1\right\|_{\mathrm{L}^2(\mathscr{B})}=\|F_1\|_{\mathscr{H}_s^{\mathrm{I}}(\mathscr{B})},$$
$$\left\|\left({}^{**}\Delta^{\mathrm{II}}\right)^{s/2} F_2\right\|_{\mathrm{L}^2(\mathscr{B})}=\|F_2\|_{\mathscr{H}_s^{\mathrm{II}}(\mathscr{B})}.\tag{10.41}$$

Therefore, the Sobolev norms $\|\cdot\|_{\mathscr{H}_s^{\mathrm{I}}(\mathscr{B})}$ and $\|\cdot\|_{\mathscr{H}_s^{\mathrm{II}}(\mathscr{B})}$ can be interpreted as the $\mathrm{L}^2(\mathscr{B})$-norms of certain generalized derivatives (i.e., in the sense of pseudodifferential operators). This is important to know, since we will also get a first minimum property for splines on the 3D ball, that is, the interpolating spline minimizes the Sobolev norm in comparison to all other interpolants. Theorem 10.8 allows us

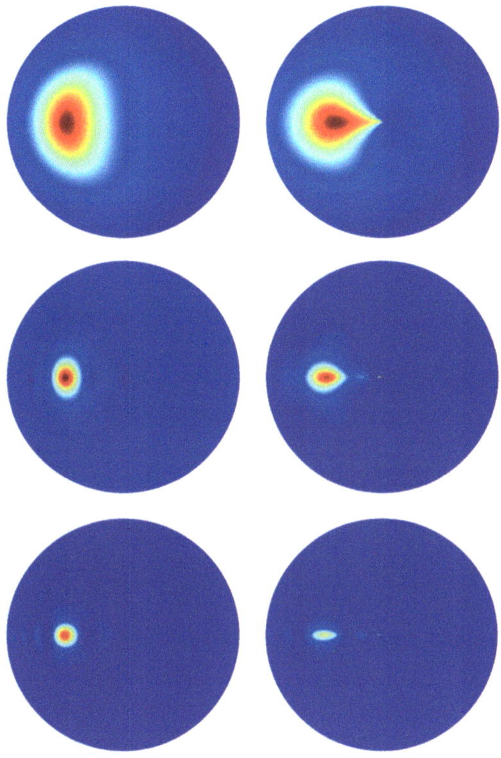

Fig. 10.2 Spline basis functions $K_{\mathscr{H}}(x_k,\cdot)$ for $x_k = (0, \frac{\beta}{2}, 0)^{\mathrm{T}}$ with $h = 0.5$ (*top*), $h = 0.8$ (*middle*), and $h = 0.9$ (*bottom*) as well as $X = \mathrm{I}$ (*left hand*) and $X = \mathrm{II}$ (*right hand*); the functions are plotted on the x_2-x_3-plane (from [54])

(as Theorem 6.12 on p. 153 did) to see a link between the first minimum property and Holladay's Theorem (Theorem 3.26 on p. 60).

Figures 10.1–10.3 show spline basis functions[2] $\mathscr{F}_x^k K_{\mathscr{H}}(x,\cdot)$ for the case $\mathscr{F}^k F := F(x_k)$. The kernels $K_{\mathscr{H}}$ correspond to the sequence

$$A_{m,n} := \begin{cases} h^{-(m+n)/2}, & \text{if } m \leq 10 \text{ and } n \leq 512 \\ 0, & \text{else} \end{cases} \tag{10.42}$$

for different choices of $h \in]0,1[$. Obviously, we get localized trial functions again, where the localization is controllable by the parameter h.

[2]For the notation, see Sect. 6.4.1 on p. 177.

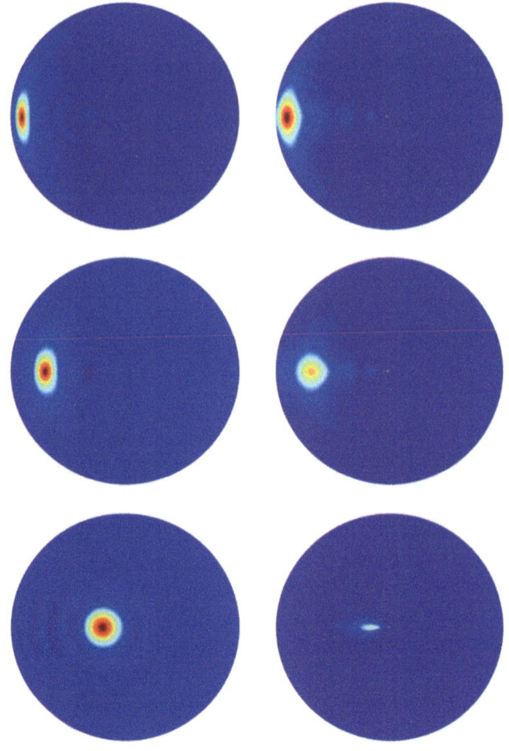

Fig. 10.3 Spline basis functions $K_{\mathscr{H}}(x_k,\cdot)$ for $h=0.8$ with $x_k=(0,0.9\beta,0)^{\mathrm{T}}$ (*top*), $x_k=(0,0.7\beta,0)^{\mathrm{T}}$ (*middle*), and $x_k=(0,0.2\beta,0)^{\mathrm{T}}$ (*bottom*) as well as $X=\mathrm{I}$ (*left hand*) and $X=\mathrm{II}$ (*right hand*); the functions are plotted on the x_2-x_3-plane (from [54])

10.2 Splines on the 3D Ball

In the following, the theoretical considerations are independent of the choice of type I or type II. We will, therefore, simply write $\mathscr{H} := \mathscr{H}((A_{m,n}),X,\mathscr{B})$ and omit the upper indices "I" and "II."

The indirect data of an unknown function $F\in\mathscr{H}$ are now assumed to be given by

$$\mathscr{F}^k F = b_k \quad \forall k=1,\ldots,N, \tag{10.43}$$

where each $\mathscr{F}^k : \mathscr{H} \to \mathbb{R}$; $k=1,\ldots,N$; is a linear and continuous functional, and $b_1,\ldots,b_N\in\mathbb{R}$ are given data. Note that the case of direct data, that is, $\mathscr{F}^k F := F(x_k)$ for $x_k\in\mathscr{B}$, is a particular case of this more general kind of data, since the evaluation functionals are linear and continuous (see the considerations after Lemma 10.3). The spline basis functions are now constructed as follows: We keep $y\in\mathscr{B}$ fixed. Then $\mathscr{H}\ni x\mapsto K_{\mathscr{H}}(x,y)\in\mathbb{R}$ is a function in $\mathscr{H}$ (see property (i) in Definition 5.13 on p. 109 and Theorem 5.14). Hence, we can apply $\mathscr{F}^k$ to this function and get a real value $\mathscr{F}^k_x K_{\mathscr{H}}(x,y)$. Since this can be done for all $y\in\mathscr{B}$, we finally end up with a function in $\mathscr{H}$ (see Theorem 6.4 on p. 148), that is, $\mathscr{B}\ni y\mapsto \mathscr{F}^k_x K_{\mathscr{H}}(x,y)\in\mathbb{R}$. This is one out of N spline basis functions.

Definition 10.9. Every function $S \in \mathscr{H}$ of the form

$$S(y) = \sum_{k=1}^{N} a_k \mathscr{F}_x^k K_{\mathscr{H}}(x,y), \quad y \in \mathscr{B}, \tag{10.44}$$

with constants $a_1,\ldots,a_N \in \mathbb{R}$ is called a **spline** in $\mathscr{H}$ (with respect to $\mathscr{F}^1,\ldots,\mathscr{F}^N$). The corresponding space of all splines is denoted by $\mathrm{Spline}((A_{m,n}),X,\mathscr{F})$, where $\mathscr{F} := (\mathscr{F}^1,\ldots,\mathscr{F}^N) \in (\mathscr{H}^*)^N$ with $\mathscr{H}^*$ being the dual space of $\mathscr{H}$.

The determination of the interpolating spline yields the system of linear equations

$$\mathscr{F}_y^l \left(\sum_{k=1}^{N} a_k \mathscr{F}_x^k K_{\mathscr{H}}(x,y) \right) = b_l \quad \forall l = 1,\ldots,N, \tag{10.45}$$

which is equivalent to

$$\sum_{k=1}^{N} a_k \mathscr{F}_y^l \mathscr{F}_x^k K_{\mathscr{H}}(x,y) = b_l \quad \forall l = 1,\ldots,N. \tag{10.46}$$

The corresponding matrix is a Gramian matrix due to Theorem 6.4:

$$\mathscr{F}_y^l \mathscr{F}_x^k K_{\mathscr{H}}(x,y) = \left\langle \mathscr{F}_x^k K_{\mathscr{H}}(x,\cdot), \mathscr{F}_y^l K_{\mathscr{H}}(y,\cdot) \right\rangle_{\mathscr{H}}. \tag{10.47}$$

Hence, this matrix is positive definite if and only if the chosen trial functions $\mathscr{F}_x^1 K_{\mathscr{H}}(x,\cdot),\ldots,\mathscr{F}_x^N K_{\mathscr{H}}(x,\cdot)$ are linearly independent. Otherwise, the matrix is singular. Furthermore, the linear independence occurs, by definition, if and only if

$$\sum_{k=1}^{N} a_k \mathscr{F}_x^k K_{\mathscr{H}}(x,\cdot) = 0 \tag{10.48}$$

only has the trivial solution $a_1 = \cdots = a_N = 0$. Equivalently, we can require that

$$\left\langle \sum_{k=1}^{N} a_k \mathscr{F}_x^k K_{\mathscr{H}}(x,\cdot), F \right\rangle_{\mathscr{H}} = 0 \quad \forall F \in \mathscr{H} \tag{10.49}$$

only has the trivial solution $a_1 = \cdots = a_N = 0$. Referring again to Theorem 6.4, we observe that (10.49) actually means that

$$\sum_{k=1}^{N} a_k \mathscr{F}^k F = 0 \quad \forall F \in \mathscr{H}. \tag{10.50}$$

In other words, the matrix above is positive definite if and only if the functionals $\mathscr{F}^1,\ldots,\mathscr{F}^N$ are linearly independent. This appears reasonable because otherwise a particular datum could be obtained by linearly combining some of the other data.

Theorem 10.10 (Existence and Uniqueness of the Interpolating Spline). *The spline interpolation problem $\mathscr{F}^k S = b_k \ \forall k = 1,\dots,N$, where $b \in \mathbb{R}^N$ is given and $S \in \mathrm{Spline}((A_{m,n}),X,\mathscr{F})$ is unknown, is uniquely solvable if and only if the functionals $\mathscr{F}^1,\dots,\mathscr{F}^N \in \mathscr{H}^*$ are linearly independent. In this case, the matrix of the system of linear equations (10.46) is positive definite.*

The spline interpolation problem is, consequently, well-posed.

Assumption 10.11. *For the rest of this chapter, we assume that $\mathscr{F}^1,\dots,\mathscr{F}^N \in \mathscr{H}^*$ are linearly independent. We write*

$$\mathscr{F} := \left(\mathscr{F}^1,\dots,\mathscr{F}^N\right). \tag{10.51}$$

Due to Theorem 6.4, most proofs of the main properties of the splines on the ball are similar to the corresponding proofs on the sphere.

Lemma 10.12. *Let $S \in \mathrm{Spline}((A_{m,n}),X,\mathscr{F})$ with $S(y) = \sum_{k=1}^N a_k \mathscr{F}_x^k K_{\mathscr{H}}(x,y)$, $y \in \mathscr{B}$, and $F \in \mathscr{H}$ be given. Then*

$$\langle S,F\rangle_{\mathscr{H}} = \sum_{k=1}^N a_k \mathscr{F}^k F. \tag{10.52}$$

Proof. The proof is rather short and uses Theorem 6.4:

$$\langle S,F\rangle_{\mathscr{H}} = \left\langle \sum_{k=1}^N a_k \mathscr{F}_x^k K_{\mathscr{H}}(x,\cdot),\, F \right\rangle_{\mathscr{H}} = \sum_{k=1}^N a_k \left\langle \mathscr{F}_x^k K_{\mathscr{H}}(x,\cdot),\, F \right\rangle_{\mathscr{H}} = \sum_{k=1}^N a_k \mathscr{F}^k F. \tag{10.53}$$

□

Theorem 10.13 (First Minimum Property). *If $b \in \mathbb{R}^N$ is a given vector and the spline $S^* \in \mathrm{Spline}((A_{m,n}),X,\mathscr{F})$ is given by $\mathscr{F}^k S^* = b_k \ \forall k = 1,\dots,N$, then S^* is the unique minimizer of*

$$\|S^*\|_{\mathscr{H}} = \min\left\{ \|F\|_{\mathscr{H}} \,\middle|\, F \in \mathscr{H} \text{ with } \mathscr{F}^k F = b_k \forall k = 1,\dots,N \right\}. \tag{10.54}$$

Proof. Let $F \in \mathscr{H}$ with $\mathscr{F}^k F = b_k \ \forall k = 1,\dots,N$. We have

$$\|F\|_{\mathscr{H}}^2 = \|F - S^* + S^*\|_{\mathscr{H}}^2 = \|F - S^*\|_{\mathscr{H}}^2 + 2\langle F - S^*, S^*\rangle_{\mathscr{H}} + \|S^*\|_{\mathscr{H}}^2. \tag{10.55}$$

If $S^* = \sum_{k=1}^N a_k^* \mathscr{F}_x^k K_{\mathscr{H}}(x,\cdot)$, then Lemma 10.12 yields

$$\langle F - S^*, S^*\rangle_{\mathscr{H}} = \sum_{k=1}^N a_k^* \underbrace{\mathscr{F}^k (F - S^*)}_{=0} = 0. \tag{10.56}$$

Hence,

$$\|F\|_{\mathscr{H}}^2 = \|F - S^*\|_{\mathscr{H}}^2 + \|S^*\|_{\mathscr{H}}^2 \geq \|S^*\|_{\mathscr{H}}^2 . \tag{10.57}$$

□

Theorem 10.14 (Second Minimum Property). *If $F \in \mathscr{H}$ is a given function and the spline $S^* \in \mathrm{Spline}((A_{m,n}),X,\mathscr{F})$ is defined by $\mathscr{F}^k S^* = \mathscr{F}^k F \; \forall k = 1,\ldots,N$, then S^* is the unique minimizer of*

$$\|F - S^*\|_{\mathscr{H}} = \min \{ \|F - S\|_{\mathscr{H}} \,|\, S \in \mathrm{Spline}\,((A_{m,n}),X,\mathscr{F}) \} . \tag{10.58}$$

Proof. We write $S = \sum_{k=1}^N a_k \mathscr{F}_x^k K_{\mathscr{H}}(x,\cdot)$ and $S^* = \sum_{k=1}^N a_k^* \mathscr{F}_x^k K_{\mathscr{H}}(x,\cdot)$ and observe that

$$\begin{aligned} \|F - S\|_{\mathscr{H}}^2 &= \|F - S^* + S^* - S\|_{\mathscr{H}}^2 \\ &= \|F - S^*\|_{\mathscr{H}}^2 + 2\,\langle F - S^*, S^* - S\rangle_{\mathscr{H}} + \|S^* - S\|_{\mathscr{H}}^2 , \end{aligned} \tag{10.59}$$

where Lemma 10.12 yields

$$\langle F - S^*, S^* - S\rangle_{\mathscr{H}} = \sum_{k=1}^N (a_k^* - a_k) \underbrace{\mathscr{F}^k (F - S^*)}_{=0} = 0. \tag{10.60}$$

Hence,

$$\|F - S\|_{\mathscr{H}}^2 = \|F - S^*\|_{\mathscr{H}}^2 + \|S^* - S\|_{\mathscr{H}}^2 \geq \|F - S^*\|_{\mathscr{H}}^2 . \tag{10.61}$$

□

Due to our previous considerations (see Theorem 10.8), we already know that the interpretation of the spline and its properties corresponds to the interpretation of the spherical spline method. For example, the interpolating spline is, in some sense, the smoothest function in comparison to all other interpolants in $\mathscr{H}$ due to the first minimum property. Moreover, the best approximation property based on the second minimum property is also valid on the ball.

Theorem 10.15 (Shannon Sampling Theorem in Spline $((A_{m,n}),X,\mathscr{F})$). *Let the functions*

$$L_k := \sum_{l=1}^N a_l^k \mathscr{F}_x^l K_{\mathscr{H}}(x,\cdot) \in \mathrm{Spline}\,((A_{m,n}),X,\mathscr{F}); \quad k = 1,\ldots,N; \tag{10.62}$$

be given[3] *by computing the coefficients $(a_l^k)_{k,l=1,\ldots,N}$ as solutions of the N systems of linear equations*

[3]Note that "k" in a_l^k is only an upper index and *not* an exponent.

$$\sum_{l=1}^{N} a_l^k \mathscr{F}_y^j \mathscr{F}_x^l K_{\mathscr{H}}(x,y) = \delta_{jk} \quad \forall\, j,k = 1,\ldots,N. \tag{10.63}$$

Then every spline $S \in \mathrm{Spline}((A_{m,n}), X, \mathscr{F})$ may be represented by the given values $\mathscr{F}^1 S, \ldots, \mathscr{F}^N S$ as

$$S(y) = \sum_{k=1}^{N} \left(\mathscr{F}^k S\right) L_k(y) \quad \forall\, y \in \mathscr{B}. \tag{10.64}$$

Proof. From (10.62) and (10.63), we conclude that

$$\mathscr{F}^j L_k = \delta_{jk} \quad \forall\, j,k = 1,\ldots,N. \tag{10.65}$$

Hence,

$$\mathscr{F}_y^j \left(\sum_{k=1}^{N} \left(\mathscr{F}^k S\right) L_k(y) \right) = \mathscr{F}^j S \quad \forall\, j = 1,\ldots,N. \tag{10.66}$$

Thus, Theorem 10.10 yields the desired result. □

Again, it is often necessary to regularize the system of linear equations.

Theorem 10.16 (Spline Approximation on the 3D Ball). *Let $b \in \mathbb{R}^N$ and $\lambda > 0$ be given. If the vector $a = (a_k)_{k=1,\ldots,N} \in \mathbb{R}^N$ is the solution of*

$$\left(\left(\mathscr{F}_y^i \mathscr{F}_x^j K_{\mathscr{H}}(x,y)\right)_{i,j=1,\ldots,N} + \lambda I_N \right) a = b, \tag{10.67}$$

where I_N is the $N \times N$-identity matrix, then the spline given by

$$S(y) := \sum_{k=1}^{N} a_k \mathscr{F}_x^k K_{\mathscr{H}}(x,y), \quad y \in \mathscr{B}, \tag{10.68}$$

is the unique minimizer of the functional

$$\mathscr{H} \ni F \mapsto \sum_{i=1}^{N} \left(b_i - \mathscr{F}^i F\right)^2 + \lambda \|F\|_{\mathscr{H}}^2 . \tag{10.69}$$

As it was the case for the previous proofs, the proof of this theorem is analogous to its spherical counterpart (in this case, it is Theorem 6.40 on p. 179). We only have to replace $F(\eta_i)$, $S(\eta_i)$, and $K_{\mathscr{H}}(\eta_i \cdot \eta_j)$ by $\mathscr{F}^i F$, $\mathscr{F}^i S$, and $\mathscr{F}_y^j \mathscr{F}_x^i K_{\mathscr{H}}(x,y)$, respectively.

The derivation of error estimates and convergence results is, however, not analogous to the spherical case because of the more general data that are allowed. Nevertheless, in [8,9], an error estimate and a convergence result could be proved.

Theorem 10.17 (Error Estimate for a Spline). *Let $F \in \mathscr{H}$ be a given function and $S^* \in \mathrm{Spline}((A_{m,n}), X, \mathscr{F})$ be the spline with $\mathscr{F}^k S^* = \mathscr{F}^k F \; \forall k = 1,\ldots,N$. Then*

$$\sup_{\substack{\mathscr{G}\in\mathscr{H}^* \\ \|\mathscr{G}\|_{\mathscr{H}^*}=1}} |\mathscr{G}F - \mathscr{G}S^*| \leq 2\Lambda \|F\|_{\mathscr{H}} \text{ and } \|F - S^*\|_{\mathscr{H}} \leq 2\sqrt{\Lambda}\|F\|_{\mathscr{H}}, \tag{10.70}$$

where the constant Λ *is given by*

$$\Lambda := \sup_{\substack{\mathscr{G}\in\mathscr{H}^* \\ \|\mathscr{G}\|_{\mathscr{H}^*}=1}} \left(\min_{\mathscr{J}\in \operatorname{span}\{\mathscr{F}^1,\ldots,\mathscr{F}^N\}} \|\mathscr{G} - \mathscr{J}\|_{\mathscr{H}^*} \right). \tag{10.71}$$

Note that Λ measures the radius of the largest "gap" in $\{\mathscr{F}^1,\ldots,\mathscr{F}^N\}$ in a similar manner as Θ_{X_N} measures the radius of the largest gap in the spherical point grid X_N (see Definition 6.35 on p. 172). Furthermore, the minimum always exists, since $\operatorname{span}\{\mathscr{F}^1,\ldots,\mathscr{F}^N\}$ is a finite-dimensional (and, consequently, closed) linear space. Moreover, Λ is bounded from above by 1, as

$$\min_{\mathscr{J}\in \operatorname{span}\{\mathscr{F}^1,\ldots,\mathscr{F}^N\}} \|\mathscr{G} - \mathscr{J}\|_{\mathscr{H}^*} \leq \|\mathscr{G}\|_{\mathscr{H}^*} = 1. \tag{10.72}$$

Proof.

(1) The first estimate:
For proving Theorem 10.17, we consider $\mathscr{G} \in \mathscr{H}^*$ with $\|\mathscr{G}\|_{\mathscr{H}^*} = 1$. There exists $\mathscr{J}_{\mathscr{G}} \in \operatorname{span}\{\mathscr{F}^1,\ldots,\mathscr{F}^N\} =: \operatorname{span}(\mathscr{F})$ such that $\|\mathscr{G} - \mathscr{J}_{\mathscr{G}}\|_{\mathscr{H}^*} \leq \Lambda$. Moreover, we observe that $\mathscr{J}_{\mathscr{G}}F = \mathscr{J}_{\mathscr{G}}S^*$, since $\mathscr{F}^kF = \mathscr{F}^kS^*$ for all $k = 1,\ldots,N$. Consequently,

$$\begin{aligned} \mathscr{G}F - \mathscr{G}S^* &= \mathscr{G}F \underbrace{- \mathscr{J}_{\mathscr{G}}F + \mathscr{J}_{\mathscr{G}}S^*}_{=0} - \mathscr{G}S^* \\ &= (\mathscr{G} - \mathscr{J}_{\mathscr{G}})(F - S^*). \end{aligned} \tag{10.73}$$

Using Theorem 6.4 and the Cauchy–Schwarz inequality, we conclude that

$$\begin{aligned} |\mathscr{G}F - \mathscr{G}S^*| &= \left|\left\langle F - S^*, (\mathscr{G} - \mathscr{J}_{\mathscr{G}})_x K_{\mathscr{H}}(x,\cdot)\right\rangle_{\mathscr{H}}\right| \\ &\leq \|F - S^*\|_{\mathscr{H}} \left\|(\mathscr{G} - \mathscr{J}_{\mathscr{G}})_x K_{\mathscr{H}}(x,\cdot)\right\|_{\mathscr{H}}. \end{aligned} \tag{10.74}$$

From the triangle inequality and the first minimum property (Theorem 10.13), we obtain

$$\|F - S^*\|_{\mathscr{H}} \leq \|F\|_{\mathscr{H}} + \|S^*\|_{\mathscr{H}} \leq 2\|F\|_{\mathscr{H}}. \tag{10.75}$$

Moreover, Theorem 6.4 says that the function $(\mathscr{G} - \mathscr{J}_{\mathscr{G}})_x K_{\mathscr{H}}(x,\cdot)$ is the "representative" of the functional $\mathscr{G} - \mathscr{J}_{\mathscr{G}} \in \mathscr{H}^*$ in the sense of the Riesz representation theorem (Theorem 2.26 on p. 26). Hence,

$$\left\|(\mathscr{G} - \mathscr{J}_{\mathscr{G}})_x K_{\mathscr{H}}(x,\cdot)\right\|_{\mathscr{H}} = \|\mathscr{G} - \mathscr{J}_{\mathscr{G}}\|_{\mathscr{H}^*} \leq \Lambda. \tag{10.76}$$

Combining (10.74), (10.75), and (10.76), we get the first part of (10.70).

(2) The second estimate:
Let $\mathscr{G} \in \mathscr{H}^*$ be chosen such that $F - S^* \in \mathscr{H}$ is its "representative" in the sense of the Riesz representation theorem (Theorem 2.26), that is, $\mathscr{G}\, G = \langle G, F - S^* \rangle_{\mathscr{H}}$ for all $G \in \mathscr{H}$ and $\|\mathscr{G}\|_{\mathscr{H}^*} = \|F - S^*\|_{\mathscr{H}}$. In particular,

$$\mathscr{G}_y K_{\mathscr{H}}(x,y) = \langle K_{\mathscr{H}}(x,\cdot), F - S^* \rangle_{\mathscr{H}} = (F - S^*)(x) \tag{10.77}$$

for all $x \in \mathscr{B}$. Furthermore, the triangle inequality and the first minimum property (Theorem 10.13) yield (again)

$$\|\mathscr{G}\|_{\mathscr{H}^*} = \|F - S^*\|_{\mathscr{H}} \le \|F\|_{\mathscr{H}} + \|S^*\|_{\mathscr{H}} \le 2\,\|F\|_{\mathscr{H}}\,. \tag{10.78}$$

If $F = S^*$, then the second part of (10.70) is trivial. Let us, therefore, assume that $\|F - S^*\|_{\mathscr{H}} > 0$ and set $\hat{\mathscr{G}} := \|\mathscr{G}\|_{\mathscr{H}^*}^{-1}\,\mathscr{G}$. Since $\hat{\mathscr{G}} \in \mathscr{H}^*$ with $\|\hat{\mathscr{G}}\|_{\mathscr{H}^*} = 1$, we finally get, using (10.77), Theorem 6.4, the first part of (10.70), and (10.78):

$$\begin{aligned}
\|F - S^*\|_{\mathscr{H}}^2 &= \langle F - S^*, F - S^* \rangle_{\mathscr{H}} \\
&= \left\langle F - S^*, \mathscr{G}_y K_{\mathscr{H}}(\cdot,y) \right\rangle_{\mathscr{H}} \\
&= \mathscr{G}\,(F - S^*) \\
&= \|\mathscr{G}\|_{\mathscr{H}^*}\, \hat{\mathscr{G}}\,(F - S^*) \\
&\le \|\mathscr{G}\|_{\mathscr{H}^*}\, 2\Lambda\, \|F\|_{\mathscr{H}} \\
&\le 4\Lambda\, \|F\|_{\mathscr{H}}^2\,.
\end{aligned} \tag{10.79}$$

□

Theorem 10.18 (Spline Convergence on the 3D Ball). *Let $\{\mathscr{F}^k \,|\, k \in \mathbb{N}\} \subset \mathscr{H}^*$ be a countable, infinite, and linearly independent system. For every $F \in \mathscr{H}$ and every $N \in \mathbb{N}$, let $S_{F,N}^* \in \mathscr{H}$ be given by*

$$\begin{aligned}
S_{F,N}^* &\in \mathrm{Spline}\left((A_{m,n}), X, \left(\mathscr{F}^1, \ldots, \mathscr{F}^N\right)\right), \\
\mathscr{F}^l\, S_{F,N}^* &= \mathscr{F}^l\, F \quad \forall l = 1, \ldots, N.
\end{aligned} \tag{10.80}$$

Then the following statements are equivalent:

(i) $\lim_{N\to\infty} |\mathscr{G}\,F - \mathscr{G}\,S_{F,N}^*| = 0\ \forall F \in \mathscr{H}\ \forall \mathscr{G} \in \mathscr{H}^*$.
(ii) $\lim_{N\to\infty} \|F - S_{F,N}^*\|_{\mathscr{H}} = 0\ \forall F \in \mathscr{H}$.
(iii) The system $\{\mathscr{F}^1, \mathscr{F}^2, \ldots\}$ is complete in $\mathscr{H}^*$, that is, if $F \in \mathscr{H}$ with $\mathscr{F}^l\, F = 0\ \forall l \in \mathbb{N}$, then $F = 0$.

Proof.

(1) (ii)⇔ (iii):
We first observe two facts:

- We have

$$\overline{\bigcup_{N\in\mathbb{N}} \mathrm{Spline}\left((A_{m,n}),X,(\mathscr{F}^1,\ldots,\mathscr{F}^N)\right)}^{\|\cdot\|_{\mathscr{H}}} = \overline{\mathrm{span}\left\{\mathscr{F}_x^k K_{\mathscr{H}}(x,\cdot)\,\middle|\, k\in\mathbb{N}\right\}}^{\|\cdot\|_{\mathscr{H}}}. \tag{10.81}$$

- Due to Theorem 6.4, $\mathscr{F}^k F = \langle F, \mathscr{F}_x^k K_{\mathscr{H}}(x,\cdot)\rangle_{\mathscr{H}}$ for all $k\in\mathbb{N}$ and all $F\in\mathscr{H}$. Hence, the completeness of $\{\mathscr{F}^k \,|\, k\in\mathbb{N}\}\subset\mathscr{H}^*$ is equivalent to the completeness of $\{\mathscr{F}_x^k K_{\mathscr{H}}(x,\cdot)\,|\,k\in\mathbb{N}\}\subset\mathscr{H}$. From Theorem 3.5 (see p. 38), we get that the latter holds if and only if $\{\mathscr{F}_x^k K_{\mathscr{H}}(x,\cdot)\,|\,k\in\mathbb{N}\}$ is closed (in the sense of the approximation theory, see Definition 3.4) in $\mathscr{H}$.

As a consequence, (iii) is equivalent to

$$\overline{\bigcup_{N\in\mathbb{N}} \mathrm{Spline}\left((A_{m,n}),X,(\mathscr{F}^1,\ldots,\mathscr{F}^N)\right)}^{\|\cdot\|_{\mathscr{H}}} = \mathscr{H}. \tag{10.82}$$

Now let (10.82) be valid. If $F\in\mathscr{H}$ and $\varepsilon>0$ are chosen, then there exists $N_0\in\mathbb{N}$ and

$$S_{N_0}\in\mathrm{Spline}\left((A_{m,n}),X,\left(\mathscr{F}^1,\ldots,\mathscr{F}^{N_0}\right)\right) \tag{10.83}$$

such that $\|F-S_{N_0}\|_{\mathscr{H}}<\varepsilon$. Hence, the second minimum property (Theorem 10.14, note that we use it twice here) yields for all $N\geq N_0$ the inequality

$$\left\|F-S_{F,N}^*\right\|_{\mathscr{H}} \leq \left\|F-S_{F,N_0}^*\right\|_{\mathscr{H}} \leq \left\|F-S_{N_0}\right\|_{\mathscr{H}} < \varepsilon. \tag{10.84}$$

This implies (ii).
Vice versa, let (ii) be valid. Then we find, for every $F\in\mathscr{H}$ and every $\varepsilon>0$, an integer $N_0\in\mathbb{N}$ such that, for all $N\geq N_0$, $\|F-S_{F,N}^*\|_{\mathscr{H}}<\varepsilon$. Since

$$S_{F,N}^*\in\mathrm{Spline}\left((A_{m,n}),X,\left(\mathscr{F}^1,\ldots,\mathscr{F}^N\right)\right), \tag{10.85}$$

we get (10.82) and, therefore, (iii).

(2) (ii) ⇒ (i):
For all $\mathscr{G}\in\mathscr{H}^*$, we have

$$\left|\mathscr{G}F-\mathscr{G}S_{F,N}^*\right| \leq \|\mathscr{G}\|_{\mathscr{H}^*}\left\|F-S_{F,N}^*\right\|_{\mathscr{H}}. \tag{10.86}$$

Hence, (ii) implies (i).

(3) (i) ⇒ (iii):
We prove this implication indirectly, that is, we assume that (iii) is false. Hence, there exists $G \in \mathscr{H}$ such that $\mathscr{F}^k G = 0$ for all $k \in \mathbb{N}$, but $G \neq 0$. From the Riesz representation theorem (Theorem 2.26 on p. 26), we know that there exists $\mathscr{G} \in \mathscr{H}^*$ such that G is its "representative." Corresponding to G, there exists a sequence of interpolating splines $(S^*_{G,N})_{N\in\mathbb{N}}$, where the spline coefficients are denoted by $(a_k^N)_{k=1,\dots,N;N\in\mathbb{N}}$, that is,

$$S^*_{G,N} = \sum_{k=1}^{N} a_k^N \mathscr{F}_x^k K_{\mathscr{H}}(x,\cdot) \tag{10.87}$$

for all $N \in \mathbb{N}$. Thus, Lemma 10.12 yields

$$\mathscr{G} S^*_{G,N} = \left\langle S^*_{G,N}, G \right\rangle_{\mathscr{H}} = \sum_{k=1}^{N} a_k^N \mathscr{F}^k G = 0 \tag{10.88}$$

for all $N \in \mathbb{N}$. As a consequence,

$$\lim_{N\to\infty} \left|\mathscr{G} G - \mathscr{G} S^*_{G,N}\right| = |\mathscr{G} G| = \langle G, G\rangle_{\mathscr{H}} = \|G\|^2_{\mathscr{H}} > 0. \tag{10.89}$$

Hence, (i) is false. □

Note that the convergence with respect to $\|\cdot\|_{\mathscr{H}}$ also implies the uniform convergence. Furthermore, the second minimum property (Theorem 10.14) also implies that the sequence $(\|F - S^*_{F,N}\|_{\mathscr{H}})_N$ is monotonically decreasing.

Finally, there also exist at least some partial analogues of Theorem 6.39 (see p. 175). These results show that the spline basis yields a localized alternative to the orthonormal basis systems. The closure results for the 3D ball were first proved in [54, 57] for the case of evaluation functionals $\mathscr{F}^k F := F(x_k)$. These theorems can also be proved for the more general data used in the case of splines.

Theorem 10.19. *Let $\{\mathscr{F}^k\}_{k\in\mathbb{N}}$ be a complete system in $\mathscr{H}^*$. Then the function system $\{\mathscr{F}_x^k K_{\mathscr{H}}(x,\cdot) \mid k \in \mathbb{N}\}$ is closed (in the sense of the approximation theory, see Definition 3.4 on p. 38) in $(\mathscr{H}, \|\cdot\|_{\mathscr{H}})$.*

Proof. Due to Theorem 6.4, the space

$$h := \overline{\operatorname{span}\{\mathscr{F}_x^k K_{\mathscr{H}}(x,\cdot) \mid k \in \mathbb{N}\}}^{\|\cdot\|_{\mathscr{H}}} \tag{10.90}$$

is a closed linear subspace of $\mathscr{H}$. Let $F \in \mathscr{H}$ and $F \perp_{\mathscr{H}} h$, that is, $\langle F, G\rangle_{\mathscr{H}} = 0$ for all $G \in h$ (see also Theorem 2.23 on p. 25). In particular, $\langle F, \mathscr{F}_x^k K_{\mathscr{H}}(x,\cdot)\rangle = 0$ for all $k \in \mathbb{N}$. Using again Theorem 6.4, we conclude that $\mathscr{F}^k F = 0$ for all $k \in \mathbb{N}$. Since the set of functionals $\mathscr{F}^k$ is complete in $\mathscr{H}^*$, only $F = 0$ is possible. Hence, $h = \mathscr{H}$. □

Whereas Theorem 10.19, which we actually already proved in the context of Theorem 10.18, is valid for both types I and II, the following theorems could (at least at present) only be proved for type I.

Theorem 10.20. *Let $\{\mathscr{F}^k\}_{k\in\mathbb{N}}$ be a complete system in $\mathscr{H}^*$, where the Sobolev space $\mathscr{H} = \mathscr{H}((A_{m,n}), \mathrm{I}, \mathscr{B})$ is given with $A_{m,n} \neq 0$ for all $m,n \in \mathbb{N}_0$. Then the system $\{\mathscr{F}_x^k K_{\mathscr{H}}(x,\cdot)\}_{k\in\mathbb{N}}$ is closed (in the sense of the approximation theory) in the Banach space $(\mathrm{C}(\mathscr{B}), \|\cdot\|_{\mathrm{C}(\mathscr{B})})$.*

Proof.

(1) $\mathscr{H}$ is dense in $\mathrm{C}(\mathscr{B})$:
Obviously, the fact that no coefficient $A_{m,n}$ may vanish, and the Sobolev Lemma (Lemma 10.3) imply that

$$g := \operatorname{span}\left\{ G^{\mathrm{I}}_{m,n,j} \,\middle|\, m,n \in \mathbb{N}_0;\ j = 1,\ldots,2n+1 \right\} \subset \mathscr{H} \subset \mathrm{C}(\mathscr{B}) , \qquad (10.91)$$

see Definition 10.1. Moreover, Theorem 9.4 on p. 253 tells us that g is a linear subspace of $\mathrm{Pol}(\mathscr{B})$, and the completeness of the orthonormal system in $\mathrm{L}^2(\mathscr{B}) \supset \mathrm{Pol}(\mathscr{B})$ (see Theorem 9.1) additionally yields that $g = \mathrm{Pol}(\mathscr{B})$.

Furthermore, note that every $F \in \mathrm{C}(\mathscr{B})$ can be continuously extended to the cube $[-\beta,\beta]^3$ (and, actually, to $\mathbb{R}^3$) by defining $\tilde{F} \in \mathrm{C}([-\beta,\beta]^3)$ with $\tilde{F}|_{\mathscr{B}} = F$ as

$$\tilde{F}(x) := \begin{cases} F(x), & \text{if } x \in \mathscr{B}, \\ F\left(\frac{x}{|x|}\beta\right), & \text{if } x \notin \mathscr{B} . \end{cases} \qquad (10.92)$$

Next, we notice that the Weierstraß approximation theorem can be generalized to multivariate functions on cuboids in $\mathbb{R}^n$. We omit the proof of this theorem and refer to [43, Corollary 6.6.4]. In particular,

$$\overline{\mathrm{Pol}\left([-\beta,\beta]^3\right)}^{\|\cdot\|_{\mathrm{C}\left([-\beta,\beta]^3\right)}} = \mathrm{C}\left([-\beta,\beta]^3\right) \qquad (10.93)$$

and, consequently,

$$\overline{\mathrm{Pol}(\mathscr{B})}^{\|\cdot\|_{\mathrm{C}(\mathscr{B})}} = \mathrm{C}(\mathscr{B}) . \qquad (10.94)$$

If we combine this result with (10.91), we get

$$\overline{\mathscr{H}}^{\|\cdot\|_{\mathrm{C}(\mathscr{B})}} = \mathrm{C}(\mathscr{B}) . \qquad (10.95)$$

(2) Back to the spline basis:
Now let $F \in \mathrm{C}(\mathscr{B})$ and $\varepsilon > 0$ be given. From the considerations above, we find $G \in \mathscr{H}$ such that $\|F - G\|_{\mathrm{C}(\mathscr{B})} < \varepsilon$, and, from Theorem 10.19, we find $H \in \operatorname{span}\{\mathscr{F}_x^k K_{\mathscr{H}}(x,\cdot) \,|\, k \in \mathbb{N}\}$ such that $\|G - H\|_{\mathscr{H}} < \varepsilon$. The latter result uses

a different norm. However, from (10.9) with $N = 0$, we obtain $\|G - H\|_{\mathrm{C}(\Omega)} \leq C\|G - H\|_{\mathscr{H}}$, where C is a constant given by the summability condition of type I. In total, we get

$$\|F - H\|_{\mathrm{C}(\mathscr{B})} \leq (1 + C)\, \varepsilon\,. \tag{10.96}$$

□

Theorem 10.21. *Let $\{\mathscr{F}^k\}_{k\in\mathbb{N}}$ be a complete system in $\mathscr{H}^*$, where the Sobolev space $\mathscr{H} = \mathscr{H}((A_{m,n}), \mathrm{I}, \mathscr{B})$ is given with $A_{m,n} \neq 0$ for all $m, n \in \mathbb{N}_0$. Then the system $\{\mathscr{F}^k_x K_{\mathscr{H}}(x,\cdot)\}_{k\in\mathbb{N}}$ is closed (in the sense of the approximation theory) in the Hilbert space $(\mathrm{L}^2(\mathscr{B}), \|\cdot\|_{\mathrm{L}^2(\mathscr{B})})$.*

Proof.

(1) $\mathrm{C}(\mathscr{B})$ is dense in $\mathrm{L}^2(\mathscr{B})$:
Every $F \in \mathrm{C}(\mathscr{B})$ can be continuously extended to an open ball $\tilde{\mathscr{B}}$ with radius $\tilde{\beta} > \beta$ by using (10.92). From [188, Theorem 3.2.2] we get that the linear space $\{\varphi \in \mathrm{C}^{(\infty)}(\tilde{\mathscr{B}}) \,|\, \operatorname{supp}\varphi \subset \tilde{\mathscr{B}} \text{ compact}\} \subset \mathrm{C}(\tilde{\mathscr{B}})$ is dense in $\mathrm{L}^2(\tilde{\mathscr{B}})$. As a consequence, $\mathrm{C}(\mathscr{B})$ is dense in $\mathrm{L}^2(\mathscr{B})$.

(2) Back to the spline basis:
Let $F \in \mathrm{L}^2(\mathscr{B})$ and $\varepsilon > 0$ be given. Due to the considerations above, there exists $G \in \mathrm{C}(\mathscr{B})$ such that $\|F - G\|_{\mathrm{L}^2(\mathscr{B})} < \varepsilon$. Moreover, Theorem 10.20 yields the existence of $H \in \operatorname{span}\{\mathscr{F}^k_x K_{\mathscr{H}}(x,\cdot) \,|\, k \in \mathbb{N}\}$ with $\|G - H\|_{\mathrm{C}(\mathscr{B})} < \varepsilon$. Finally, (2.28) on p. 20 shows that $\|G - H\|_{\mathrm{L}^2(\mathscr{B})} \leq \|G - H\|_{\mathrm{C}(\mathscr{B})} \sqrt{\frac{4}{3}\pi\beta^3}$. Consequently, we get

$$\|F - H\|_{\mathrm{L}^2(\mathscr{B})} \leq \left(1 + \sqrt{\frac{4}{3}\pi\beta^3}\right) \varepsilon\,. \tag{10.97}$$

□

We show the result of an application of the spline method to the inversion of heterogeneous gravitational data. Above South America, data (of the gravitational potential) at a denser grid and at a lower altitude (slightly above the surface) were used. These data were combined with global data (of the second radial derivative of the gravitational potential) at a coarser grid and a higher altitude (200 km). From these heterogeneous data, the spline in Fig. 10.4 was calculated (for further details, see [135]). To obtain a unique solution, a harmonicity constraint was applied, that is, $X = \mathrm{I}$ and $A_{m,n} = 0$ for $m > 0$. We see that the higher resolution of the data over South America is reflected by the spline. Note that low-altitude data are better for an inversion than high-altitude data.

The spline method presented here was extended to the case where the domain is an ellipsoid of revolution in [2, 3]. Moreover, a related version of the spline method presented here was developed for an inverse problem of medical imaging in [59].

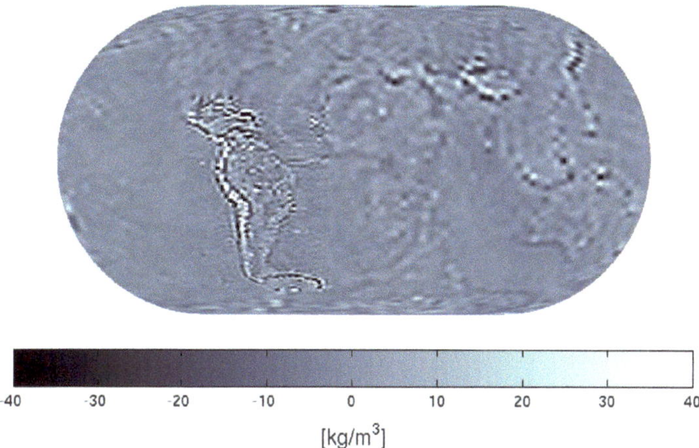

Fig. 10.4 Result of a spline-based inversion of heterogeneous gravitational data for mass density variations near the Earth's surface: The resolution of the spline is locally adapted to the data structure and quality. Since the data are better and denser over South America, the spline is automatically equipped with a higher resolution there. The image is from [135]

10.3 Questions for Understanding

- How are Sobolev spaces defined on $\mathscr{B}$?
- In correspondence to the two different basis systems, two summability conditions are distinguished on the ball. Which of them do you consider to be more difficult to satisfy? Give examples for summable sequences!
- What do the Sobolev spaces on the ball have in common with the spherical Sobolev spaces? Which difference exists?
- For the splines on a ball, we allowed more general data in comparison to the construction of spherical splines. In which sense? What is the reason for this generalization? Why is the data type used for spherical splines a particular case of the data type used for splines on a ball?
- What are the main features of the constructed splines on a ball? Which of them are analogous to features of spherical splines?

Chapter 11
Wavelets for Inverse Problems on the 3D Ball

We present here one particular wavelet method, which was developed by the author. This is certainly not the only wavelet method for tomographic problems on the 3D ball. There exist alternatives, where at least [60, 175, 176] should be mentioned here.

11.1 The Problem

We will solve problems of the following class.

Problem 11.1. There exists a measurable set $D \subset \mathbb{R}^m$, a complete subspace $\mathscr{Y} \subset \mathrm{L}^2(D, \mathbb{R}^q)$, $q \in \mathbb{N}$, and an operator $T : \mathrm{L}^2(\mathscr{B}) \to \mathscr{Y}$ with a known singular value decomposition:

$$TF = \sum_{m,n=0}^{\infty} \sum_{j=1}^{2n+1} \tau_{m,n} \left\langle F, G_{m,n,j} \right\rangle_{\mathrm{L}^2(\mathscr{B})} b_{m,n,j} \quad \forall F \in \mathrm{L}^2(\mathscr{B}), \tag{11.1}$$

where the (not necessarily pairwise distinct) functions $b_{m,n,j}$ are elements of $\mathscr{Y}$ such that $\{b_{m,n,j}\}_{m,n \in \mathbb{N}_0, \tau_{m,n} \neq 0; j=1,\ldots,2n+1}$ is complete in $\mathscr{Y}$ and

$$\left\langle b_{m,n,j}, b_{\mu,\nu,\iota} \right\rangle_{\mathrm{L}^2(D,\mathbb{R}^q)} = \delta_{m\mu} \delta_{n\nu} \delta_{j\iota}, \quad \text{if } \tau_{m,n} \tau_{\mu,\nu} \neq 0. \tag{11.2}$$

In other words, the system of all $b_{m,n,j}$ for $m, n \in \mathbb{N}_0$ with $\tau_{m,n} \neq 0$ and $j = 1, \ldots, 2n+1$ is an orthonormal basis of the Hilbert space $(\mathscr{Y}, \langle \cdot, \cdot \rangle_{\mathrm{L}^2(D,\mathbb{R}^q)})$. Moreover, we assume that $\{\tau_{m,n}\}_{m,n \in \mathbb{N}_0}$ is bounded, that is, T is continuous. The inverse problem is now: Given $g \in \mathscr{Y}$, find $f \in \mathrm{L}^2(\mathscr{B})$ such that $Tf = g$.

Example 11.2. For example, if $T : \mathrm{L}^2(\mathscr{B}) \to \mathrm{L}^2(\sigma\Omega)$ with $\sigma > \beta$ is the operator which maps a mass density distribution $\rho \in \mathrm{L}^2(\mathscr{B})$ to the corresponding gravitational potential at a spherical satellite orbit with radius σ, that is,

V. Michel, *Lectures on Constructive Approximation*, Applied and Numerical Harmonic Analysis, DOI 10.1007/978-0-8176-8403-7_11,

$$(T\rho)(y) = \int_{\mathscr{B}} \frac{\rho(x)}{|x-y|}\,\mathrm{d}x, \quad |y| = \sigma, \tag{11.3}$$

then (see [130])

$$(T\rho)(\sigma\eta) = \sum_{n=0}^{\infty} \frac{4\pi}{2n+1} \left(\frac{\beta}{\sigma}\right)^n \sqrt{\frac{\beta^3}{2n+3}} \sum_{j=1}^{2n+1} \langle \rho, G^{\mathrm{I}}_{0,n,j} \rangle_{\mathrm{L}^2(\mathscr{B})} \frac{1}{\sigma} Y_{n,j}(\eta), \, \eta \in \Omega, \tag{11.4}$$

that is,

$$\tau_{m,n} = \delta_{m0} \frac{4\pi}{2n+1} \left(\frac{\beta}{\sigma}\right)^n \sqrt{\frac{\beta^3}{2n+3}}, \tag{11.5}$$

$$b_{0,n,j} = \frac{1}{\sigma} Y_{n,j}\left(\frac{\cdot}{\sigma}\right), \tag{11.6}$$

$$D = \sigma\Omega, \tag{11.7}$$

$$\mathscr{Y} = \mathrm{L}^2(\sigma\Omega), \tag{11.8}$$

$$q = 1, \tag{11.9}$$

and $b_{m,n,j}$ may be chosen arbitrarily for $m > 0$. This is the inverse gravimetric problem, which we already mentioned above. There also exist expansions for the case that derivatives of the potential are given. For a detailed discussion, see [130]. In such cases, it can occur that the functions $b_{m,n,j}$ are not scalar but vectorial or tensorial.

In [126, 127, 129] a wavelet method for the inverse gravimetric problem (Example 11.2) was developed. We will follow here the generalized approach in [128, 130, 132, 134], where inverse problems of the kind of Problem 11.1 are treated. Note also that we can set $T = \mathrm{Id}$, $\tau_{m,n} = 1 \forall m,n$, $b_{m,n,j} = G_{m,n,j} \forall m,n,j$ in order to discuss the approximation of $F \in \mathrm{L}^2(\mathscr{B})$ from direct data (this case is also briefly described in [131]). Moreover, note that an Approximate Identity on the 3D ball based on locally supported kernels was constructed in [4].

Furthermore, it should be mentioned that an extension to the case where a singular value decomposition of the operator T is unknown was developed in [125]. Note that the inverse problem can be ill-posed. This is, in particular, the case, if

$$\lim_{N\to\infty} \sup_{m+n\leq N} |\tau_{m,n}| = 0, \tag{11.10}$$

which is the case of a compact operator. Then $(\tau_{m,n}^{-1})_{m,n}$ is unbounded,[1] and the sequence $(T^{-1}b_{m,n,j})_{m,n,j}$ is unbounded as well, though $\|b_{m,n,j}\|_{\mathrm{L}^2(D,\mathbb{R}^q)} = 1$ for all m,n,j.

11.2 Convolutions

We adapt now the concept of convolutions introduced in Definition 3.31 (see p. 67). Note that every $G_{m,n,j}$ is scalar, but the functions $b_{m,n,j}$ can be scalar, vectorial, or tensorial.[2]

Definition 11.3. Let $K \in \mathrm{L}^2(\mathscr{B} \times D, \mathbb{R}^q)$ and $g \in \mathrm{L}^2(D,\mathbb{R}^q)$ be given. Then, we define the **convolution** of K and g by

$$(K * g)(x) := \int_D K(x,y) \cdot g(y)\,\mathrm{d}y\,, \quad x \in \mathscr{B}. \tag{11.11}$$

Note that $K(x,y) \cdot g(y)$ is here a Euclidean inner product.

Theorem 11.4. *If $K \in \mathrm{L}^2(\mathscr{B} \times D, \mathbb{R}^q)$ and $g \in \mathrm{L}^2(D, \mathbb{R}^q)$, then $K * g \in \mathrm{L}^2(\mathscr{B}, \mathbb{R})$.*

Proof. The $\mathrm{L}^2(\mathscr{B})$-norm of $K * g$ can be estimated by using the Cauchy–Schwarz inequality as follows:

$$\begin{aligned}
\int_{\mathscr{B}} ((K * g)(x))^2\,\mathrm{d}x &= \int_{\mathscr{B}} \left(\int_D K(x,y) \cdot g(y)\,\mathrm{d}y \right)^2 \mathrm{d}x \\
&\le \int_{\mathscr{B}} \int_D |K(x,y)|_{\mathbb{R}^q}^2\,\mathrm{d}y \int_D |g(y)|_{\mathbb{R}^q}^2\,\mathrm{d}y\,\mathrm{d}x \\
&= \|g\|_{\mathrm{L}^2(D,\mathbb{R}^q)}^2 \, \|K\|_{\mathrm{L}^2(\mathscr{B}\times D,\mathbb{R}^q)}^2 < +\infty\,.
\end{aligned} \tag{11.12}$$

□

Theorem 11.5. *Let $g \in \mathscr{Y}$ be given and $K \in \mathrm{L}^2(\mathscr{B} \times D, \mathbb{R}^q)$ be of the form*

$$K(x,y) = \sum_{m,n=0}^{\infty} \sum_{j=1}^{2n+1} K^{\wedge}(m,n)\, G_{m,n,j}(x)\, b_{m,n,j}(y), \quad x \in \mathscr{B}, y \in D, \tag{11.13}$$

where $K^{\wedge}(m,n) = 0$, if $\tau_{m,n} = 0$. Then

---

[1] For this example, we assume that $\tau_{m,n} \neq 0$ for all $m,n \in \mathbb{N}_0$.

[2] In the case of tensorial functions, that is, $b_{m,n,j} : D \to \mathbb{R}^{3\times 3}$, we can use the isomorphism of $\mathbb{R}^{3\times 3}$ and $\mathbb{R}^9$ and interpret this case as $q = 9$.

$$K * g = \sum_{m,n=0}^{\infty} \sum_{j=1}^{2n+1} K^{\wedge}(m,n) \langle g, b_{m,n,j} \rangle_{\mathrm{L}^2(D,\mathbb{R}^q)} G_{m,n,j} \tag{11.14}$$

in the sense of $\mathrm{L}^2(\mathscr{B})$.

Before we prove Theorem 11.5, let us note that the functions

$$\mathscr{B} \times D \ni (x,y) \mapsto G_{m,n,j}(x)\, b_{m,n,j}(y) \tag{11.15}$$

are orthonormal in $\mathrm{L}^2(\mathscr{B} \times D, \mathbb{R}^q)$ (at least for those (m,n,j) with $\tau_{m,n} \neq 0$), since Fubini's theorem yields

$$\begin{aligned}
&\int_{\mathscr{B}\times D} G_{m,n,j}(x)\, b_{m,n,j}(y)^{\mathrm{T}} G_{\mu,\nu,\iota}(x)\, b_{\mu,\nu,\iota}(y)\, \mathrm{d}(x,y) \\
&= \underbrace{\int_{\mathscr{B}} G_{m,n,j}(x)\, G_{\mu,\nu,\iota}(x)\, \mathrm{d}x}_{=\delta_{m\mu}\delta_{n\nu}\delta_{j\iota}} \int_D b_{m,n,j}(y)^{\mathrm{T}} b_{\mu,\nu,\iota}(y)\, \mathrm{d}y \\
&= \delta_{m\mu}\delta_{n\nu}\delta_{j\iota} \underbrace{\left\| b_{m,n,j} \right\|^2_{\mathrm{L}^2(D,\mathbb{R}^q)}}_{=1,\text{ if } \tau_{m,n}\neq 0 \text{ due to (11.2)}} .
\end{aligned} \tag{11.16}$$

However, this system is not complete for the following reason: Let us assume, without loss of generality, that $\tau_{0,0} \neq 0$ and $\tau_{1,0} \neq 0$. Then we get [using (11.2)]

$$\begin{aligned}
&\int_{\mathscr{B}\times D} G_{0,0,1}(x) b_{1,0,1}(y)^{\mathrm{T}} G_{m,n,j}(x)\, b_{m,n,j}(y)\, \mathrm{d}(x,y) \\
&= \underbrace{\int_{\mathscr{B}} G_{0,0,1}(x)\, G_{m,n,j}(x)\, \mathrm{d}x}_{=\delta_{m0}\delta_{n0}\delta_{j1}} \int_D b_{1,0,1}(y)^{\mathrm{T}} b_{m,n,j}(y)\, \mathrm{d}y \\
&= \langle b_{1,0,1}, b_{0,0,1} \rangle_{\mathrm{L}^2(D,\mathbb{R}^q)} \\
&= 0 \quad \forall m,n \in \mathbb{N}_0\ \forall j \in \{1,\ldots,2n+1\},
\end{aligned} \tag{11.17}$$

although (11.2) also implies that $b_{1,0,1} \neq b_{0,0,1}$.

Nevertheless, the orthonormality alone suffices to conclude that

$$\|K\|^2_{\mathrm{L}^2(\mathscr{B}\times D,\mathbb{R}^q)} = \sum_{m,n=0}^{\infty} (2n+1) \left(K^{\wedge}(m,n)\right)^2, \tag{11.18}$$

provided that K is given by (11.13).

Now let us prove Theorem 11.5.

Proof. Since $K \in \mathrm{L}^2(\mathscr{B} \times D, \mathbb{R}^q)$ and $g \in \mathscr{Y} \subset \mathrm{L}^2(D, \mathbb{R}^q)$, the Fourier expansions of $K(x,\cdot)$ and g converge strongly in $\mathrm{L}^2(D, \mathbb{R}^q)$ (for almost every $x \in \mathscr{B}$). Hence, we get [using (11.2) and the completeness of $\{b_{m,n,j}\}_{\tau_{m,n}\neq 0}$]

$$
\begin{aligned}
(K*g)(x) &= \langle K(x,\cdot), g\rangle_2 \\
&= \left\langle \sum_{m,n=0}^{\infty} \sum_{j=1}^{2n+1} K^{\wedge}(m,n)\, G_{m,n,j}(x)\, b_{m,n,j}, \sum_{\substack{\mu,\nu=0\\ \tau_{\mu,\nu}\neq 0}}^{\infty} \sum_{\iota=1}^{2\nu+1} \langle g, b_{\mu,\nu,\iota}\rangle_2\, b_{\mu,\nu,\iota} \right\rangle_2 \\
&= \sum_{\substack{m,n,\mu,\nu=0\\ \tau_{\mu,\nu}\neq 0}}^{\infty} \sum_{j=1}^{2n+1} \sum_{\iota=1}^{2\nu+1} K^{\wedge}(m,n) \langle g, b_{\mu,\nu,\iota}\rangle_2\, G_{m,n,j}(x) \langle b_{m,n,j}, b_{\mu,\nu,\iota}\rangle_2 \\
&= \sum_{m,n=0}^{\infty} \sum_{j=1}^{2n+1} K^{\wedge}(m,n) \langle g, b_{m,n,j}\rangle_2\, G_{m,n,j}(x), \quad x \in \mathscr{B}, \qquad (11.19)
\end{aligned}
$$

in the sense of $\mathrm{L}^2(\mathscr{B})$, where we used the abbreviation $\langle\cdot,\cdot\rangle_2 := \langle\cdot,\cdot\rangle_{\mathrm{L}^2(D,\mathbb{R}^q)}$. □

Theorem 11.5 is an analogue of Corollary 7.6 on p. 188, which played an important role in the spherical wavelet theory.

In the following, we will present a linear wavelet theory for the regularization of the inverse problem $Tf = g$. "Linear" means that we do without the use of dual wavelets—in contrast to a bilinear theory as we had it for the spherical case. A bilinear theory for $Tf = g$ is presented in [128, 130]. In the numerical implementations, there is no significant difference between both concepts.

11.3 Scaling Functions

Definition 11.6. We define a **scaling function** [with respect to the considered inverse problem (Problem 11.1)] as a family of functions Φ_J on $\mathscr{B}\times D$ of the form

$$
\Phi_J(x,y) = \sum_{m,n=0}^{\infty} \sum_{j=1}^{2n+1} \Phi_J^{\wedge}(m,n)\, G_{m,n,j}(x)\, b_{m,n,j}(y); \quad x\in\mathscr{B}, y \in D; \qquad (11.20)
$$

if the coefficients $\Phi_J^{\wedge}(m,n)$ satisfy the following conditions[3]:

(SB1) $\Phi_J^{\wedge}(m,n) = 0$ for all $(m,n)\in\mathbb{N}_0^2$ with $\tau_{m,n} = 0$ and all $J\in\mathbb{N}_0$.
(SB2) $|\Phi_J^{\wedge}(m,n)| \le |\Phi_{J+1}^{\wedge}(m,n)|$ for all $J,m,n\in\mathbb{N}_0$.
(SB3) $\lim_{J\to\infty} \Phi_J^{\wedge}(m,n) = \tau_{m,n}^{-1}$ for all $(m,n)\in\mathbb{N}_0^2$ with $\tau_{m,n}\neq 0$.
(SB4) $\sum_{m,n=0}^{\infty} n(\Phi_J^{\wedge}(m,n))^2 < +\infty$ for all $J\in\mathbb{N}_0$.

Note that (SB4) implies that $\Phi_J \in \mathrm{L}^2(\mathscr{B}\times D)$ for all $J\in\mathbb{N}_0$ due to (11.18). The main property of a scaling function is the fact that it can be used to calculate arbitrarily good but stable approximations to the (possibly instable) exact solution, that is, it provides us with a regularization of the inverse problem. We will prove this in two steps.

[3] "SB" refers here to "scaling function" and "ball."

Theorem 11.7 (Approximation of the Solution). *Let the functions $f \in \mathrm{L}^2(\mathscr{B})$ and $g \in \mathrm{L}^2(D, \mathbb{R}^q)$ satisfy $Tf = g$ and $\langle f, G_{m,n,j} \rangle_{\mathrm{L}^2(\mathscr{B})} = 0$ for all (m,n,j) with $\tau_{m,n} = 0$. Moreover, let $\{\Phi_J\}_{J \in \mathbb{N}_0}$ be a corresponding scaling function. Then*

$$\lim_{J \to \infty} \|\Phi_J * g - f\|_{\mathrm{L}^2(\mathscr{B})} = 0. \tag{11.21}$$

If $T = \mathrm{Id}$ (and, consequently, $f = g$), then Theorem 11.7 says that the scaling function yields an Approximate Identity.

Proof. Note that $Tf = g$ guarantees that $g \in \mathscr{Y}$ due to (11.1) and the requirements on $\{b_{m,n,j}\}_{m,n \in \mathbb{N}_0;\, j=1,\ldots,2n+1}$. We can now use Theorem 11.5 to conclude that

$$\Phi_J * g = \sum_{m,n=0}^{\infty} \sum_{j=1}^{2n+1} \Phi_J^{\wedge}(m,n) \left\langle g, b_{m,n,j} \right\rangle_{\mathrm{L}^2(D,\mathbb{R}^q)} G_{m,n,j} \tag{11.22}$$

in the sense of $\mathrm{L}^2(\mathscr{B})$. Moreover, the solution f of $Tf = g$ under the constraints of the theorem is uniquely given by

$$f = \sum_{\substack{m,n=0 \\ \tau_{m,n} \neq 0}}^{\infty} \sum_{j=1}^{2n+1} \tau_{m,n}^{-1} \left\langle g, b_{m,n,j} \right\rangle_{\mathrm{L}^2(D,\mathbb{R}^q)} G_{m,n,j} \tag{11.23}$$

in the sense of $\mathrm{L}^2(\mathscr{B})$. Hence, the Parseval identity in combination with condition (SB1) yields

$$\|\Phi_J * g - f\|_{\mathrm{L}^2(\mathscr{B})}^2 = \sum_{\substack{m,n=0 \\ \tau_{m,n} \neq 0}}^{\infty} \sum_{j=1}^{2n+1} \left(\Phi_J^{\wedge}(m,n) - \tau_{m,n}^{-1} \right)^2 \left\langle g, b_{m,n,j} \right\rangle_{\mathrm{L}^2(D,\mathbb{R}^q)}^2 . \tag{11.24}$$

Furthermore, conditions (SB2) and (SB3) allow us to conclude that

$$0 \leq \left(\Phi_J^{\wedge}(m,n) - \tau_{m,n}^{-1} \right)^2 \leq \left(\left| \Phi_J^{\wedge}(m,n) \right| + \left| \tau_{m,n}^{-1} \right| \right)^2 \leq 4 \tau_{m,n}^{-2}. \tag{11.25}$$

Since the solvability of the problem $Tf = g$ (actually, the fact that $\|f\|_{\mathrm{L}^2(\mathscr{B})} < +\infty$) yields the convergence of the series

$$\sum_{\substack{m,n=0 \\ \tau_{m,n} \neq 0}}^{\infty} \sum_{j=1}^{2n+1} \tau_{m,n}^{-2} \left\langle g, b_{m,n,j} \right\rangle_{\mathrm{L}^2(D,\mathbb{R}^q)}^2 < +\infty, \tag{11.26}$$

we see that the series in (11.24) is uniformly convergent with respect to $J \in \mathbb{N}_0$. Hence, condition (SB3) implies

$$\lim_{J\to\infty} \|\Phi_J * g - f\|^2_{\mathrm{L}^2(\mathscr{B})} = \sum_{\substack{m,n=0\\ \tau_{m,n}\neq 0}}^{\infty} \sum_{j=1}^{2n+1} \lim_{J\to\infty} \left(\Phi_J^\wedge(m,n) - \tau_{m,n}^{-1}\right)^2 \langle g, b_{m,n,j}\rangle^2_{\mathrm{L}^2(D,\mathbb{R}^q)} = 0. \tag{11.27}$$

□

Theorem 11.8 (Stability of the Approximation). *The mapping*

$$R_J : \mathscr{Y} \ni g \mapsto \Phi_J * g \in \mathrm{L}^2(\mathscr{B}) \tag{11.28}$$

is continuous.

Proof. Obviously, R_J is linear. Therefore, we show that the operator is bounded. From Theorem 11.5 and the Parseval identity, we get

$$\begin{aligned} \|R_J g\|^2_{\mathrm{L}^2(\mathscr{B})} &= \sum_{m,n=0}^{\infty} \left(\Phi_J^\wedge(m,n)\right)^2 \sum_{j=1}^{2n+1} \langle g, b_{m,n,j}\rangle^2_{\mathrm{L}^2(D,\mathbb{R}^q)} \\ &\leq \left(\sup_{m,n\in\mathbb{N}_0} \Phi_J^\wedge(m,n)^2\right) \|g\|^2_{\mathrm{L}^2(D,\mathbb{R}^q)}, \end{aligned} \tag{11.29}$$

where the supremum is finite due to condition (SB4). □

This shows us the principle of a regularization, which is as follows: If we know that the mapping $g \mapsto T^{-1}g$ is discontinuous, we construct an alternative class of continuous mappings $g \mapsto \Phi_J * g$ such that the sequence $(\Phi_J * g)_J$ tends to $T^{-1}g$. In practice, there is usually an optimal J^*, after which the approximation deteriorates.[4]

Also, the scaling functions which we have here provide us with a multiresolution analysis.

Definition 11.9. Let $\{\Phi_J\}_{J\in\mathbb{N}_0}$ be a scaling function. Then the spaces

$$V_J := \left\{\Phi_J * g \,\middle|\, g \in T\left(\mathrm{L}^2(\mathscr{B})\right)\right\}, \quad J \in \mathbb{N}_0, \tag{11.30}$$

are called the corresponding **scale spaces**.

[4]This is, for example, caused by the following fact: Typically, g is noisy, where the relative contribution of the noise to the Fourier coefficients $\langle g, b_{m,n,j}\rangle_{\mathrm{L}^2(D,\mathbb{R}^q)}$ increases with increasing degree m or n. On the one hand, we get that $\Phi_J^\wedge(m,n)$ tends to zero as $m\to\infty$ or $n\to\infty$ from (SB4). On the other hand, (SB2) and (SB3) require that $|\Phi_J^\wedge(m,n)|$ increases with increasing scale J and approaches the (absolute value of the) reciprocal of the singular value. The former causes that Fourier coefficients corresponding to high degrees are equipped with small factors due to the convolution $\Phi_J * g$—which is good due to the noisy character of these coefficients. The latter, however, causes that this smoothing effect gets weaker and weaker with increasing scale J such that finally the noise is not sufficiently attenuated any more. An optimal scale J^* corresponds to a trade-off between a notable attenuation and a low approximation error.

Note that we only convolve here the scaling function with right-hand sides g for which a solution exists.

Theorem 11.10 (Multiresolution Analysis). *Let $\{\Phi_J\}_{J\in\mathbb{N}_0}$ be a scaling function. Then the corresponding scale spaces $\{V_J\}_{J\in\mathbb{N}_0}$ constitute a multiresolution analysis, that is,*

(i) For all $J \in \mathbb{N}_0$,

$$V_0 \subset \cdots \subset V_J \subset V_{J+1} \subset \cdots \subset \mathrm{L}^2(\mathscr{B}). \tag{11.31}$$

(ii) The following property holds:

$$\overline{\bigcup_{J\in\mathbb{N}_0} V_J}^{\|\cdot\|_{\mathrm{L}^2(\mathscr{B})}} = \left\{ f \in \mathrm{L}^2(\mathscr{B}) \,\middle|\, \langle f, G_{m,n,j}\rangle_{\mathrm{L}^2(\mathscr{B})} = 0, \text{ if } \tau_{m,n} = 0 \right\}. \tag{11.32}$$

Proof.

(1) Part (i):
From Theorem 11.4, we already know that $V_J \subset \mathrm{L}^2(\mathscr{B})$ for all $J \in \mathbb{N}_0$. Now let $\Phi_J * g$ with $g \in T(\mathrm{L}^2(\mathscr{B}))$ be an arbitrary element of V_J. We define $h \in \mathscr{Y}$ by its Fourier coefficients

$$\langle h, b_{m,n,j}\rangle_{\mathrm{L}^2(D,\mathbb{R}^q)} := \begin{cases} \frac{\Phi_J^\wedge(m,n)}{\Phi_{J+1}^\wedge(m,n)} \langle g, b_{m,n,j}\rangle_{\mathrm{L}^2(D,\mathbb{R}^q)}, & \text{if } \Phi_{J+1}^\wedge(m,n) \neq 0,\\ 0, & \text{else.} \end{cases} \tag{11.33}$$

Indeed, $h \in \mathscr{Y} \subset \mathrm{L}^2(D,\mathbb{R}^q)$ because $0 \leq \Phi_J^\wedge(m,n)^2 \leq \Phi_{J+1}^\wedge(m,n)^2$ due to condition (SB2) and, consequently,

$$\sum_{m,n=0}^{\infty} \sum_{j=1}^{2n+1} \langle h, b_{m,n,j}\rangle^2_{\mathrm{L}^2(D,\mathbb{R}^q)} \leq \sum_{\substack{m,n=0\\ \tau_{m,n}\neq 0}}^{\infty} \sum_{j=1}^{2n+1} \langle g, b_{m,n,j}\rangle^2_{\mathrm{L}^2(D,\mathbb{R}^q)} = \|g\|^2_{\mathrm{L}^2(D,\mathbb{R}^q)} < +\infty, \tag{11.34}$$

where we used condition (SB1) and (11.33).
Moreover, there also exists a solution $f \in \mathrm{L}^2(\mathscr{B})$ of $Tf = h$. Such a solution is, for example, given by

$$\langle f, G_{m,n,j}\rangle_{\mathrm{L}^2(\mathscr{B})} := \begin{cases} \tau_{m,n}^{-1} \langle h, b_{m,n,j}\rangle_{\mathrm{L}^2(D,\mathbb{R}^q)}, & \text{if } \tau_{m,n} \neq 0,\\ 0, & \text{else.} \end{cases} \tag{11.35}$$

The verification of $Tf = h$ is easy [see (11.1)]. Furthermore, $f \in \mathrm{L}^2(\mathscr{B})$, because

$$\begin{aligned} \sum_{m,n=0}^{\infty} \sum_{j=1}^{2n+1} \langle f, G_{m,n,j}\rangle^2_{\mathrm{L}^2(\mathscr{B})} &= \sum_{\substack{m,n=0\\ \tau_{m,n}\neq 0}}^{\infty} \sum_{j=1}^{2n+1} \tau_{m,n}^{-2} \langle h, b_{m,n,j}\rangle^2_{\mathrm{L}^2(D,\mathbb{R}^q)} \\ &\leq \sum_{\substack{m,n=0\\ \tau_{m,n}\neq 0}}^{\infty} \sum_{j=1}^{2n+1} \tau_{m,n}^{-2} \langle g, b_{m,n,j}\rangle^2_{\mathrm{L}^2(D,\mathbb{R}^q)}, \end{aligned} \tag{11.36}$$

and $g \in T(\mathrm{L}^2(\mathscr{B}))$ [see also (11.26)]. Hence, $\Phi_{J+1} * h \in V_{J+1}$. The rest is analogous to the spherical case (see Theorem 7.22 on p. 207): Theorem 11.5 yields

$$\begin{aligned}
\Phi_{J+1} * h &= \sum_{\substack{m,n=0 \\ \Phi_{J+1}^{\wedge}(m,n)\neq 0}}^{\infty} \sum_{j=1}^{2n+1} \Phi_{J+1}^{\wedge}(m,n) \left\langle h, b_{m,n,j} \right\rangle_{\mathrm{L}^2(D,\mathbb{R}^q)} G_{m,n,j} \\
&= \sum_{\substack{m,n=0 \\ \Phi_{J+1}^{\wedge}(m,n)\neq 0}}^{\infty} \sum_{j=1}^{2n+1} \Phi_{J}^{\wedge}(m,n) \left\langle g, b_{m,n,j} \right\rangle_{\mathrm{L}^2(D,\mathbb{R}^q)} G_{m,n,j} \\
&= \Phi_J * g
\end{aligned} \tag{11.37}$$

in the sense of $\mathrm{L}^2(\mathscr{B})$, where $\Phi_J^{\wedge}(m,n) = 0$ for all $(m,n) \in \mathbb{N}_0^2$ with $\Phi_{J+1}^{\wedge}(m,n) = 0$ due to condition (SB2). Thus, $V_J \subset V_{J+1}$.

(2) Part (ii):
Let $f \in \mathrm{L}^2(\mathscr{B})$ with $\langle f, G_{m,n,j}\rangle_{\mathrm{L}^2(\mathscr{B})} = 0$ for all $(m,n) \in \mathbb{N}_0^2$ with $\tau_{m,n} = 0$ and all $j = 1,\ldots,2n+1$ be arbitrary. Then $g := Tf \in T(\mathrm{L}^2(\mathscr{B}))$ and $(\Phi_J * g)_{J\in\mathbb{N}_0}$ is a sequence in $\bigcup_{J\in\mathbb{N}_0} V_J$, which converges to f in the sense of $\mathrm{L}^2(\mathscr{B})$ according to Theorem 11.7. □

11.4 Wavelets

In correspondence to the choice of a linear theory (i.e., we use Φ_J and not $\Phi_J^{(2)}$, which would anyway not make any sense here), we do not distinguish primal and dual wavelets anymore. The wavelet theory is here very simple.

Definition 11.11. Let $\{\Phi_J\}_{J\in\mathbb{N}_0}$ be a given scaling function. Then the corresponding wavelet $\{\Psi_J\}_{J\in\mathbb{N}_0\cup\{-1\}}$ is defined by

$$\Psi_J := \Phi_{J+1} - \Phi_J \quad \forall J \in \mathbb{N}_0, \qquad \Psi_{-1} := \Phi_0. \tag{11.38}$$

We immediately get the following result.

Theorem 11.12 (Scale-Step Property). *Let $\{\Phi_J\}_{J\in\mathbb{N}_0}$ be a scaling function and $\{\Psi_J\}_{J\in\mathbb{N}_0}$ be the corresponding wavelet. If $f \in \mathrm{L}^2(\mathscr{B})$ and $g \in \mathrm{L}^2(D, \mathbb{R}^q)$ such that $Tf = g$ and $\langle f, G_{m,n,j}\rangle_{\mathrm{L}^2(\mathscr{B})} = 0$ for all (m,n,j) with $\tau_{m,n} = 0$, then*

$$\Phi_{J_2} * g = \Phi_{J_1} * g + \sum_{j=J_1}^{J_2-1} \Psi_j * g\,, \tag{11.39}$$

$$f = \Phi_{J_1} * g + \sum_{j=J_1}^{\infty} \Psi_j * g\,, \tag{11.40}$$

$$f = \sum_{j=-1}^{\infty} \Psi_j * g \tag{11.41}$$

for all $J_1, J_2 \in \mathbb{N}_0$ with $J_1 < J_2$, where the latter two identities hold in the sense of $\mathrm{L}^2(\mathscr{B})$. Moreover, the ***detail spaces***

$$W_J := \left\{ \Psi_J * g \,\middle|\, g \in T\left(\mathrm{L}^2(\mathscr{B})\right) \right\} \tag{11.42}$$

satisfy

$$V_{J_2} = V_{J_1} + \sum_{j=J_1}^{J_2-1} W_j \tag{11.43}$$

for all $J_1, J_2 \in \mathbb{N}_0$ with $J_1 < J_2$.

Proof. Only one part is nontrivial: the inclusion

$$V_J + W_J \subset V_{J+1} \quad \forall J \in \mathbb{N}_0, \tag{11.44}$$

which yields (11.43) by induction, since the opposite inclusion is trivial. Let $g_1, g_2 \in T(\mathrm{L}^2(\mathscr{B}))$, that is, there exist $f_1, f_2 \in \mathrm{L}^2(\mathscr{B})$ such that $Tf_1 = g_1$ and $Tf_2 = g_2$. Hence, $\Phi_J * g_1$ and $\Psi_J * g_2$ are arbitrary elements of V_J and W_J, respectively.
Now, let $h, h_1, h_2 \in \mathscr{Y}$ be defined by $h := h_1 + h_2$ and

$$\left\langle h_1, b_{m,n,j} \right\rangle_{\mathrm{L}^2(D,\mathbb{R}^q)} := \begin{cases} \frac{\Phi_J^\wedge(m,n)}{\Phi_{J+1}^\wedge(m,n)} \left\langle g_1, b_{m,n,j} \right\rangle_{\mathrm{L}^2(D,\mathbb{R}^q)}, & \text{if } \Phi_{J+1}^\wedge(m,n) \neq 0 \\ 0, & \text{else,} \end{cases}$$

$$\left\langle h_2, b_{m,n,j} \right\rangle_{\mathrm{L}^2(D,\mathbb{R}^q)} := \begin{cases} \frac{\Phi_{J+1}^\wedge(m,n) - \Phi_J^\wedge(m,n)}{\Phi_{J+1}^\wedge(m,n)} \left\langle g_2, b_{m,n,j} \right\rangle_{\mathrm{L}^2(D,\mathbb{R}^q)}, & \text{if } \Phi_{J+1}^\wedge(m,n) \neq 0 \\ 0 & \text{else} \end{cases}$$

for all $m, n \in \mathbb{N}_0$ and all $j \in \{1, \ldots, 2n+1\}$.
Note that $\langle h_k, b_{m,n,j} \rangle_{\mathrm{L}^2(D,\mathbb{R}^q)} = 0 \; \forall k = 1,2$, if $\tau_{m,n} = 0$, due to condition (SB1) in Definition 11.6. Moreover, condition (SB2) in the same definition also yields that, indeed, $h_k \in \mathscr{Y} \subset \mathrm{L}^2(D, \mathbb{R}^q) \; \forall k = 1,2$ (and, consequently, $h \in \mathscr{Y} \subset \mathrm{L}^2(D, \mathbb{R}^q)$), since

$$\left| \left\langle h_k, b_{m,n,j} \right\rangle_{\mathrm{L}^2(D,\mathbb{R}^q)} \right| \leq 2 \left| \left\langle g_k, b_{m,n,j} \right\rangle_{\mathrm{L}^2(D,\mathbb{R}^q)} \right| \tag{11.45}$$

for all $k \in \{1,2\}$, $m, n \in \mathbb{N}_0$, and all $j \in \{1, \ldots, 2n+1\}$.
Furthermore, if we set

$$\left\langle F_k, G_{m,n,j} \right\rangle_{\mathrm{L}^2(\mathscr{B})} := \begin{cases} \tau_{m,n}^{-1} \left\langle h_k, b_{m,n,j} \right\rangle_{\mathrm{L}^2(D,\mathbb{R}^q)}, & \text{if } \tau_{m,n} \neq 0, \\ 0 & \text{else} \end{cases} \tag{11.46}$$

for all $k \in \{1,2\}$, $m, n \in \mathbb{N}_0$, and all $j \in \{1, \ldots, 2n+1\}$, then (11.45) and the fact that $Tf_k = g_k$ for all $k \in \{1,2\}$ imply the inequality

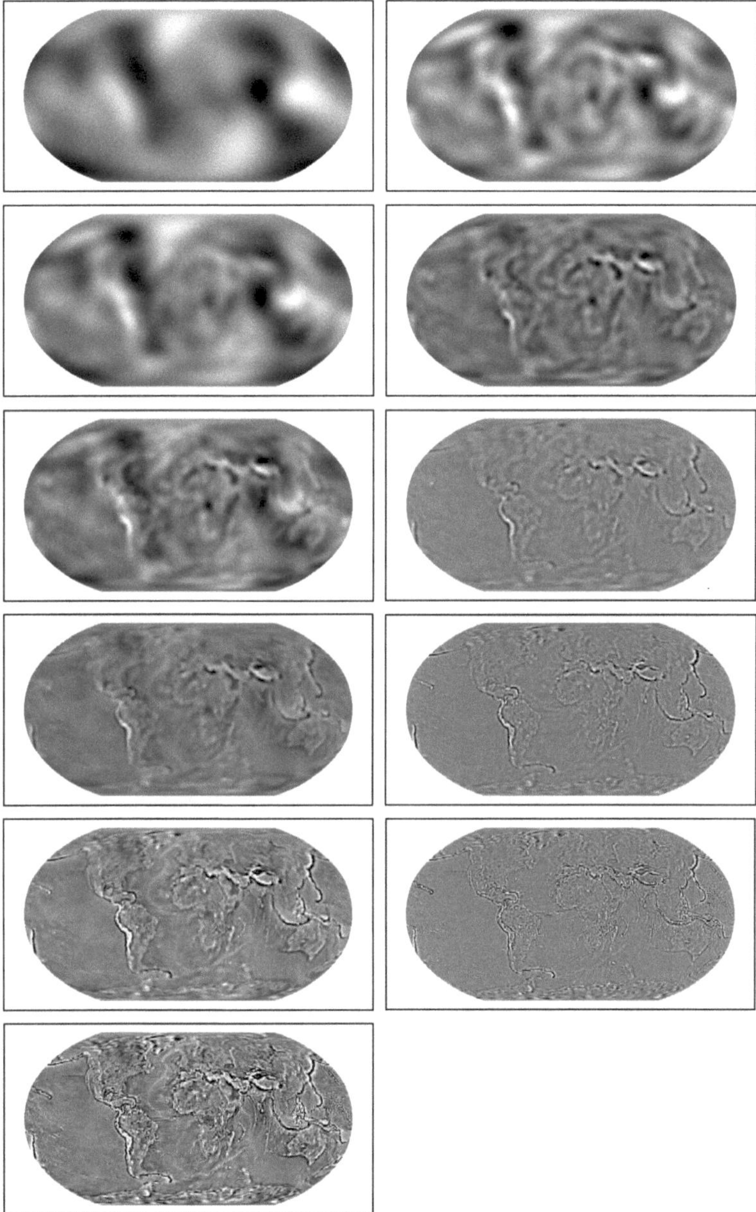

Fig. 11.1 $\Phi_J * \frac{\partial^2 V}{\partial r^2}$ (*left-hand column*) and $\Psi_J * \frac{\partial^2 V}{\partial r^2}$ (*right-hand column*) for $J = 3, \ldots, 8$ (*top* to *bottom*), where V is the gravitational potential from EGM96 (see [110]): The *left-hand column* shows approximations of the harmonic density variations near the surface of the Earth at different scales, and the *right-hand column* shows the details corresponding to the scale steps (from [130])

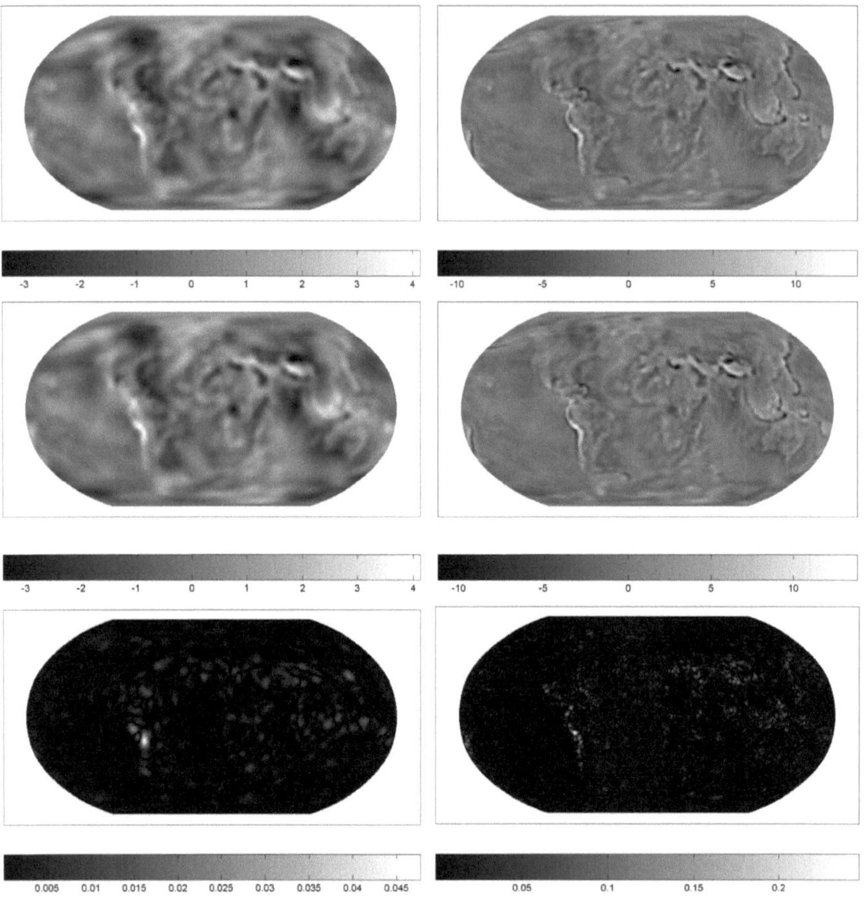

Fig. 11.2 $\Phi_J * \frac{\partial^2 V}{\partial r^2}$ (*top*, like in the left-hand column of Fig. 11.1 but with different color scales), $\Phi_J * \frac{\partial^2 \tilde{V}}{\partial r^2}$ (*middle*), where the perturbed function $\frac{\partial^2 \tilde{V}}{\partial r^2}$ was obtained by adding 10 % artificial noise to $\frac{\partial^2 V}{\partial r^2}$, and $|\Phi_J * \frac{\partial^2 V}{\partial r^2} - \Phi_J * \frac{\partial^2 \tilde{V}}{\partial r^2}|$ (*bottom*) for $J = 5$ (*left hand*) and $J = 6$ (*right hand*). The physical unit of the *color bar* is kg m^{-3} (from [130])

$$\begin{aligned}
\|F_k\|^2_{\mathrm{L}^2(\mathscr{B})} &= \sum_{\substack{m,n=0\\ \tau_{m,n}\neq 0}}^{\infty} \sum_{j=1}^{2n+1} \tau_{m,n}^{-2} \left\langle h_k, b_{m,n,j}\right\rangle^2_{\mathrm{L}^2(D,\mathbb{R}^q)} \\
&\leq 4 \sum_{\substack{m,n=0\\ \tau_{m,n}\neq 0}}^{\infty} \sum_{j=1}^{2n+1} \tau_{m,n}^{-2} \left\langle g_k, b_{m,n,j}\right\rangle^2_{\mathrm{L}^2(D,\mathbb{R}^q)} \\
&\leq 4 \|f_k\|^2_{\mathrm{L}^2(\mathscr{B})} < +\infty
\end{aligned} \tag{11.47}$$

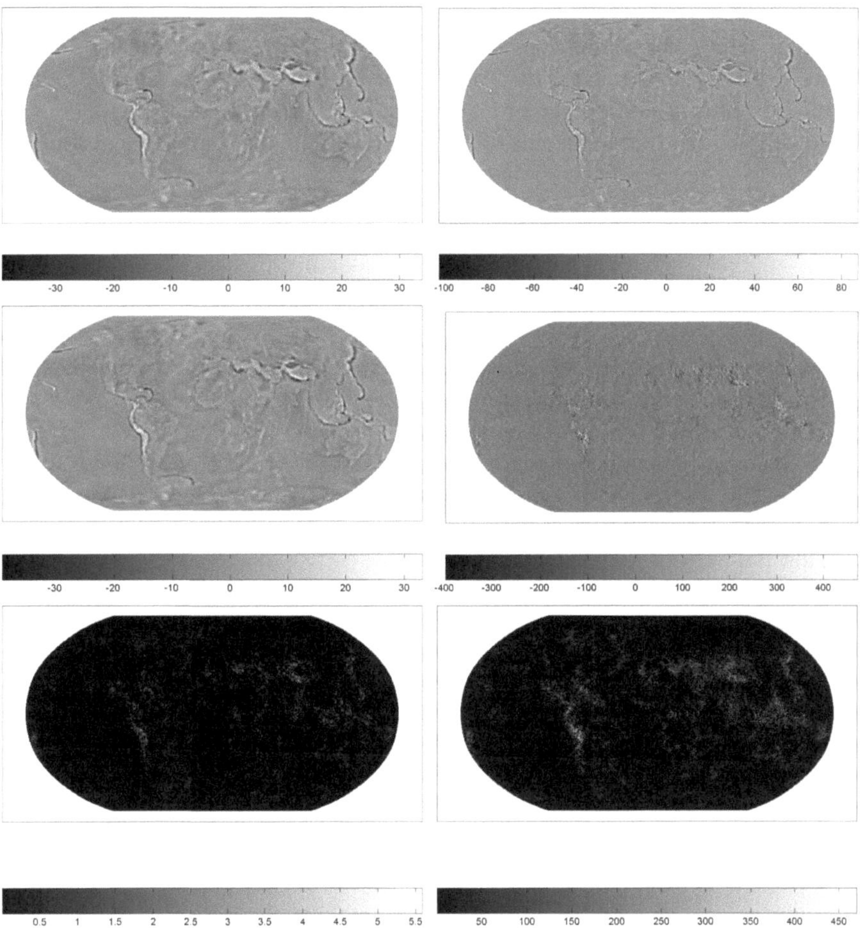

Fig. 11.3 $\Phi_J * \frac{\partial^2 V}{\partial r^2}$ (*top*, like in the left-hand column of Fig. 11.1 but with different color scales), $\Phi_J * \frac{\partial^2 \tilde{V}}{\partial r^2}$ (*middle*), where the perturbed function $\frac{\partial^2 \tilde{V}}{\partial r^2}$ was obtained by adding 10 % artificial noise to $\frac{\partial^2 V}{\partial r^2}$, and $|\Phi_J * \frac{\partial^2 V}{\partial r^2} - \Phi_J * \frac{\partial^2 \tilde{V}}{\partial r^2}|$ (*bottom*) for $J = 7$ (*left hand*) and $J = 8$ (*right hand*). Note that the low-pass filter at scale 8 is too weak such that the noise dominates the result. The occurrence of such an effect at large scales is typical for ill-posed inverse problems. The physical unit of the color bar is kg m^{-3} (from [130])

for all $k \in \{1,2\}$. Hence, $h_1, h_2 \in T(\mathrm{L}^2(\mathscr{B}))$ and, consequently, $h \in T(\mathrm{L}^2(\mathscr{B}))$. Finally, Theorem 11.5 yields

$$V_{J+1} \ni \Phi_{J+1} * h = \Phi_J * g_1 + \underbrace{(\Phi_{J+1} - \Phi_J)}_{=\Psi_J} * g_2. \tag{11.48}$$

□

Figures 11.1–11.3 show the result of a regularization of the inverse gravimetric problem, where the second radial derivative of the gravitational potential of the Earth is given 200 km above the Earth's surface. The results are approximations to the harmonic solution of the inverse problem, which are plotted at a sphere with radius 0.999β. For further details of the numerical implementation, see [130] (note that $\tilde{\Psi}_J * \Psi_J$ in the paper corresponds to Ψ_J in this book).

11.5 Questions for Understanding

- What kind of inverse problems can be solved by the considered wavelet method?
- How did we define convolutions on the ball?
- Why is the result of a convolution still an L^2-function?
- All kinds of convolutions have in common that a particular property holds for the Fourier transform/coefficients of a convolution. Which property is this (on the ball and on the sphere)?
- How did we define scaling functions?
- Give examples for scaling functions in the case of the inverse gravimetric problem!
- What do you know about the following keywords: multiresolution analysis, wavelets, scale-step property?
- A linear wavelet theory was preferred for the inverse problems. Why? In the case of the bilinear spherical wavelets, there is a refinement equation. What is the analogue in a linear wavelet theory?
- The wavelet method is a regularization. What does this mean and why is this the case here?

Chapter 12
The Regularized Functional Matching Pursuit

12.1 The Idea

The spline and wavelet methods presented so far have several advantages, which are connected to the localization of the used (reproducing) kernel functions. However, some drawbacks remain. These are as follows:

The choice of the centers of the "hat functions" is not flexible enough. In the case of the wavelet method, a quadrature grid is needed. In the case of splines, the grid of centers is immediately connected to the grid of data, and we cannot use a sparser grid of centers due to the algorithm. If we have, however, a nowadays realistic number of data between 10^4 and 10^6, then a numerical inversion of the spline matrix becomes impossible. Moreover, we are limited to a preliminary choice of basis functions, that is, we have to decide in the very beginning which scaling function we use or which reproducing kernel we use (and the latter means, that all "hats" have the same "width"). Furthermore, the $G_{m,n,j}$-functions, which are ideal to cover global trends, cannot be incorporated in the expansion of the solution.

In the Euclidean setting, novel approaches which are called, for example, greedy algorithms, matching pursuits, sparse regularization, or compressed sensing, have become popular for several years (see, e.g., [14, 16, 25, 29, 37, 41, 42, 45, 115, 152, 181–183]). As usual, methods which perform well in the Euclidean setting do not automatically work on the sphere or the ball. Hence, the knowledge of constructive approximation on the 3D ball is needed to construct a novel algorithm that provides us with properties that are analogous to those known for the Euclidean methods. In [18, 54, 57], such a method is presented. This novel approach, which is called the Regularized Functional Matching Pursuit (RFMP), yields promising first results and needs to be investigated and enhanced in the future research. It appears not to be reasonable to give a one-by-one copy of the paper [57] here, which gives a concise description of the current state of the art. Instead, we will briefly explain the main ideas of the algorithm.

V. Michel, *Lectures on Constructive Approximation*, Applied and Numerical Harmonic Analysis, DOI 10.1007/978-0-8176-8403-7_12,

1. The algorithm is iterative: we start with $F_0 := 0$ and consecutively determine the summands of the expansion, that is, given $F_n = \sum_{k=1}^{n} \alpha_k d_k$, find $\alpha_{n+1} d_{n+1}$.
2. Whereas α_k is a constant real number, d_k is an element of a preliminarily chosen subset $\mathscr{D} \subset \mathrm{L}^2(\mathscr{B})$, which is called a dictionary. This set is typically overcomplete. It consists, for example, of orthonormal functions $G_{m,n,j}$ and "hat" functions $K_{\mathscr{H}_1}(x_k,\cdot),\ldots,K_{\mathscr{H}_r}(x_k,\cdot)$ originating from different Sobolev spaces $\mathscr{H}_1,\ldots,\mathscr{H}_r$ (i.e., we use "hats" with different "widths"). This enables us to represent different features of the solution by trial functions of different kinds.
3. One novel feature is that we do not decide from the very beginning which expansion functions d_k are used. The algorithm will only select some of the elements of the dictionary for the expansion. More precisely, the new summand $\alpha_{n+1} d_{n+1}$ is chosen such that the functional

$$\left\|y - \mathscr{F}\left(F_n + \alpha_{n+1} d_{n+1}\right)\right\|_{\mathbb{R}^l}^2 + \lambda \left\|F_n + \alpha_{n+1} d_{n+1}\right\|_{\mathrm{L}^2(\mathscr{B})}^2 \tag{12.1}$$

is minimized, where $y \in \mathbb{R}^l$ is the given data vector, $\mathscr{F} : \mathrm{L}^2(\mathscr{B}) \to \mathbb{R}^l$ is an operator representing the (discretized) inverse problem[1], and λ is the regularization parameter. One currently open question in the research is whether a different norm in the penalty term $\lambda\|F_n + \alpha_{n+1} d_{n+1}\|_{\mathrm{L}^2(\mathscr{B})}^2$ can yield improvements. Numerical experiments, however, indicate that norms which are commonly used in the Euclidean setting do not yield satisfactory results here.
4. A joint inversion of different kinds of data is possible. This is, in particular, important for the determination of a model of the Earth's interior due to the null spaces associated to each data type.

Figure 12.1 shows the result of an inversion of 25,440 gravitational data[2] by the RFMP with $\lambda = 4.6416$. The dictionary consists of the functions $G_{0,n,j}^{\mathrm{I}}$ for $n = 3,\ldots,8$; $j = 1,\ldots,2n+1$ and (normalized versions of) $K_{\mathscr{H}_i}(x_k,\cdot)$ for three different Sobolev spaces $\mathscr{H}_i$ and an equiangular grid $\{x_k\}_k$ of 39,800 points, that is, there are approximately 120,000 functions available in the dictionary to choose from. The left-hand image of the figure shows $F_{20{,}000}$ (i.e., the algorithm was truncated after 20,000 functions were selected from the dictionary). We see typical mass density anomalies in South America (like the Andes) and its vicinity (like the Caribbean and parts of the Mid-Atlantic Ridge). The right-hand image shows the locations x_k of the centers of those "hat" functions $K_{\mathscr{H}_i}(x_k,\cdot)$, which were selected from $\mathscr{D}$ to build $F_{20{,}000}$. Clearly, the algorithm uses a lot of basis functions in regions where the solution requires a high resolution, whereas the grid of chosen centers is sparse in areas where the solution has a low detail structure. This is one of the main features of the novel method.

[1] For instance, $\mathscr{F}F$ represents the gravitational potential associated to the mass density function F and measured at l points in space.

[2] This number is far beyond the limit for a stable inversion of the corresponding spline matrix.

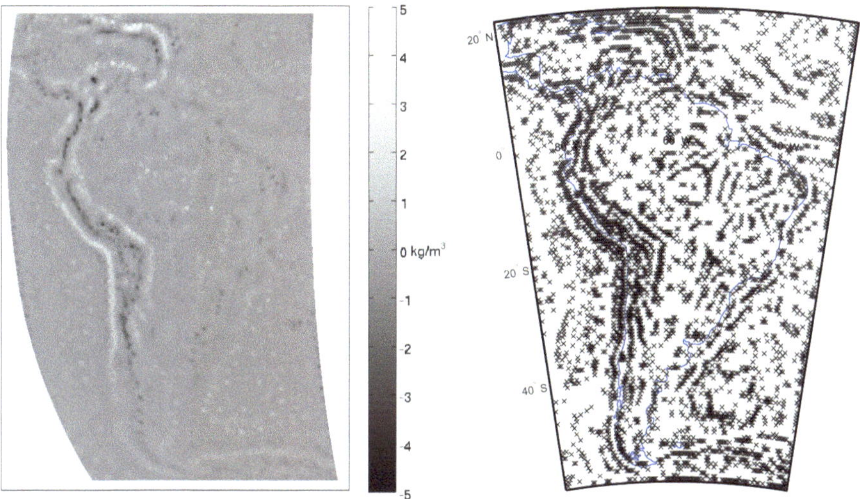

Fig. 12.1 Result of the RFMP for the inversion of gravitational data (*left hand*) and grid of "hat" functions chosen by the algorithm (*right hand*): the higher the local detail structure of the solution, the more basis functions the algorithm installs in the corresponding area (from [57])

For further numerical results, including the identification of climate-based (water) mass transports, see [54–57].

12.2 Questions for Understanding

- What are the advantages and disadvantages of Fourier, spline, and wavelet methods?
- What is the idea of the RFMP? What kinds of problems can the RFMP tackle that the previous methods cannot tackle?
- Which practical examples of inverse problems on the ball can you imagine? What are possible difficulties connected to them?

References

1. Abramowitz, M., Stegun, I.A.: Handbook of Mathematical Functions with Formulas, Graphs, and Mathematical Tables. Dover, New York (1972)
2. Akhtar, N.: A multiscale harmonic spline interpolation method for the inverse spheroidal gravimetric problem. Ph.D. thesis, University of Siegen, Department of Mathematics, Geomathematics Group. Shaker, Aachen (2009)
3. Akhtar, N., Michel, V.: Reproducing Kernel based splines for the regularization of the inverse ellipsoidal gravimetric problem. Appl. Anal. (2011). Accepted for publication, pre-published online via doi:10.1080/00036811.2011.590479
4. Akram, M.: Constructive approximation on the 3-dimensional ball with focus on locally supported kernels and the Helmholtz decomposition. Ph.D. thesis, University of Kaiserslautern, Department of Mathematics, Geomathematics Group. Shaker, Aachen (2009)
5. Akram, M., Amina, I., Michel, V.: A study of differential operators for particular complete orthonormal systems on a 3D ball. Int. J. Pure Appl. Math. **73**, 489–506 (2011)
6. Alfeld, P., Neamtu, M., Schumaker, L.L.: Fitting scattered data on sphere-like surfaces using spherical splines. J. Comput. Appl. Math. **73**, 5–43 (1996)
7. Amann, H., Escher, J.: Analysis III, 2nd edn. Birkhäuser, Basel (2008)
8. Amirbekyan, A.: The application of reproducing kernel based spline approximation to seismic surface and body wave tomography: theoretical aspects and numerical results. Ph.D. thesis, University of Kaiserslautern, Department of Mathematics, Geomathematics Group (2007). www.kluedo.ub.uni-kl.de/volltexte/2007/2103/pdf/ThesisAbel.pdf
9. Amirbekyan, A., Michel, V.: Splines on the three-dimensional ball and their application to seismic body wave tomography. Inverse Probl. **24**, 1–25 (2008)
10. Antoine, J.P., Demanet, L., Jacques, L., Vandergheynst, P.: Wavelets on the sphere: implementations and approximations. Appl. Comput. Harm. Anal. **13**, 177–200 (2002)
11. Antoine, J.P., Vandergheynst, P.: Wavelets on the 2-sphere: A group-theoretic approach. Appl. Comput. Harm. Anal. **7**, 1–30 (1999)
12. Ballani, L., Engels, J., Grafarend, E.W.: Global base functions for the mass density in the interior of a massive body (Earth). Manuscr. Geodaet. **18**, 99–114 (1993)
13. Bäni, W.: Wavelets: Eine Einführung für Ingenieure. Oldenburg, München (2002)
14. Barron, A.R., Cohen, A., Dahmen, W., DeVore, R.A.: Approximation and learning by greedy algorithms. Ann. Stat. **36**, 64–94 (2008)
15. Beckmann, J., Mhaskar, H.N., Prestin, J.: Quadrature formulas for integration of multivariate trigonometric polynomials on spherical triangles. Int. J. Geomath. **3**, 119–138 (2012)
16. Berg, A.P., Mikhael, W.B.: A survey of mixed transform techniques for speech and image coding. In: Proceedings of the 1999 IEEE International Symposium on Circuits and Systems, vol. 4, pp. 106–109 (1999)

V. Michel, *Lectures on Constructive Approximation*, Applied and Numerical Harmonic Analysis, DOI 10.1007/978-0-8176-8403-7,

17. Berkel, P.: Multiscale methods for the combined inversion of normal mode and gravity variations. Ph.D. thesis, University of Kaiserslautern, Department of Mathematics, Geomathematics Group. Shaker, Aachen (2009)
18. Berkel, P., Fischer, D., Michel, V.: Spline multiresolution and numerical results for joint gravitation and normal mode inversion with an outlook on sparse regularisation. Int. J. Geomath. **1**, 167–204 (2011)
19. Berkel, P., Michel, V.: On mathematical aspects of a combined inversion of gravity and normal mode variations by a spline method. Math. Geosci. **42**, 795–816 (2010)
20. Blatter, C.: Wavelets: Eine Einführung. Vieweg, Braunschweig (1998)
21. Blick, C., Freeden, W.: Spherical spline application to radio occultation data. J. Geodetic Sci **1**, 379–395 (2011)
22. Bogdanova, I., Vandergheynst, P., Antoine, J.P., Jacques, L., Morvidone, M.: Stereographic wavelet frames on the sphere. Appl. Comput. Harm. Anal. **19**, 223–252 (2005)
23. Böhme, M., Potts, D.: A fast algorithm for filtering and wavelet decomposition on the sphere. Electron. Trans. Numer. Anal. **16**, 70–93 (2003)
24. Chambodut, A., Panet, I., Mandea, M., Diament, M., Holschneider, M., Jamet, O.: Wavelet frames: An alternative to spherical harmonic representation of potential fields. Geophys. J. Int. **163**, 875–899 (2005)
25. Chen, S.S., Donoho, D.L., Saunders, M.A.: Atomic decomposition by basis pursuit. SIAM Rev. **43**, 129–159 (2001)
26. Chihara, T.S.: An Introduction to Orthogonal Polynomials. Gordon and Breach, New York (1978)
27. Chui, C.K.: An Introduction to Wavelets. Academic, San Diego (1992)
28. Cohen, A.: Numerical Analysis of Wavelet Methods. Elsevier, Amsterdam (2003)
29. Coifman, R., Meyer, Y., Wickerhauser, V.: Adapted wave form analysis; waveletpackets and applications. In: ICIAM 91, Proceedings of the Second International Conference on Industrial and Applied Mathematics, pp. 41–50 (1992)
30. Coifman, R., Wickerhauser, V.: Entropy-based algorithms for best basis selection. IEEE Trans. Inform. Theory **38**, 713–718 (1992)
31. Conrad, M., Prestin, J.: Multiresolution on the sphere. In: Iske, A., Quak, E., Floater, M.S. (eds.) Summer School Lecture Notes on Principles of Multiresolution in Geometric Modelling, pp. 165–202, Munich (2001)
32. Cooley, J.W., Tukey, J.W.: An algorithm for the machine calculation of complex Fourier series. Math. Comp. **19**, 297–301 (1965)
33. Cui, J., Freeden, W.: Equidistribution on the sphere. SIAM J. Sci. Comput. **18**, 595–609 (1997)
34. Dahlen, F.A., Simons, F.J.: Spectral estimation on a sphere in geophysics and cosmology. Geophys. J. Int. **174**, 774–807 (2008)
35. Dahlen, F.A., Tromp, J.: Theoretical Global Seismology. Princeton University Press, Princeton (1998)
36. Dahlke, S., Dahmen, W., Schmitt, E., Weinreich, I.: Multiresolution analysis and wavelets on S^2 and S^3. Numer. Func. Anal. Opt. **16**, 19–41 (1995)
37. Dahlke, S., Fornasier, M., Raasch, T.: Multilevel preconditioning and adaptive sparse solution of inverse problems. Math. Comput. **81**, 419–446 (2009)
38. Dahlke, S., Steidl, G., Teschke, G.: Coorbit spaces and Banach frames on homogeneous spaces with applications to the sphere. Adv. Comput. Math. **21**, 147–180 (2004)
39. Dahlke, S., Steidl, G., Teschke, G.: Frames and coorbit theory on homogeneous spaces with a special guidance on the sphere. J. Fourier Anal. Appl. **13**, 387–403 (2007)
40. Daubechies, I.: Ten Lectures on Wavelets. SIAM, Philadelphia (1992)
41. Daubechies, I., Defrise, M., DeMol, C.: An iterative thresholding algorithm for linear inverse problems with sparsity constraint. Commun. Pur. Appl. Math. **57**, 1413–1457 (2004)
42. Daubechies, I., Fornasier, M., Loris, I.: Accelerated projected gradient method for linear inverse problems with sparsity constraints. J. Fourier Anal. Appl. **14**, 764–792 (2008)
43. Davis, P.J.: Interpolation and Approximation. Dover, New York (1975)

44. Deuflhard, P.: On algorithms for the summation of certain special functions. Computing **17**, 37–48 (1975)
45. DeVore, R.A.: Nonlinear approximation. Acta Numerica **7**, 51–150 (1998)
46. Driscoll, J.R., Healy, R.M.: Computing Fourier transforms and convolutions on the 2-sphere. Adv. Appl. Math. **15**, 202–250 (1994)
47. Dufour, H.M.: Fonctions orthogonales dans la sphère. résolution théorique du problème du potentiel terrestre. B. Geod. **51**, 227–237 (1977)
48. Dunkl, C.F., Xu, Y.: Orthogonal polynomials of several variables. In: Encyclopedia of Mathematics and its Applications. Cambridge University Press, Cambridge (2001)
49. Engl, H.W., Grever, W.: Using the L-curve for determining optimal regularization parameters. Numer. Math. **69**, 25–31 (1994)
50. Fasshauer, G.E., Schumaker, L.L.: Scattered data fitting on the sphere. In: Dæhlen, M., Lyche, T., Schumaker, L.L. (eds.) Mathematical Methods for Curves and Surfaces II, pp. 117–166. Vanderbilt University Press, Nashville, TN (1998)
51. Feinerman, R.P., Newman, D.J.: Polynomial Approximation. The Williams and Wilkins Company, Baltimore (1974)
52. Fengler, M.J., Freeden, W., Kohlhaas, A., Michel, V., Peters, T.: Wavelet modelling of regional and temporal variations of the Earth's gravitational potential observed by GRACE. J. Geodesy **81**, 5–15 (2007)
53. Fengler, M.J., Michel, D., Michel, V.: Harmonic spline-wavelets on the 3-dimensional ball and their application to the reconstruction of the Earth's density distribution from gravitational data at arbitrarily shaped satellite orbits. Z. Angew. Math. Mech. **86**, 856–873 (2006)
54. Fischer, D.: Sparse regularization of a joint inversion of gravitational data and normal mode anomalies. Ph.D. thesis, University of Siegen, Department of Mathematics, Geomathematics Group (2011). Verlag Dr. Hut, München
55. Fischer, D., Michel, V.: How to combine spherical harmonics and localized bases for regional gravity modelling and inversion. In: Siegen Preprints on Geomathematics, vol. 8. University of Siegen, Germany (2012, Preprint)
56. Fischer, D., Michel, V.: Inverting GRACE gravity data for local climate effects. In: Siegen Preprints on Geomathematics, vol. 9. University of Siegen, Germany (2012, Preprint)
57. Fischer, D., Michel, V.: Sparse regularization of inverse gravimetry — case study: spatial and temporal mass variations in South America. Inverse Probl. **28** (2012). 065012
58. Fletcher, N.H., Rossing, T.D.: The Physics of Musical Instruments, 2nd edn. Springer, New York (1998)
59. Fokas, A.S., Hauk, O., Michel, V.: Electro-magneto-encephalography for the three-shell model: numerical implementation via splines for distributed current in spherical geometry. Inverse Probl. **28** (2012). 035009 (28 pp)
60. Fornasier, M., Pitolli, F.: Adaptive iterative thresholding algorithms for magnetoencephalography (MEG). J. Comput. Appl. Math. **221**, 386–395 (2008)
61. Freeden, W.: On approximation by harmonic splines. Manuscr. Geodaet. **6**, 193–244 (1981)
62. Freeden, W.: On spherical spline interpolation and approximation. Math. Methods Appl. Sci. **3**, 551–575 (1981)
63. Freeden, W.: Multiscale Modelling of Spaceborne Geodata. B G Teubner. Stuttgart, Leipzig (1999)
64. Freeden, W., Gerhards, C.: Poloidal and toroidal field modeling in terms of locally supported vector wavelets. Math. Geosci. **42**, 817–838 (2010)
65. Freeden, W., Gervens, T., Schreiner, M.: Tensor spherical harmonics and tensor spherical splines. Manuscr. Geodaet. **19**, 70–100 (1994)
66. Freeden, W., Gervens, T., Schreiner, M.: Constructive Approximation on the Sphere with Applications to Geomathematics. Oxford University Press, Oxford (1998)
67. Freeden, W., Mayer, C.: Wavelets generated by layer potentials. Appl. Comput. Harm. Anal. **14**, 195–237 (2003)

68. Freeden, W., Michel, V.: Constructive approximation and numerical methods in geodetic research today—an attempt at a categorization based on an uncertainty principle. J. Geodesy **73**, 452–465 (1999)
69. Freeden, W., Michel, V.: Multiscale Potential Theory (with Applications to Geoscience). Birkhäuser, Boston (2004)
70. Freeden, W., Michel, V.: Orthogonal zonal, tesseral and sectorial wavelets on the sphere for the analysis of satellite data. Adv. Comput. Math. **21**, 181–217 (2004)
71. Freeden, W., Michel, V., Nutz, H.: Satellite-to-satellite tracking and satellite gravity gradiometry (advanced techniques for high-resolution geopotential field determination). J. Eng. Math. **43**, 19–56 (2002)
72. Freeden, W., Nutz, H.: Satellite gravity gradiometry as tensorial inverse problem. Int. J. Geomath. **2**, 177–218 (2011)
73. Freeden, W., Schneider, F.: Regularization wavelets and multiresolution. Inverse Probl. **14**, 225–243 (1998)
74. Freeden, W., Schreiner, M.: Orthogonal and non-orthogonal multiresolution analysis, scale discrete and exact fully discrete wavelet transform on the sphere. Constr. Appr. **14**, 493–515 (1998)
75. Freeden, W., Schreiner, M.: Spherical Functions of Mathematical Geosciences, a Scalar, Vectorial, and Tensorial Setup. Springer, Berlin (2009)
76. Freeden, W., Windheuser, U.: Earth's gravitational potential and its MRA approximation by harmonic singular integrals. Z. Angew. Math. Mech. **75**, 633–634 (1995)
77. Freeden, W., Windheuser, U.: Spherical wavelet transform and its discretization. Adv. Comput. Math. **5**, 51–94 (1996)
78. Freeden, W., Windheuser, U.: Combined spherical harmonic and wavelet expansion—a future concept in Earth's gravitational determination. Appl. Comput. Harm. Anal. **4**, 1–37 (1997)
79. Gerhards, C.: Spherical decompositions in a global and local framework: theory and application to geomagnetic modeling. Int. J. Geomath. **1**, 205–256 (2011)
80. Gerhards, C.: Spherical multiscale methods in terms of locally supported wavelets: theory and application to geomagnetic modeling. Ph.D. thesis, University of Kaiserslautern, Department of Mathematics, Geomathematics Group (2011). Verlag Dr. Hut, München
81. Gledhill, J.A.: Aeronomic effects of the South Atlantic anomaly. Rev. Geophys. **14**, 173–187 (1976)
82. Göttelmann, J.: Locally supported wavelets on the sphere. Z. Angew. Math. Mech. **78**, 919–920 (1998)
83. Goupillaud, P., Grossmann, A., Morlet, J.: Cycle-octave and related transforms in seismic signal analysis. Geoexploration **23**, 85–102 (1984/85)
84. Greville, T.N.E.: Introduction to spline functions. In: Greville, T.N.E. (ed.) Theory and Applications of Spline Functions, pp. 1–35. Academic, New York (1969)
85. Gronwall, T.: On the degree of convergence of Laplace series. Trans. Am. Math. Soc. **15**, 1–30 (1914)
86. Haar, A.: Zur Theorie der orthogonalen Funktionen-Systeme. Math. Ann. **69**, 331–371 (1910)
87. Hansen, P.C.: Analysis of discrete ill-posed problems by means of the L-curve. SIAM Rev. **34**, 561–580 (1992)
88. Hansen, P.C.: The L-curve and its use in the numerical treatment of inverse problems. In: Johnston, P. (ed.) Computational Inverse Problems in Electrocardiology, pp. 119–142. WIT Press, Southampton (2000)
89. Hebinger, G., Michel, V., Richter, M., Simon, A.: Speech Recognition Support of Assisted Living. Schriften zur Funktionalanalysis und Geomathematik **40** (2008)
90. Heirtzler, J.R.: The future of the South Atlantic anomaly and implications for radiation damage in space. J. Atmos. Sol.-Terr. Phy. **64**, 1701–1708 (2002)
91. Heiskanen, W.A., Moritz, H.: Physical Geodesy, Reprint. Institute of Physical Geodesy, Technical University Graz/Austria (1981)

92. Hesse, K., Sloan, I.H., Womersly, R.S.: Numerical integration on the sphere. In: Freeden, W., Nashed, M.Z., Sonar, T. (eds.) Handbook of Geomathematics, pp. 1187–1219. Springer, Heidelberg (2010)
93. Heuser, H.: Funktionalanalysis, 3rd edn. B G Teubner, Stuttgart (1992)
94. Hobson, E.W.: The Theory of Spherical and Ellipsoidal Harmonics. Chelsea, New York (1965)
95. Holschneider, M.: Continuous wavelet transforms on the sphere. J. Math. Phys. **37**, 4156–4165 (1996)
96. Holschneider, M., Chambodut, A., Mandea, M.: From global to regional analysis of the magnetic field on the sphere using wavelet frames. Phys. Earth Planet. In. **135**, 107–124 (2003)
97. Holschneider, M., Iglewska-Nowak, I.: Poisson wavelets on the sphere. J. Fourier Anal. Appl. **13**, 405–419 (2007)
98. Johnston, I.: Measured Tones. The Interplay of Physics and Music. Institute of Physics Publishing, Bristol (1989)
99. Jones, F.: Lebesgue Integration on Euclidean Spaces. Jones and Bartlett Publishers, Boston (1993)
100. Keiner, J., Prestin, J.: A Fast Algorithm for Spherical Basis Approximation. In: Govil, N.K., Mhaskar, H.N., Mohapatra, R.N., Nashed, Z., Szabados, J. (eds.) Frontiers in Interpolation and Approximation, pp. 259–286. Chapman & Hall/CRC, Boca Raton (2006)
101. Kellogg, O.D.: Foundations of Potential Theory. Springer, Berlin (1967)
102. Kress, R.: Numerical Analysis. Springer, New York (1998)
103. Kufner, A., John, O., Fučík, S.: Function Spaces. Noordhoff International Publishing, Leyden (1977)
104. Kunis, S., Potts, D.: Fast spherical Fourier algorithms. J. Comput. Appl. Math. **161**, 75–98 (2003)
105. Lai, M.J., Shum, C.K., Baramidze, V., Wenston, P.: Triangulated spherical splines for geopotential reconstruction. J. Geodesy **83**, 695–708 (2009)
106. Laín Fernández, N.: Optimally space-localized band-limited wavelets on $\mathbb{S}^{q-1}$. J. Comput. Appl. Math. **199**, 68–79 (2007)
107. Landau, H.J., Pollak, H.O.: Prolate spheroidal wave functions, Fourier analysis and uncertainty—II. Bell Syst. Tech. J. **40**, 65–84 (1961)
108. Lang, S.: Undergraduate Analysis, 2nd edn. Springer, New York (2001)
109. Le Gia, Q.T., Mhaskar, H.N.: Localized linear polynomial operators and quadrature formulas on the sphere. SIAM J. Numer. Anal. **47**, 440–466 (2008)
110. Lemoine, F.G., Smith, D.E., Kunz, L., Smith, R., Pavlis, E.C., Pavlis, N.K., Klosko, S.M., Chinn, D.S., Torrence, M.H., Williamson, R.G., Cox, C.M., Rachlin, K.E., Wang, Y.M., Kenyon, S.C., Salman, R., Trimmer, R., Rapp, R.H., Nerem, R.S.: The development of the NASA GSFC and NIMA joint geopotential model. In: Proceedings of the International Symposium on Gravity, Geoid, and Marine Geodesy (GRAGEOMAR 1996), The University of Tokyo. Springer (1996)
111. Li, T.H.: Multiscale representation and analysis of spherical data by spherical wavelets. SIAM J. Sci. Comput. **21**, 924–953 (1999)
112. Louis, A.K., Maaß, P., Rieder, A.: Wavelets: Theory and Applications. Wiley, Chichester (1997)
113. Magnus, W., Oberhettinger, F., Soni, R.P.: Formulas and Theorems for the Special Functions of Mathematical Physics. Springer, Berlin (1966)
114. Mallat, S.: A Wavelet Tour of Signal Processing, 3rd edn. Academic, Burlington (2009)
115. Mallat, S.G., Zhang, Z.: Matching pursuits with time-frequency dictionaries. IEEE Trans. Signal Process. **41**, 3397–3415 (1993)
116. Masters, G., Richards-Dinger, K.: On the efficient calculation of ordinary and generalized spherical harmonics. Geophys. J. Int. **135**, 307–309 (1998)
117. Maus, S., Rother, M., Hemant, K., Stolle, C., Lühr, H., Kuvshinov, A., Olsen, N.: Earth's lithospheric magnetic field determined to spherical harmonic degree 90 from CHAMP satellite measurements. Geophys. J. Int. **164**, 319–330 (2006)

118. Maus, S., Rother, M., Holme, R., Lühr, H., Olsen, N., Haak, V.: First scalar magnetic anomaly map from CHAMP satellite data indicates weak lithospheric field. Geophys. Res. Lett. **29**, 47–1 to 47–4 (2002)
119. McShane, E.J.: Integration. Princeton University Press, Princeton (1974)
120. Mhaskar, H.N.: Local quadrature formulas on the sphere. J. Complex. **20**, 753–772 (2004)
121. Mhaskar, H.N.: Local quadrature formulas on the sphere, II. In: Neamtu M., Saff, E.B. (eds.) Advances in Constructive Approximation, pp. 333–344. Nashboro Press, Brentwood (2004)
122. Mhaskar, H.N., Narcowich, F.J., Prestin, J., Ward, J.D.: Polynomial frames on the sphere. Adv. Comput. Math. **13**, 387–403 (2000)
123. Mhaskar, H.N., Narcowich, F.J., Ward, J.D.: Spherical Marcinkiewicz–Zygmund inequalities and positive quadrature. Math. Comput. **70**, 1113–1130 (2000)
124. Mhaskar, H.N., Prestin, J.: Polynomial frames: a fast tour. In: Chui, C.K., Neamtu, M., Schumaker, L.L. (eds.) Approximation Theory XI: Gatlinburg 2004, pp. 101–132. Nashboro Press, Brentwood (2004)
125. Michel, D.: Framelet based multiscale operator decomposition. Ph.D. thesis, University of Kaiserslautern, Department of Mathematics, Geomathematics Group. Shaker, Aachen (2006)
126. Michel, V.: A wavelet based method for the gravimetry problem. In: Freeden, W. (ed.) Progress in Geodetic Science, Proceedings of the Geodetic Week, pp. 283–298. Shaker, Aachen (1998)
127. Michel, V.: A multiscale method for the gravimetry problem: theoretical and numerical aspects of harmonic and anharmonic modelling. Ph.D. thesis, University of Kaiserslautern, Department of Mathematics, Geomathematics Group. Shaker, Aachen (1999)
128. Michel, V.: A multiscale approximation for operator equations in separable Hilbert spaces—case study: reconstruction and description of the Earth's interior, Habilitation thesis. Shaker, Aachen (2002)
129. Michel, V.: Scale continuous, scale discretized and scale discrete harmonic wavelets for the outer and the inner space of a sphere and their application to an inverse problem in geomathematics. Appl. Comput. Harm. Anal. **12**, 77–99 (2002)
130. Michel, V.: Regularized wavelet-based multiresolution recovery of the harmonic mass density distribution from data of the Earth's gravitational field at satellite height. Inverse Probl. **21**, 997–1025 (2005)
131. Michel, V.: Wavelets on the 3-dimensional ball. Proc. Appl. Math. Mech. **5**, 775–776 (2005)
132. Michel, V.: Tomography—problems and multiscale solutions. In: Freeden, W., Nashed, M.Z., Sonar, T. (eds.) Handbook of Geomathematics, pp. 949–972. Springer, Heidelberg (2010)
133. Michel, V.: Optimally localized approximate identities on the 2-sphere. Numer. Func. Anal. Opt. **32**, 877–903 (2011)
134. Michel, V., Fokas, A.S.: A unified approach to various techniques for the non-uniqueness of the inverse gravimetric problem and wavelet-based methods. Inverse Probl. **24** (2008). 045019 (25 pp)
135. Michel, V., Wolf, K.: Numerical aspects of a spline-based multiresolution recovery of the harmonic mass density out of gravity functionals. Geophys. J. Int. **173**, 1–16 (2008)
136. Mikhlin, S.G.: Mathematical Physics, an Advanced Course. North-Holland Publishing Company, Amsterdam (1970)
137. Mohlenkamp, M.J.: A fast transform for spherical harmonics. J. Fourier Anal. Appl. **5**, 159–184 (1999)
138. Müller, C.: Über die ganzen Lösungen der Wellengleichung. Math. Ann. **124**, 235–264 (1952)
139. Müller, C.: Spherical Harmonics. Springer, Berlin (1966)
140. Müller, C.: Foundations of the Mathematical Theory of Electromagnetic Waves. Springer, Berlin (1969)
141. Narcowich, F.J., Petrushev, P., Ward, J.D.: Localized tight frames on spheres. SIAM J. Math. Anal. **38**, 574–594 (2006)
142. Narcowich, F.J., Ward, J.D.: Nonstationary wavelets on the m-sphere for scattered data. Appl. Comput. Harm. Anal. **3**, 324–336 (1996)

143. Nievergelt, Y.: Wavelets Made Easy. Birkhäuser, Boston (1999)
144. Nikiforov, A.F., Uvarov, V.B.: Special Functions of Mathematical Physics—A Unified Introduction with Applications. Birkhäuser, Basel (1988). Translated from the Russian by R. P. Boss
145. Olson, H.F.: Music, Physics and Engineering, 2nd edn. Dover, New York (1967)
146. Pavlis, N.K., Holmes, S.A., Kenyon, S.C., Factor, J.K.: An Earth gravitational model to degree 2160: EGM2008. Presentation given at the 2008 European Geosciences Union General Assembly held in Vienna, Austria, 13–18 Apr 2008. http://earth-info.nga.mil/GandG/wgs84/gravitymod/egm2008/NPavlis&al_EGU2008.ppt
147. Plato, R.: Numerische Mathematik kompakt, 4th edn. Vieweg + Teubner, Wiesbaden (2010)
148. Potts, D., Steidl, G., Tasche, M.: Kernels of spherical harmonics and spherical frames. In: Fontanella, F., Jetter, K., Laurent, P.J. (eds.) Advanced Topics in Multivariate Approximation, pp. 287–301. World Scientific, Singapore (1996)
149. Prestin, J., Rosca, D.: On some cubature formulas on the sphere. J. Approx. Theory **142**, 1–19 (2006)
150. Protter, M.H., Morrey, C.B.: A First Course in Real Analysis, 2nd edn. Springer, New York (1977)
151. Purucker, M.E., Dyment, J.: Satellite magnetic anomalies related to seafloor spreading in the South Atlantic ocean. Geophys. Res. Lett. **27**, 2765–2768 (2000)
152. Qian, S., Chen, D.: Signal representation using adaptive normalized Gaussian functions. Signal Process. **36**, 1–11 (1994)
153. Quarteroni, A., Sacco, R., Saleri, F.: Numerical Mathematics, 2nd edn. Springer, Berlin (2007)
154. Reigber, C., Balmino, G., Schwintzer, P., Biancale, R., Bode, A., Lemoine, J-M., König, R., Loyer, S., Neumayer, H., Marty, J-C., Barthelmes, F., Perosanz, F., Zhu, S.Y.: A high-quality global gravity field model from CHAMP GPS tracking data and accelerometry (EIGEN-1S). Geophys. Res. Lett. **29**, 37–1 to 37–4 (2002)
155. Renardy, M., Rogers, R.C.: An Introduction to Partial Differential Equations. Springer, New York (1996)
156. Renka, R.J.: Interpolation of data on the surface of a sphere. ACM T. Math. Software **10**, 417–436 (1984)
157. Reuter, R.: Über Integralformeln der Einheitssphäre und harmonische Splinefunktionen. Ph.D. thesis, Veröff. Geod. Inst. RWTH Aachen, RWTH Aachen, vol. 33 (1982)
158. Riley, K.F., Hobson, M.P., Bence, S.J.: Mathematical Methods for Physics and Engineering, 4th edn. Cambridge University Press, Cambridge (2008)
159. Rivlin, T.J.: An Introduction to the Approximation of Functions. Blaisdell Publishing Company, Waltham (1969)
160. Robin, L.: Fonctions Sphérique de Legendre et Fonctions Sphéroidale, vol. 1. Gauthier-Villars, Paris (1957)
161. Robin, L.: Fonctions Sphérique de Legendre et Fonctions Sphéroidale, vol. 2. Gauthier-Villars, Paris (1958)
162. Robin, L.: Fonctions Sphérique de Legendre et Fonctions Sphéroidale, vol. 3. Gauthier-Villars, Paris (1959)
163. Sard, A.: Linear Approximation. American Mathematical Society, Providence (1963)
164. Schaeben, H., Bernstein, S., Hielscher, R., Beckmann, J., Keiner, J., Prestin, J.: High resolution texture analysis with spherical wavelets. Mater. Sci. Forum **495–497**, 245–254 (2005)
165. Schmidt, M., Fengler, M., Mayer-Gürr, T., Eicker, A., Kusche, J., Sánchez, L., Han, S-C.: Regional gravity modeling in terms of spherical base functions. J. Geodesy **81**, 17–38 (2007)
166. Schneider, F.: Inverse problems in satellite geodesy and their approximate solution by splines and wavelets. Ph.D. thesis, University of Kaiserslautern, Geomathematics Group. Shaker, Aachen (1997)
167. Schoenberg, I.J.: On best approximations of linear operators. Nederl. Akad. Wetensch. Proc. Ser. A **67**, 155–163 (1964)

168. Schreiner, M.: On a new condition for strictly positive definite functions on spheres. Proc. Am. Math. Soc. **125**, 531–539 (1997)
169. Schröder, P., Sweldens, W.: Spherical wavelets: efficiently representing functions on the sphere. In: SIGGRAPH'95 Proceedings of the 22nd Annual Conference on Computer Graphics and Interactive Techniques pp. 161–172. ACM, New York (1995)
170. Schwarz, H.R.: Numerical Analysis: A Comprehensive Introduction. Wiley, Chichester (1989)
171. Sethares, W.A.: Tuning, Timbre, Spectrum, Scale. Springer, London (2005)
172. Simons, F.J.: Slepian functions and their use in signal estimation and spectral analysis. In: Freeden, W., Nashed, M.Z., Sonar, T. (eds.) Handbook of Geomathematics, pp. 891–923. Springer, Heidelberg (2010)
173. Simons, F.J., Dahlen, F.A.: Spherical Slepian functions and the polar gap in geodesy. Geophys. J. Int. **166**, 1039–1061 (2006)
174. Simons, F.J., Dahlen, F.A., Wieczorek, M.A.: Spatiospectral concentration on a sphere. SIAM Rev. **48**, 504–536 (2006)
175. Simons, F.J., Loris, I., Brevdo, E., Daubechies, I.C.: Wavelets and wavelet-like transforms on the sphere and their application to geophysical data inversion. Proc. SPIE **8138** (2011). 81380X
176. Simons, F.J., Loris, I., Nolet, G., Daubechies, I.C., Voronin, S., Judd, J.S., Vetter, P.A., Charléty, J., Vonesch, C.: Solving or resolving global tomographic models with spherical wavelets, and the scale and sparsity of seismic heterogeneity. Geophys. J. Int. **187**, 969–988 (2011)
177. Slepian, D.: Prolate spheroidal wave functions, Fourier analysis and uncertainty—IV: extensions to many dimensions; generalized prolate spheroidal functions. Bell Syst. Tech. J. **43**, 3009–3057 (1964)
178. Slepian, D., Pollak, H.O.: Prolate spheroidal wave functions, Fourier analysis and uncertainty—I. Bell Syst. Tech. J. **40**, 43–63 (1961)
179. Sloan, I.H., Womersley, R.S.: Extremal systems of points and numerical integration on the sphere. Adv. Comput. Math. **21**, 107–125 (2004)
180. Szegö, G.: Orthogonal Polynomials, vol. XXIII, 14th edn. AMS Colloquium Publications, Providence (1975)
181. Temlyakov, V.N.: Greedy algorithms and m-term approximation. J. Approx. Theor. **98**, 117–145 (1999)
182. Temlyakov, V.N.: Greedy algorithms with regard to multivariate systems with special structure. Constr. Approx. **16**, 399–425 (1999)
183. Temlyakov, V.N.: Nonlinear methods of approximation. Found. Comput. Math. **3**, 33–107 (2003)
184. Triebel, H.: Interpolation Theory, Function Spaces, Differential Operators. Johann Ambrosius Barth Verlag, Heidelberg (1995)
185. Trim, D.: Calculus. Prentice Hall, Scarborough (1993)
186. Tscherning, C.C.: Isotropic reproducing kernels for the inner of a sphere or spherical shell and their use as density covariance functions. Math. Geol. **28**, 161–168 (1996)
187. Tygert, M.: Fast algorithms for spherical harmonic expansions II. J. Comput. Phys. **227**, 4260–4279 (2008)
188. Voigt, A., Wloka, J.: Hilberträume und elliptische Differentialoperatoren. Bibliographisches Institut, Mannheim (1975)
189. Walnut, D.F.: An Introduction to Wavelet Analysis. Birkhäuser, Boston (2002)
190. Walter, W.: Einführung in die Potentialtheorie. Bibliographisches Institut, Mannheim (1971)
191. Walter, W.: Analysis 2, 3rd edn. Springer, Berlin (1992)
192. Wang, Z., Dahlen, F.A.: Spherical-spline parameterization of three-dimensional Earth models. Geophys. Res. Lett. **22**, 3099–3102 (1995)
193. Wang, Z.X., Guo, D.R.: Special Functions. World Scientific, Singapore (1989)
194. Weinreich, I.: A construction of C^1-wavelets on the two-dimensional sphere. Appl. Comput. Harm. Anal. **10**, 1–26 (2001)

195. Werner, J.: Numerische Mathematik I: Lineare und nichtlineare Gleichungssysteme, Interpolation, numerische Integration. Vieweg, Braunschweig, Wiesbaden (1992)
196. Wesfried, E., Wickerhauser, M.V.: Adapted local trigonometric transforms and speech processing. IEEE Trans. Signal Process. **41**, 3596–3600 (1993)
197. Wickerhauser, M.V.: INRIA lectures on wavelet packet algorithms. In: Minicourse lecture notes. INRIA, Rocquencourt (1991)
198. Wieczorek, M.A., Simons, F.J.: Localized spectral analysis on the sphere. Geophys. J. Int. **162**, 655–675 (2005)
199. Wieczorek, M.A., Simons, F.J.: Minimum-variance spectral analysis on the sphere. J. Fourier Anal. Appl. **13**, 665–692 (2007)
200. Windheuser, U.: Sphärische Wavelets: Theorie und Anwendung in der Physikalischen Geodäsie. Ph.D. thesis, University of Kaiserslautern, Geomathematics Group (1995)
201. Wojtaszczyk, P.: A Mathematical Introduction to Wavelets. Cambridge University Press, Cambridge (1997)
202. Wood, A.: The Physics of Music. University Paperbacks, London (1962)
203. WWW: http://cddis.nasa.gov/926/egm96/egm96.html
204. WWW: http://earth-info.nga.mil/gandg/wgs84/gravitymod/egm2008/
205. WWW: http://www.csr.utexas.edu/grace
206. Xu, Y., Cheney, E.W.: Strictly positive definite functions on spheres. Proc. Am. Math. Soc. **116**, 977–981 (1992)
207. Yosida, K.: Functional Analysis, 6th edn. Classics in Mathematics. Springer, Berlin (1995)
208. Zeidler, E. (ed.): Teubner-Taschenbuch der Mathematik, originally from I.N. Bronstein and K.A. Semendjajew. Teubner, Leipzig (1996)

Index

V. Michel, *Lectures on Constructive Approximation*, Applied and Numerical Harmonic Analysis, DOI 10.1007/978-0-8176-8403-7,

D

E

Y

Z

Applied and Numerical Harmonic Analysis

A.I. Saichev and W.A. Woyczyński: *Distributions in the Physical and Engineering Sciences* (ISBN 978-0-8176-3924-2)

R. Tolimieri and M. An: *Time-Frequency Representations* (ISBN 978-0-8176-3918-1)

G.T. Herman: *Geometry of Digital Spaces* (ISBN 978-0-8176-3897-9)

A. Procházka, J. Uhlíř, P.J.W. Rayner, and N.G. Kingsbury: *Signal Analysis and Prediction* (ISBN 978-0-8176-4042-2)

J. Ramanathan: *Methods of Applied Fourier Analysis* (ISBN 978-0-8176-3963-1)

A. Teolis: *Computational Signal Processing with Wavelets* (ISBN 978-0-8176-3909-9)

W.O. Bray and C.V. Stanojević: *Analysis of Divergence* (ISBN 978-0-8176-4058-3)

G.T Herman and A. Kuba: *Discrete Tomography* (ISBN 978-0-8176-4101-6)

J.J. Benedetto and P.J.S.G. Ferreira: *Modern Sampling Theory* (ISBN 978-0-8176-4023-1)

A. Abbate, C.M. DeCusatis, and P.K. Das: *Wavelets and Subbands* (ISBN 978-0-8176-4136-8)

L. Debnath: *Wavelet Transforms and Time-Frequency Signal Analysis* (ISBN 978-0-8176-4104-7)

K. Gröchenig: *Foundations of Time-Frequency Analysis* (ISBN 978-0-8176-4022-4)

D.F. Walnut: *An Introduction to Wavelet Analysis* (ISBN 978-0-8176-3962-4)

O. Bratteli and P. Jorgensen: *Wavelets through a Looking Glass* (ISBN 978-0-8176-4280-8)

H.G. Feichtinger and T. Strohmer: *Advances in Gabor Analysis* (ISBN 978-0-8176-4239-6)

O. Christensen: *An Introduction to Frames and Riesz Bases* (ISBN 978-0-8176-4295-2)

L. Debnath: *Wavelets and Signal Processing* (ISBN 978-0-8176-4235-8)

J. Davis: *Methods of Applied Mathematics with a MATLAB Overview* (ISBN 978-0-8176-4331-7)

G. Bi and Y. Zeng: *Transforms and Fast Algorithms for Signal Analysis and Representations* (ISBN 978-0-8176-4279-2)

J.J. Benedetto and A. Zayed: *Sampling, Wavelets, and Tomography* (ISBN 978-0-8176-4304-1)

E. Prestini: *The Evolution of Applied Harmonic Analysis* (ISBN 978-0-8176-4125-2)

O. Christensen and K.L. Christensen: *Approximation Theory* (ISBN 978-0-8176-3600-5)

L. Brandolini, L. Colzani, A. Iosevich, and G. Travaglini: *Fourier Analysis and Convexity* (ISBN 978-0-8176-3263-2)

W. Freeden and V. Michel: *Multiscale Potential Theory* (ISBN 978-0-8176-4105-4)

O. Calin and D.-C. Chang: *Geometric Mechanics on Riemannian Manifolds* (ISBN 978-0-8176-4354-6)

J.A. Hogan and J.D. Lakey: *Time-Frequency and Time-Scale Methods* (ISBN 978-0-8176-4276-1)

C. Heil: *Harmonic Analysis and Applications* (ISBN 978-0-8176-3778-1)

K. Borre, D.M. Akos, N. Bertelsen, P. Rinder, and S.H. Jensen: *A Software-Defined GPS and Galileo Receiver* (ISBN 978-0-8176-4390-4)

T. Qian, V. Mang I, and Y. Xu: *Wavelet Analysis and Applications* (ISBN 978-3-7643-7777-9)

G.T. Herman and A. Kuba: *Advances in Discrete Tomography and Its Applications* (ISBN 978-0-8176-3614-2)

Applied and Numerical Harmonic Analysis (Cont'd)

M.C. Fu, R.A. Jarrow, J.-Y. J. Yen, and R.J. Elliott: *Advances in Mathematical Finance* (ISBN 978-0-8176-4544-1)

O. Christensen: *Frames and Bases* (ISBN 978-0-8176-4677-6)

P.E.T. Jorgensen, K.D. Merrill, and J.A. Packer: *Representations, Wavelets, and Frames* (ISBN 978-0-8176-4682-0)

M. An, A.K. Brodzik, and R. Tolimieri: *Ideal Sequence Design in Time-Frequency Space* (ISBN 978-0-8176-4737-7)

B. Luong: *Fourier Analysis on Finite Abelian Groups* (ISBN 978-0-8176-4915-9)

S.G. Krantz: *Explorations in Harmonic Analysis* (ISBN 978-0-8176-4668-4)

G.S. Chirikjian: *Stochastic Models, Information Theory, and Lie Groups, Volume 1* (ISBN 978-0-8176-4802-2)

C. Cabrelli and J.L. Torrea: *Recent Developments in Real and Harmonic Analysis* (ISBN 978-0-8176-4531-1)

M.V. Wickerhauser: *Mathematics for Multimedia* (ISBN 978-0-8176-4879-4)

P. Massopust and B. Forster: *Four Short Courses on Harmonic Analysis* (ISBN 978-0-8176-4890-9)

O. Christensen: *Functions, Spaces, and Expansions* (ISBN 978-0-8176-4979-1)

J. Barral and S. Seuret: *Recent Developments in Fractals and Related Fields* (ISBN 978-0-8176-4887-9)

O. Calin, D. Chang, K. Furutani, and C. Iwasaki: *Heat Kernels for Elliptic and Sub-elliptic Operators* (ISBN 978-0-8176-4994-4)

C. Heil: *A Basis Theory Primer* (ISBN 978-0-8176-4686-8)

J.R. Klauder: *A Modern Approach to Functional Integration* (ISBN 978-0-8176-4790-2)

J. Cohen and A. Zayed: *Wavelets and Multiscale Analysis* (ISBN 978-0-8176-8094-7)

D. Joyner and J.-L. Kim: *Selected Unsolved Problems in Coding Theory* (ISBN 978-0-8176-8255-2)

J.A. Hogan and J.D. Lakey: *Duration and Bandwidth Limiting* (ISBN 978-0-8176-8306-1)

G. Chirikjian: *Stochastic Models, Information Theory, and Lie Groups, Volume 2* (ISBN 978-0-8176-4943-2)

G. Kutyniok and D. Labate: *Shearlets* (ISBN 978-0-8176-8315-3)

P.G. Casazza and G. Kutyniok: *Finite Frames* (ISBN 978-0-8176-8372-6)

T. Andrews, R. Balan, J. Benedetto, W. Czaja, and K. Okoudjou: *Excursions in Harmonic Analysis, Volume 1* (ISBN 978-0-8176-8375-7)

T. Andrews, R. Balan, J. Benedetto, W. Czaja, and K. Okoudjou: *Excursions in Harmonic Analysis, Volume 2* (ISBN 978-0-8176-8378-8)

D. Mitrea, I. Mitrea, M. Mitrea, and S. Monniaux: *Groupoid Metrization Theory: With Applications to Analysis on Quasi-Metric Spaces and Functional Analysis* (ISBN 978-0-8176-8396-2)

V. Michel: *Lectures on Constructive Approximation* (ISBN 978-0-8176-8402-0)

For a fully up-to-date list of ANHA titles, visit ***http://www.springer.com/series/4968?detailsPage=titles*** *or* ***http://www.springerlink.com/content/t7k8lm/****.*

If you have any concerns about our products,
you can contact us on
ProductSafety@springernature.com

In case Publisher is established outside the EU,
the EU authorized representative is:
Springer Nature Customer Service Center GmbH
Europaplatz 3, 69115 Heidelberg, Germany

Printed by Libri Plureos GmbH
in Hamburg, Germany